RF Design Basics

John Fielding, ZS5JF

Radio Society of Great Britain

Published by the Radio Society of Great Britain, 3 Abbey Court, Priory Business Park, Bedford, MK44 3WH

First published 2007

Digitally reprinted 2012 onwards

ISBN: 9781-9050-8625-2

Publisher's note

The opinions expressed in this book are those of the author and not necessarily those of the RSGB. While the information presented is believed to be correct, the author, the publisher and their agents cannot accept responsibility for consequences arising from any inaccuracies or omissions.

Production: Mark Allgar, M1MPA
Cover design: Dorotea Vizer, M3VZR
Editing, typography and design: Steve Telenius-Lowe, 9M6DXX

Printed in Great Britain by 4Edge Ltd of Hockley, Essex

Contents

Introduction

DESIGNING and constructing radio frequency circuits for the amateur can be a very rewarding part of the hobby. However, a lack of understanding the full problem and the physical limitations might have us pursuing an unproductive course with disappointing results. If the basics are clearly understood it is far easier to spot the errors and potential pitfalls. One way is to learn more about the subject and the best way is to read some books or articles on the subject. Unfortunately no one textbook covers the entire subject in detail. There are many books written on certain specialised aspects of radio frequency design but to possess all of these would incur a considerable expense. In many cases only about 10% of the textbook carries information on the aspect we need to read up on. For those who do not have access to a good reference library, we need to purchase a wide variety of books in the hope of finding all the information we require.

The other problem is that most textbooks are intended for degree students and the level they aspire to attain is often far higher than the average person - unless they have an engineering degree - is fully able to comprehend. The assumption made by the textbook writer is that the reader already has an intimate knowledge of the basics. Another major problem is that once a book is published it immediately becomes out of date, such is the advancement in the technology today, but the basics will always remain the same. So if you pick up a textbook written five years ago it will already be out of date in certain aspects of technology.

Another problem with some of the amateur publications is that some contain information that simply is not true and actually defies the accepted laws of physics. Unfortunately the average amateur is not aware of this and a 'myth' or 'urban legend' is formed about a particular subject. This does untold damage and has amateurs pursuing illogical paths because the information is plain wrong. One of these bogus subjects is VSWR and what it really means. Fortunately we have people like Walter Maxwell, W2DU, whose excellent book *Reflections* published by the ARRL debunks a lot of this nonsense. If you only buy one other book, this is highly recommended reading.

To try to get around these problems I decided to put together this book for the amateur reader who only has a simple understanding of the physics involved. The amateur in most countries today has to pass a written technical examination such as the Radio Amateurs Examination (RAE) which is nearly always a multiple choice paper. In multiple choice papers a lot of very small questions are asked about the whole of the subject. This explores the general knowledge of the examinee but does not test them to any great degree. It is similar to the driving test in many countries. Passing the driving test does not mean they are a good driver, it only determines if they are safe to be let loose on the roads on their own to gain practical experience and acquire further skills towards becoming a good driver. The same argument holds for the RAE, the amateur radio hobby is for 'the self training in the science of electronics and wireless communications', or words to that effect.

In olden days to become an amateur, and obtain a transmitting licence, different countries applied different standards. In some it was necessary to construct some item, such as a simple receiver or transmitter and then this was sent to the examining authority, normally the Post Office, for perusal. Often the licensee had to explain the circuit, how the device worked and demonstrate its operation to the satisfaction of the authority. Most countries required a Morse test to establish if the amateur was competent in wireless telegraphy. This largely precluded the average man in the street from taking up the hobby.

Happily those draconian days are mostly a thing of the past, but there were some good points about the testing methods. Nowadays anyone can sit and pass the RAE with little difficulty; even if they have no prior contact with electronics or physics. (I had a student some years ago who was a retired bank manager 74 years of age who passed at the first attempt.) Although this new system allows far more people to enter the hobby, which is a good thing - we always need new blood - it arguably has reduced the general level of technical standards and this is why there is such a thirst for good books covering the amateur hobby.

When I set about compiling this book I had certain goals. One was to eliminate, as far as possible, frightening mathematical formulae and rather to explain it in simple language with examples to clarify the topic. Inevitably one cannot totally divorce formulae from the subject but I have tried as far as possible only to include the ones you will immediately need. If you require the full formulae there are plenty of other publications that delve deeper in the subject and the RSGB and the ARRL publish many which make good reading at a more advanced stage. One such book is an ARRL publication by Wes Hayward, W7ZOI, *Introduction to Radio Frequency Design*, but when I purchased my copy I was amazed at the difficulty of understanding some of it - and I was a radio frequency designer of 30 years experience. That was the initial impetus to write something more appropriate for the 'average amateur'.

My college professor once told me that it wasn't essential to be able to remember all the formulae, but you should remember which book contains it when you need it.

One thing I have done for many years is to write simple software in *Basic* for my own use to reduce the chore of getting to a solution. Included with this book is software that contains many programs to do the necessary complex calculations. These I use almost every day because often finding the correct formulae and the scientific calculator buried under the paper on my desk simply takes too long. The other thing I have used for the last 10 years or so is analytical software to solve complex circuit designs. I could do this with just a scientific calculator but this takes hours of number crunching, whereas the software does the same thing in a few minutes. The software I am now using is the *Ansoft Student Version*, which is free and can be downloaded from the Ansoft website. Prior to this I mostly used the ARRL Radio Designer software, which sadly is now no longer available.

My background as a licensed amateur started in 1972 in the UK, although from the age of nine I was building crystal sets and other more complicated receivers. At the time of obtaining my licence I was employed by the British Post Office as an investigation officer dealing with radio and television interference. One of my duties was to perform amateur inspections on a periodic basis. In those days a newly licensed amateur was visited and an inspection made to see they had the necessary basic items of test equipment, wave meter, frequency meter, dummy load etc, and that the logbook was up to date. This would be repeated every year, in theory, although in practice it was sometimes two or three years between inspections because of the other workloads. One of the good things about this process was that it built up a bond between the GPO and the local amateurs, and we soon got to know who could be trusted and which ones we needed to keep a closer watch on. These also were the days of radio pirates and keeping in close touch with the local amateur population gave us many leads of where to look for pirate stations. Being also an amateur strengthened this bond and amateurs were more likely to divulge information to me than to a 'government agent', which is how most amateurs regarded GPO officials.

Since those days I have progressed through industry holding many positions in the development and research departments of large companies, mostly military contractors, where I have worked on several large design projects involving radio frequency design and allied fields. Today I am semi-retired and operate a consulting company as well as design and manufacture a wide range of radio frequency and automotive electronics products. Today my main interest is Radio Astronomy and my recent book *Amateur Raio Astronomy*, published by the RSGB, was my first attempt at writing.

Being an amateur employed as a radio frequency design engineer in a laboratory is a bit like a busman's holiday; the pay may not have been great but the 'toys' (test equipment) were fantastic! Having at hand the most up to date instruments and working with complex equipment was a pleasure, even if it involved long hours at times to meet deadlines. Nowadays I have to rely on other people's lab equipment when I need to make accurate measurements. One project I am currently working on is a radio astronomy receiver for 1.42GHz for a Hydrogen Line spectral analysis project. This involves the Durban University of Technology and is a project to develop a low cost Small Radio Telescope (SRT) for schools.

No one person is entirely responsible for a book. After the writer passes the draft to the publisher a whole team behind the scenes takes over. Part of this is further editing and the second component is the conversion into a format for Desk-Top-Publishing, so by the time the book reaches the bookshelves it has changed from the original, sometimes quite a bit!

I should like to thank some people for help in putting this book together. Originally it was intended to be a joint effort between another local amateur, Stuart McPherson, ZR5SD, and myself. Stuart is the Director of Electronics at the Durban University of Technology and supervises the RF and Microwave Communications courses for which I am the external moderator and external consultant. Unfortunately Stuart was too busy to write the intended chapters. However, he graciously allowed me to use his published work to compile some of the chapters. Stuart also proof-read the relevant technical chapters and offered suggestions for the content.

The other person who rendered invaluable assistance is Dr Brian Austin, G0GSF, who is ex-ZS6BKW of Wits University, South Africa, and recently retired from the University of Liverpool. Brian is a recognised expert in transmission line theory, electromagnetic radiation and antenna theory. Brian was very helpful in correcting some aspects and suggesting ways to improve the technical content..

To the RSGB and Mark Allgar, M1MPA; Mike Dennison, G3XDV, and Steve Telenius-Lowe, 9M6DXX, who encouraged me to put the book together and assisted in its production, I should like to offer my gratitude. Without this organisation and its people this book would never have seen the light of day.

Finally, I cannot forget the part my wife Penny played in keeping me fed and watered whilst putting it all together, and offering sympathy when it wasn't going right!

I hope you find this book useful.

John Fielding, ZS5JF

Chapter 1

Tuned circuits

VARIOUS different types of tuned circuits are used in radio frequency design. Some are parallel resonant and used to make band-pass filters and trap filters, others are series tuned and use to make notch filters. All the different types use inductors and capacitors connected in a *network* to satisfy the requirement.

However, any inductor or a capacitor is not in itself a pure component. This gives rise to spurious responses, as the capacitor contains a small inherent inductance and inductors contain an inherent capacitance. There is no such thing as an *ideal component* and the two types each contain an inherent loss resistance that degrades the performance. At low frequencies the effect the loss resistance, parasitic inductance or capacitance have is generally small, but at VHF and UHF the effects are very noticeable.

REACTANCE (X)

The opposition to the flow of alternating current (AC) in a capacitor or inductor is known as *reactance* and is measured in ohms; it is the AC equivalent of the resistance to direct current (DC) flow. The reactance of an inductor increases with increasing frequency and the reactance of a capacitor decreases with increasing frequency.

The formulae to calculate reactance in a capacitor or inductor are:

$$X_C = \frac{1}{2\pi fC} \quad \text{capacitive reactance in } \Omega$$

$$X_L = 2\pi fL \quad \text{inductive reactance in } \Omega$$

where:
f is the frequency in Hertz
C is the capacitor value in Farads
L is the inductance value in Henrys

At resonance $X_C = X_L$

For example, at 10MHz a capacitance of 100pF has a reactance of 160Ω, an inductance of 1μH has a reactance of 62.8Ω.

At 100MHz the 100pF capacitor's reactance falls to 16Ω and the 1μH inductor reactance rises to 628Ω.

USEFUL REACTANCE TO COMPONENT VALUE FORMULAE

When a reactance needs to be converted to the value of capacitance or inductance the following formulae are useful.

To convert from a reactance to capacitor value in pF when the frequency (f) is in MHz we use:

$$C = \frac{159150}{(f \times X_C)} \quad \text{pF}$$

To convert to inductance we use:

$$L = 0.15915 \frac{X_L}{F} \quad \mu\text{H}$$

Example: A reactance of 100Ω at 10MHz is equivalent to a capacitor of ≈160pF or an inductance of ≈1.6μH.

QUALITY FACTOR - 'Q'

The parasitic series loss resistance R_S in a capacitor or inductor prevents it from being an ideal component. When working at lower frequencies (below ≈10MHz) this is not of too much concern and the effect the loss resistance has is small. But at frequencies above about 10MHz the effect is quite noticeable. To quantify the 'goodness' of a component we use a notation of 'Q'. Q is simply the ratio of the reactance to the loss resistance and is a number with no units. If the loss resistance is low the Q will be high; conversely if the loss resistance is high the Q will be low.

$$Q = X / R_S$$

where X is the reactance of either an inductor or capacitor in Ohms and R_S is the series loss resistance.

For example, a 100pF capacitor at 10MHz has a reactance of 160Ω and a Q of 1000, what is the loss resistance? R_S is 160/1000 = 0.16Ω. The value of loss resistance generally increases with increasing frequency in a more or less linear manner. Hence, the 100pF capacitor at 100MHz would have a reactance of 16Ω and the value of R_S is now, say, 1Ω. The Q at 100MHz would only be 16. This is unlikely to happen in practice and Q values of 1000 or more are common with good quality RF capacitors, hence R_S is generally quite a low value. Some types used for microwave applications have Q values of greater than 5000.

For an inductor the same effect applies and the predominant factor in the value of R_S is the skin resistance or core loss if the inductor is wound on a ferrite core. A typical 1μH molded choke would have a Q of about 50 at 30MHz and a reactance of 188Ω. Hence, the effective value of R_S is 188/50 = 3.76Ω. At 300MHz the reactance rises to 1880Ω, assuming the value of R_S remains the same the Q is now 500. This is a general trend with inductors, the Q rises with increasing frequency until the value of R_S due to the skin effect or core loss becomes dominant and then Q falls steeply to less than the lower frequency value. In practice R_S also rises with frequency but not as quickly as a linear function. The DC resistance might be as little as 50mΩ but the bulk of R_S is made up of the skin resistance and core loss, especially with very small diameter wire. Hence, for low loss coils the use of a large diameter wire with an air-wound core reduces the value of R_S and the DC resistance. For high frequency applications the effect of skin resistance means that the RF current is only flowing in the outer layer of the conductor and hence the solid wire can be replaced with a thin wall tube, such as copper.

When the inductor is wound on a ferrite core the magnitude of the DC flowing in the inductor also has some effect. When the DC flowing is low the core loss is also low, but if the combination of the DC and AC flowing in the winding approaches the non-linear magnetisation curve of the ferrite material the losses rise rapidly. The type of ferrite used also has a marked effect on the potential Q. If a low frequency ferrite is used the Q will degrade above a certain frequency. For the highest Q at high frequencies the *permeability* of the ferrite material needs to be low. Unfortunately a low permeability ferrite requires more turns to achieve the required inductance.

At low frequencies a low value of inductance will have a low Q and it will increase with frequency to some higher frequency where the inductor becomes self-resonant. The highest Q is when the parasitic winding capacitance resonates with the inductance to form a parallel resonant circuit. When this occurs the inductor appears to be a very high resistance. Similarly if an external capacitor is placed in series with the inductor and the network connected across a source of RF voltage, as the frequency is increased at some high frequency the network will become series resonant and the only component left is R_S. This is a *series-tuned trap* and exhibits very low impedance to the driving signal. At resonance the voltage across the network will fall close to zero, determined by the value of R_S.

SERIES TO PARALLEL CONVERSION - MATHEMATICAL MAYHEM!

One of the things RF designers tackle daily is the above topic. In keeping with the principles we adopted for this book, we are not even going to try to cover the frightening formulae because it is so confusing and you will be more mystified than before. If you want them you can find them in more advanced books such as the RSGB or ARRL *Handbooks*.

However, a very simple explanation is concerned with the loss resistance of an inductor or capacitor. We have explained that for a high Q we need an inductor with a low series resistance, R_S. In series to parallel conversion, and vice versa, we mathematically juggle formula to convert between series and parallel resistance. If the series loss resistance is small, the equivalent parallel case gives a high value resistance we call R_P. For a high Q we need a low series resistance or a high parallel resistance. Hence, when we 'de-Q' a parallel tuned circuit, to broaden the bandwidth or lower the value of R_P by placing a damping resistor in parallel, we are doing the same as placing a low value resistor in series with either the inductor or capacitor. Sometimes it is more convenient to place the resistor physically in parallel with the parallel network and at other times we choose to place the resistor in series with the inductor for stability reasons. As my late father used to say, *"it's six of one and half a dozen of the other"*, they both achieve the same end.

> Because we will need to perform the conversions in later chapters software which tackles this has been included on the supporting website. You simply need to run the program and input the required numbers and the operating frequency. Out come the converted parameters. We will be using this in the chapter on solid state power amplifiers and elsewhere in the book.

REJECTER CIRCUITS

The series and parallel tuned network can be used to reject signals. If a parallel-tuned trap is placed in series with a source of RF voltage and a load it will serve to 'notch out' a narrow band of frequencies around the centre frequency. Similarly if the series tuned trap is also placed in series between the voltage source and load it will pass signals only around its resonant frequency. This is the basis of band-pass filters and other types.

The depth and bandwidth of the notch obtainable

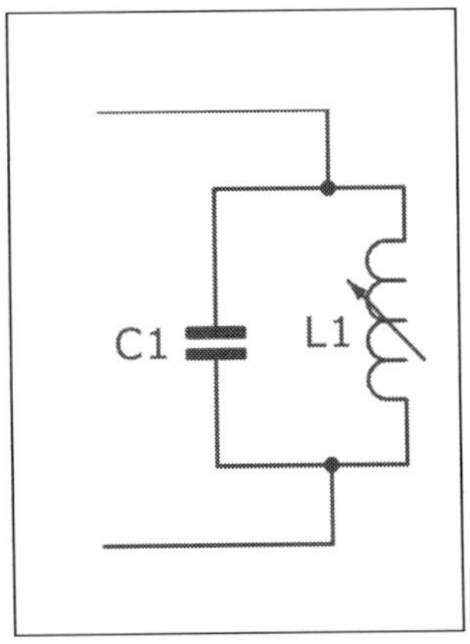

Fig 1.1. Parallel resonant network

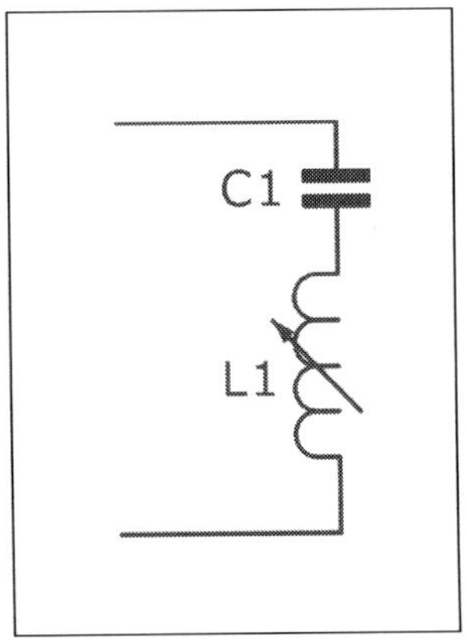

Fig 1.2. Series resonant network

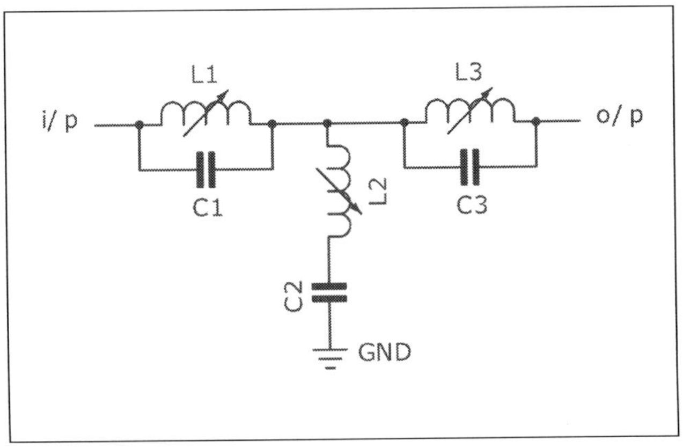

Fig 1.3. Multi section trap

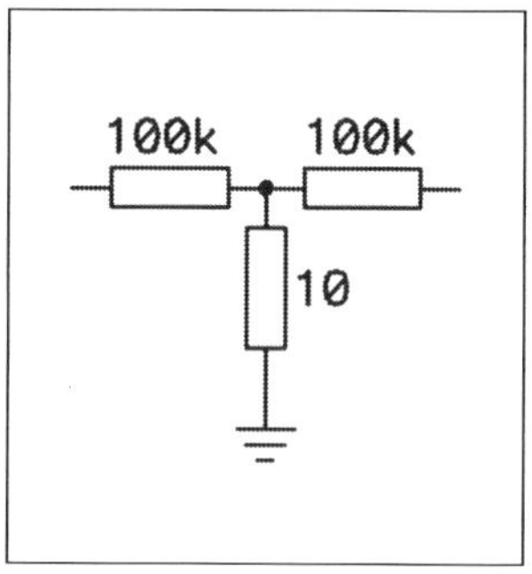

Fig 1.4. Equivalent network at resonance

is directly related to the Q of the inductor and capacitor, for a deep notch the Q of both need to be high.

With sufficient Q for the inductors and capacitors this circuit can produce over 90dB of rejection to a signal. The equivalent circuit at resonance is shown in **Fig 1.4**.

MEASUREMENT OF Q

The Q value is measured with an instrument known as a 'Q-meter' or 'Circuit Magnification Meter'. The best known instrument was manufactured by Marconi Instruments and could measure the Q of inductors or capacitors between 1kHz and 300MHz with an accuracy of ±10% for Q values up to 1200. As well as being able to measure Q it could also indirectly measure the value of an inductor or capacitor. The measuring instrument contains a precision air-variable capacitor with a readout accuracy of ±0.05pF. By finding the resonant frequency of the inductor and then noting the tuning capacitor value and frequency required to resonate the circuit the inductance can be calculated to within ±0.1%.

Similarly when an unknown capacitor needs to be measured we use an arbitrary inductor and the unknown capacitor connected in parallel and set the air variable to either zero capacitance or some low value, say 10pF. Adjusting the signal generator frequency we can find the resonant frequency. Having found the resonant frequency we remove the unknown capacitor and dial in capacitance on the air variable until the network resonates again. The amount of capacity we need to add is the *effective value* of the unknown capacitor at that frequency. At high frequencies often the capacitor marked value and the effective value differ because of the self-inductance of the capacitor or its lead length.

Unfortunately production of the Marconi Q-meter came to an end in the late 1970s, but they sometimes appear on the surplus market. If you find one snap it up quickly, but make sure it is working correctly first and it has the handbooks, they can be very difficult to repair and calibrate!

My Marconi Q-meter was made in 1963. There were two signal generators available; one covered 40kHz to 50MHz and the other 20MHz to 300MHz. **Fig 1.5** shows the 20 to 300MHz generator. The right hand side is the instrument itself; the unit on the left is the signal generator to provide the signal.

EQUIVALENT CIRCUIT

The inductor and capacitor can be modeled with the parasitic components to arrive at the physical item. If an inductor is examined we find that the equivalent circuit is as shown in **Fig 1.6**.

The component L is the actual inductor. The component C_P is the parallel capacitance due to inter-winding coupling. R_S is the series resistance due to copper losses, skin effect and core loss at high frequency. When the frequency is very high the capacitance C_P resonates with the inductance and the inductor behaves like a parallel tuned circuit. This exhibits high impedance, represented by R_D. If an external capacitor is placed in series with the inductor it can cause the inductor to become series resonant. When this occurs the reactance of L and C-series (CS) cancel out and the only component left is R_S. In a well-constructed inductor R_S will be low. So the maximum current is limited only by the value of R_S.

The capacitor also has the same sort of parasitic components. The predominant ones are R_S and the lead inductance L. The material the capacitor is made from also alters the value of L. In effect there are two series inductances in a capacitor. One is inherent in the manufacturing process; the second is the termination lead out wires. The lead-out wires have an inductance of about 1nH per mm. If the lead-out wires are long, the lead inductance partially cancels some of the capacity and lowers the series resonant frequency (SRF). Often when using a capacitor to decouple a supply rail to RF we can utilise this effect to make the capacitor appear as short-circuit (limited by the value of R_S) at the frequency we need to decouple. In effect we have made a series resonant trap at the frequency.

A factor that many people fail to realise is that once the capacitor has passed the self-resonant frequency it no longer behaves like a capacitor. Above the SRF the

Fig 1.5. Marconi Instruments 'Q-meter'

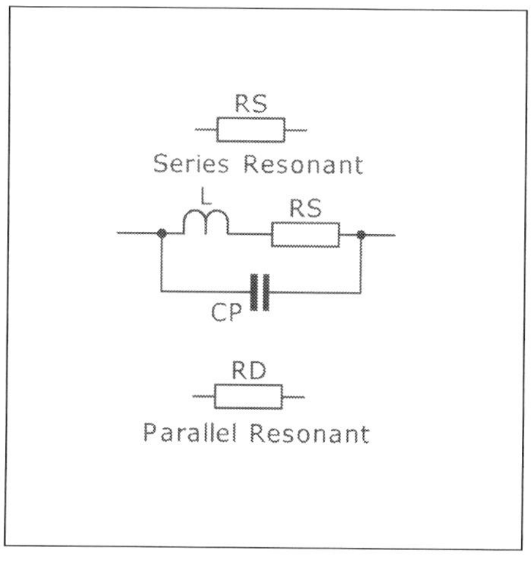

Fig 1.6. Equivalent circuit of an inductor

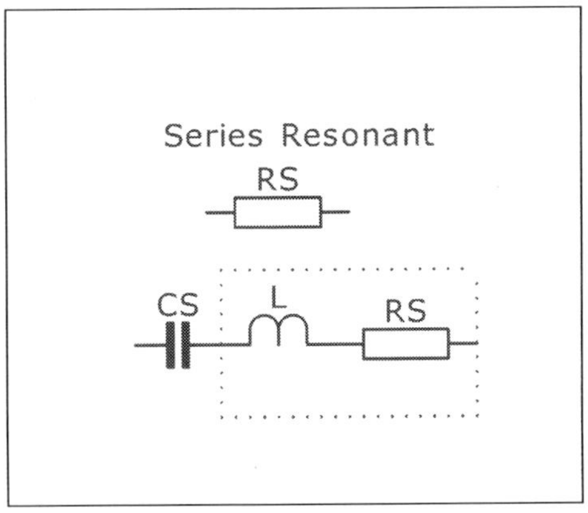

Fig 1.7. Inductor series resonant with external capacitor

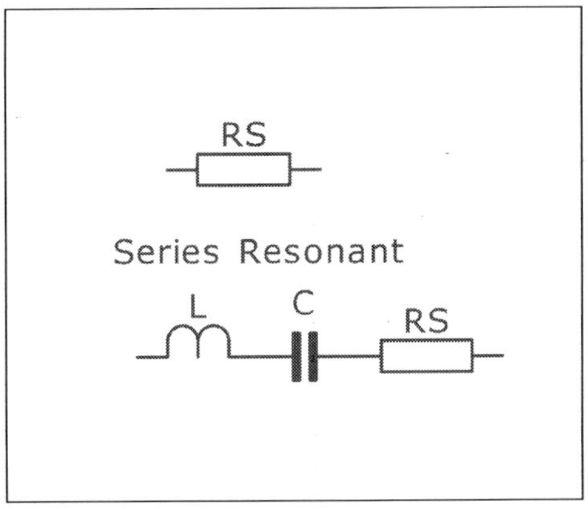

Fig 1.8. Equivalent circuit of a capacitor

component behaves like an inductance and the inductive reactance will continue increasing for increasing frequency. Hence, the decoupling or coupling capacitor starts to become less effective as the frequency rises. At some very high frequency it appears close to an open circuit. A 100nF disc ceramic capacitor is *not* a low reactance at 144MHz!

The inductor also possesses the same effect. Above the self-resonant frequency it behaves like a capacitor and progressively tends towards a short circuit at some high frequency.

One way of preventing this is to use what is known as *dynamic decoupling*. In this, the single large value capacitor is replaced with several smaller value ones all connected in parallel. Because paralleling inductance reduces the inductance and capacitance increases with paralleling several capacitors, the decoupling is much more effective. In a 2m transmitter, for example, we might use the following capacitors across the supply rail to ground. 10μF electrolytic, 100nF ceramic disc, 10nF ceramic multi-layer plate, 1nF ceramic multi-layer and 100pF ceramic multi-layer. This provides a good low-reactance decoupling from low frequencies up to at least 1GHz. The 10μF electrolytic is required to prevent low frequency instability of the very high gain the VHF devices can develop at a few tens of kilohertz.

Surface mount capacitors have very low self-inductance because of the absence of lead-out wires. A typical 0805 (2mm x 1.2mm) ceramic multi-layer capacitor has an inherent inductance of only about 0.4nH when correctly terminated on a printed board. This means they behave like capacitors up to very high frequencies, as high as several GHz for most types in common use. Hence, the values we use to decouple a supply rail need to be carefully selected. In general the reactance of a capacitor when used as a decoupling or coupling capacitor needs to have a value of at least 10 times less than the point we are trying to decouple or couple. In a 50Ω system the reactance of the capacitor needs to be less than 5Ω to ensure we do not introduce extra loss to the signal.

The same criteria apply to inductors. The reactance needs to be at least 10 times higher than the point we are trying to connect to. If the need is for an inductor to connect to a 50Ω point in a circuit, to pass the supply current to a transistor for example, the reactance needs to be at least 500Ω for the same reason.

RESONANCE

As has already been mentioned, resonance occurs when the value of X_C and X_L are the same value. In a parallel resonant network at resonance the reactive components cancel and we are left with infinite resistance, in theory. In practice because the components have some loss resistance the resistance is high but not as high as it could be with infinite Q components. To demonstrate the resonance effect the graph in **Fig 1.9** is useful.

The point on the graph where the inductive and capacitive reactance cross (Xr) is the resonant frequency, F_r. It does not matter if this is a series connected circuit or a parallel connected circuit, if the same capacitor and inductance are used the frequency for both connection schemes will be the same. In fact electrically there is no difference between the two. This can be demonstrated by examining the properties of the two different types.

We will start with the series network. At resonance the current flowing is a maximum and is only limited by the loss resistance R_{St} which is the sum of R_S for the capacitor and inductor as each has R_S incorporated in them. Assume that R_{St} = 1Ω for the two combined components. If the network is connected across an AC source of 10V rms the current flowing is 10A. If the loss resistance were 0.1Ω the current would be 100A. In **Fig 1.10** the arrow denotes the RF current flowing.

Now bend the two ends of the series network so that they are joined together. This is now a parallel tuned network, but it is really a series network as well. If we visualise the circulating current flowing around the loop of the inductor and capacitor at resonance it will be the same value as when connected originally. In other words, although a parallel resonant network presents a very high resistance to an external circuit, it has a very high circulating current flowing internally. The higher the Q of the components the higher the current will be.

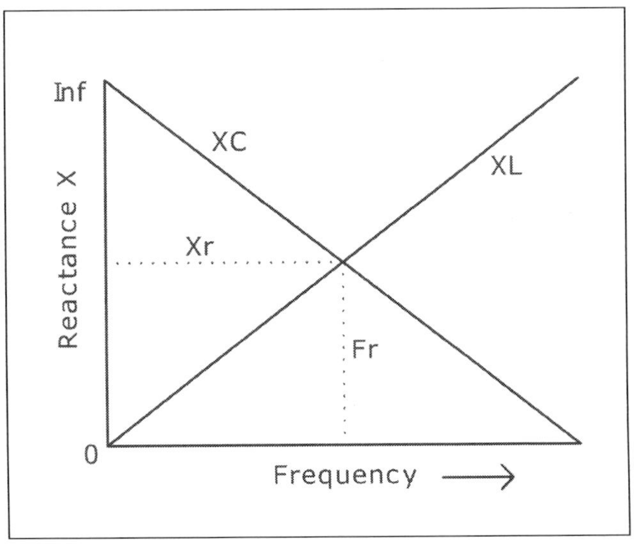

Fig 1.9. Reactance versus frequency plot

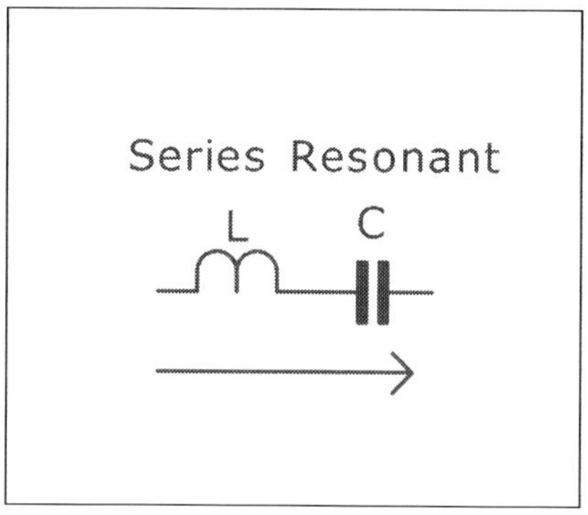

Fig 1.10. Series resonant circuit

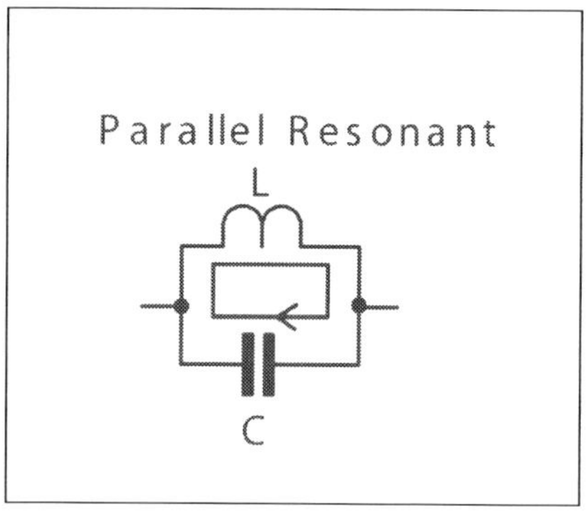

Fig 1.11. Parallel resonant circuit

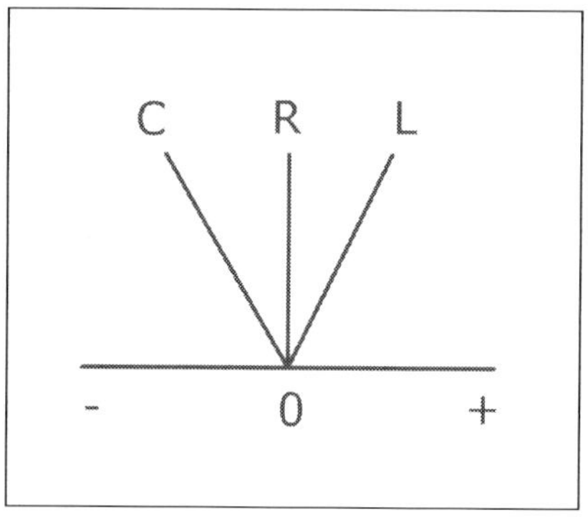

Fig 1.12. Vector diagram of resonance

You will note that we used the word *resistance* when the circuit is resonant in the parallel connection mode. This was deliberate because it really is a pure resistance and we can prove this by measuring the phase angle between the voltage and current. In a pure resistance the phase angle is 0°. However, now consider what happens when the signal generator driving this high resistance network is moved slightly in frequency. Suppose we reduce the frequency by 1%. The condition now is that the reactances do not exactly cancel, the capacitive reactance is slightly more than the inductive reactance. So the circuit no longer behaves like a pure resistance but a small value of capacitance. Hence, we cannot call it a resistance any more as it is a complex item made up of reactance and resistance. Hence, this is now *impedance*, still with numeric values stated in ohms just like a reactance or resistance.

If we now change the signal generator frequency so that it is 1% higher than the resonant point the opposite effect occurs. The inductive reactance is now slightly greater than the capacitive reactance and so the circuit behaves as a very low value of inductance with impedance. This property is used extensively in matching to complex impedance devices such as transistors. By making either the inductor or capacitor variable over a small range we can swing from net capacitive to net inductive quite simply.

The other fact we can measure is the amplitude of the AC voltage across the network at resonance. When the circuit is parallel resonant the AC voltage will be a maximum. At frequencies lower or higher the AC voltage will be less. This is the basis of the band-pass filter.

For those who find this difficult to visualise, a simple analogy to a watch can explain it. Imagine the second hand of your watch. It rotates through 360° every minute. If we call 12 o'clock the point where it is pointing directly upwards and 9 o'clock where it is pointing to the left and 3 o'clock where it is pointing to the right we have covered 180° of the dial. The 9 o'clock point corresponds to some low frequency and 3 o'clock corresponds to some high frequency. 12 o'clock hence corresponds to the centre frequency of the resonant circuit. At this point the network is purely resistive. At 11 o'clock we are below the resonant frequency by, say 1% and at 1 o'clock we are same amount above the centre frequency. As the second hand sweeps around when it gets to 9 o'clock the frequency is far below resonance and the phase angle is a large negative value. When the second hand sweeps around to 1 o'clock the frequency is above resonance and the phase angle is positive. At 3 o'clock the frequency is far above resonance and the phase angle is a large positive value. This representation is known as a *vector diagram* (see **Fig 1.12**) and the lines represent the second hand of the watch as it sweeps around the dial. Below resonance the network is capacitive and above it is inductive.

The other thing we can see from this simple vector diagram is the amplitude of the voltage across the resonant network. The height of the tip of the second hand is voltage above the horizontal baseline. At 9 o'clock the amplitude is low and so is the 3 o'clock point. But at 12 o'clock the tip of the second hand is at its highest point. (The tip of the second hand moves along an arc of constant radius.) This corresponds with the resonant point. Hence, as we move the frequency from low to high, passing through the resonant point, the voltage will rise to maximum at the resonant frequency and then fall away as we move further away in an upwards direction. In the case of the series resonant network it is current and not voltage that varies with frequency.

DYNAMIC RESISTANCE

When a capacitor and inductor are connected in parallel they will resonate at some frequency. At resonance the inductive reactance and capacitance are equal in numerical value but opposite in sign, so they will cancel out. Theoretically, if both the components were pure (no R_S) the result would be an infinite resistance appearing across the network. In practice this never happens but the effective resistance appearing across a parallel resonant network can be very high. The term we use for this is the *dynamic resistance* (R_D). This is the apparent resistance value in Ohms across the network. If the Q of the two components is high then R_D will also be high.

The formula to calculate R_D is:

$$R_D = \frac{X^2}{R_S}$$

where:
X^2 is the reactance at resonance squared
R_S is the total series loss resistance in Ohms.

The potential value of R_D depends on the value of the components used to achieve parallel resonance and the Q of these. This brings about another factor often quoted, the *L-C ratio*.

L-C RATIO

This is a bit of a misnomer because for resonance both reactances need to be numerically equal. In general, for best R_D the inductor value needs to be large and the capacity value small. A couple of examples should clarify the facts.

A parallel tuned circuit for 10MHz is required with the highest possible value for R_D using practical components. We will select a suitable variable inductor with a Q of 100 and then find the capacitor value required to resonate at 10MHz. At 10MHz with a ≈2µH inductor the capacitor required is ≈120pF. Using a 5µH inductor, also with a Q of 100, requires a capacitor of 51pF.

The values of R_D for these two cases are:

12.6kΩ for the 2µH inductor
31.3kΩ for the 5µH inductor.

So it is clear that for maximum R_D the inductor value needs to be large. In semiconductor matching, especially to low noise devices like GaAsFets, the optimum situation is to use the inherent gate capacity to resonate the inductor, this requires a large value of inductance and often the Q it is possible to attain, especially at UHF and above, limits the value of R_D.

The value of R_D is ultimately limited by the Q of the inductor as the capacitor Q is normally much greater than the inductor. In cases where we have a parallel resonant network and need to match to some specific impedance, for example the collector of a transistor, often the available inductor types leave us with a R_D that is too high. A simple solution to this is to shunt the network with a resistor to lower the R_D. The two resistors act exactly like two normal resistors in parallel. If for example we needed a collector load of ≈6kΩ at 10MHz, a 2µH inductor and a 120pF capacitor will give 12.6kΩ . If we now connect a 12kΩ resistor across the network we arrive at the required impedance. In effect what we have done is to lower the Q of the inductor by the shunt resistor. If a resistor is not a suitable method we would need to find the inductor and capacitor values and inductor Q required to give the correct R_D, a messy and complicated process; a resistor is a much cheaper and quicker solution.

If we now re-run the calculations for a higher Q inductor we can see the potential R_D we might achieve. Winding the 5µH inductor on a ferrite toroid can attain a Q value of about 250 maximum at 10MHz. This would yield a value of approx 78kΩ. If we increase the inductor value to 7.5µH the capacitor to resonate becomes 33pF. Assuming the inductor Q is also 250 the value becomes ≈120kΩ. The problem with increasing the inductor value is that we need to take into account any circuit stray capacitance. If this is significant we need to lower the inductance so we have a sensible value of tuning capacitance to work with. Often the inductance can be made variable over a small percentage to allow fine tuning, or a variable capacitor is another method.

CONSTANT R NETWORK

In some circuits we need a network that terminates a node with a constant low VSWR but also containing some selectivity. A typical application is when a double-balanced mixer is used in a receiver to mix down to an IF.

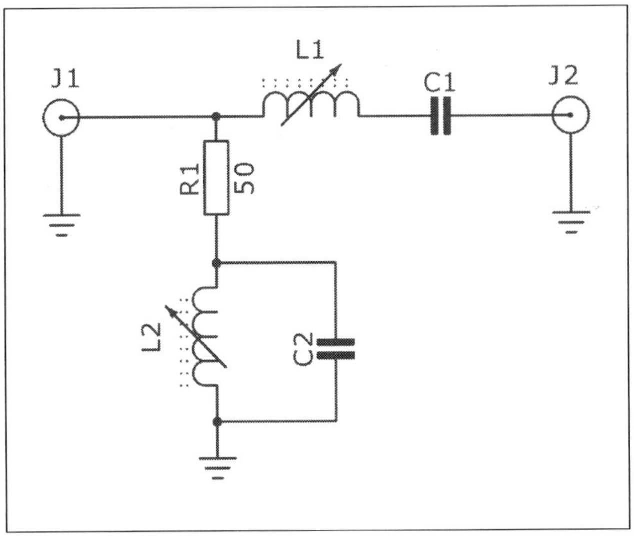

Fig 1.13. Constant R network

The mixer IF output port contains several frequencies apart from the IF. These are the LO and its harmonics generated in the mixer, the RF signal, the IF image frequency and all the other mixing products. All of these need to be terminated in a low reactive load to prevent reflections causing intermodulation products. If the IF port is not terminated in a wideband manner the reactive network used to select the IF signal will reflect power back into the mixer. (The way that mixer manufacturers characterise the mixer is in a circuit with large attenuator values to ensure a pure 50Ω load on all ports, this is a lossy method and one we would wish to avoid in practice.)

A network called a Constant R network performs this wideband termination. A typical Constant R circuit is shown in **Fig 1.13**.

The network consists of two tuned circuits. One is series resonant to pass the required IF, formed by L1 and C1. The second tuned circuit is a parallel resonant network also at the IF, formed by L2 and C2. Between these two is a load resistor R1. L2 and C2 appear as high impedance to the IF and 'disconnect' the load resistor from ground and hence they have no effect at the IF. L1 and C1 form the frequency selective path for the IF signal to the following IF filter or amplifier stages and appear as low impedance at the IF. At frequencies removed from the IF the parallel resonant network approaches low impedance and hence serves to ground the load resistor. This terminates the IF port of the mixer to frequencies other than the IF. The mixer IF port connects to J1 and the following stage to J2.

The frequency response of the network for a 21.4MHz IF is shown in **Fig 1.14**.

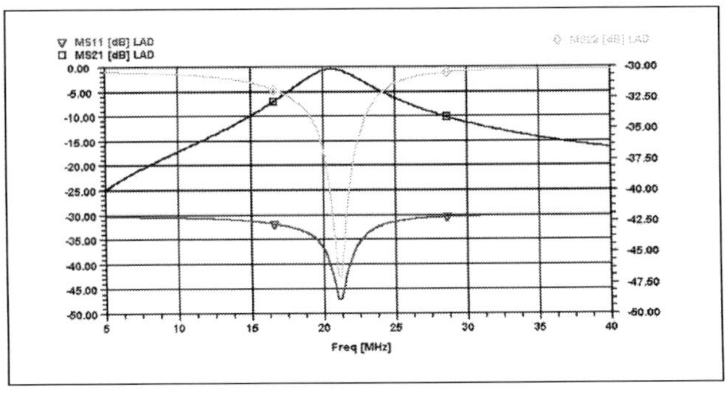

Fig 1.14. Frequency response of Constant R network

Fig 1.15. Typical low power RF inductors

The plot shows the IF insertion loss as MS21 (top solid trace). This is broadly peaked at the IF. The input VSWR is shown by MS11 (bottom solid trace), which remains below -30dB over a wide frequency range and dips to a minimum at the IF. The output match to 50Ω is shown by the trace of MS22 (light grey trace), at frequencies other than the IF it is severely mismatched, but well matched at the IF.

RF INDUCTORS

We will require RF inductors for different circuits. If the signal level is low and we require moderate Q values we can often utilise commercial ready-wound coils wound on formers or the types made with shielding cans. For some applications we need more Q than a commercial coil can attain and then a good option is to wind it on a ferrite core. A selection of inductors is shown in **Fig 1.15**.

The small inductor on the left is made by Toko RCL and is a 7mm square can 11mm tall with a ferrite-adjusting slug. The two larger ones are also made by Toko but in the larger 10mm square can and about 15mm tall, the inside of one is shown where the inductor is encapsulated in polyethylene. These also have ferrite slug adjusters. The Toko coils typically have a Q of 80 - 120 with an adjustment range of ±20%. The coil on the right is a homemade type wound on a ferrite toroid and has a Q of ≈200 at 30MHz. Adjustment of the inductance is by squeezing together the turns or spreading them further apart. An adjustment range of ±15% is normally possible.

For higher power applications we often need to use air-cored coils to attain the high Q values. Examples of two are shown in **Fig 1.16**. These are wound on a ceramic former which has a spiral groove to space the wire correctly. The former diameter is 30mm. The left-hand coil is a pi tank inductor wound with 16SWG bare copper wire for a low power transmitter. The right-hand coil is a RFC for a VHF power amplifier using valves and is wound with 20SWG enamel copper wire.

Fig 1.16. High power RF inductors

Where very high power and hence low loss is required, the inductor can be wound with either thick wire or thin-wall copper tube. **Fig 1.17** shows a typical power amplifier inductor wound from 5mm diameter copper tube with the ends hammered flat for attaching with screws passing through the ends. The coil is 50mm internal diameter. At 50MHz this inductor has a Q of ≈600 and was used as the pi tank inductor for a high power 6m amplifier.

Fig 1.17. Typical high power inductor

RF CAPACITORS

When high Q is essential we also need to select suitable capacitors. These may be fixed capacitors or variable types. The power and voltage level required determine the optimum type to use. For circuits built on printed circuit boards the types also vary. Some types are shown in the next series of pictures.

The popular ceramic plate types are at the bottom of the picture in **Fig 1.18** and vary in value from 0.68pF up to several nF. The two larger ones are 500V types with 5mm lead spacing, the smaller types are 50V or 63V rating with 2.5mm lead spacing. Above are some high current capacitors, on the left are clamped mica for transmitters with a working voltage of 250V; these are approx 6mm square. The larger one is a similar type for 350V. The small capacitors at the right are ATC 100B monolithic ceramic, with a working voltage of 50V. These are 0.1in square (the 100 refers to 100mil = 2.5mm) and have very low self-inductance making them suitable up to microwaves. These types are popular when constructing high power solid state amplifiers or low pass harmonic filters.

Fig 1.18. Fixed capacitors

Fig 1.19. Low power variable capacitors

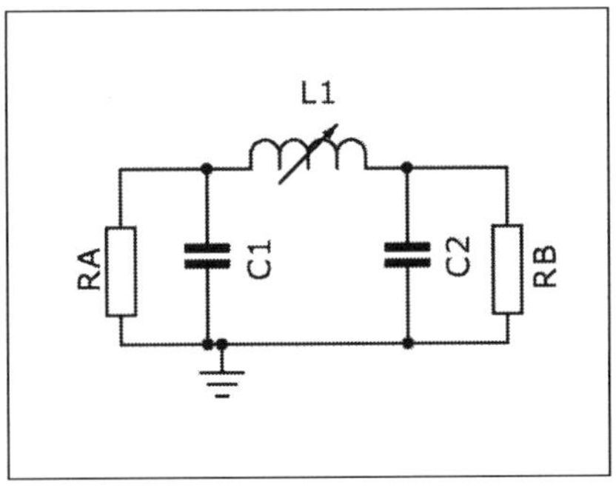

Fig 1.21. π network

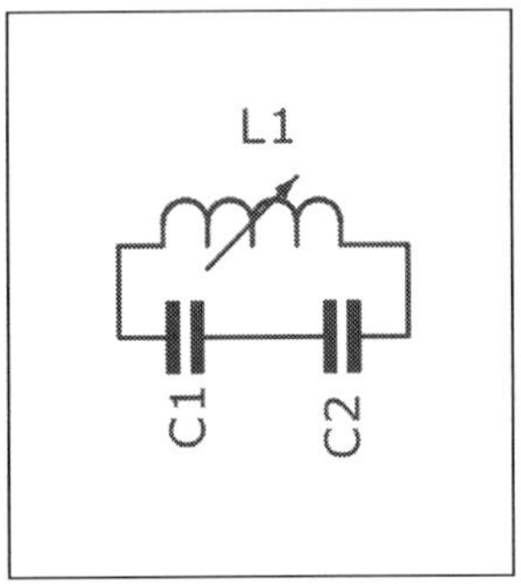

Fig 1.22. Equivalent pi network circuit

The types shown in **Fig 1.19** are designed for printed circuit board mounting. The extreme left hand one is an air variable with a 100V rating. The next two are tubular trimmers. The first is a ceramic tube with an adjusting screw to vary the capacitance. The middle capacitor is a Johannson microwave piston trimmer suitable for use up to 6GHz with a Q of over 2000. The right-hand types are a plastic film trimmer and a ceramic trimmer.

The large capacitor at the top of **Fig 1.20** is a two-gang air variable for an HF power amplifier pi network loading capacitor. The smaller types are used for similar duties. The bottom right capacitor is a split-stator dual capacitor for a push-pull amplifier. For high RF voltages the spacing between the plates needs to be wide.

Fig 1.20. A selection of air variable capacitors for transmitters

PI (π) NETWORK

The pi network is one of the most common types used. It finds applications in transistor and valve amplifiers, antenna tuning units as well as other applications. The pi network gets its name because the circuit looks like the Greek letter π.

In a pi network we have two capacitors and one inductor. The reactance values of the two capacitors and the inductor define the transformation ratio of the network and the loaded Q. The π network is essentially a low-pass network so it attenuates the harmonics in a power amplifier that uses this network. For the simplest case it is a 3-pole low-pass filter.

The circuit of a π network is shown in **Fig 1.21**. The two impedances to be transformed are shown as RA and RB. The values of capacitors C1 and C2 will be different if the values of RA and RB are different. For the end with the higher impedance the value of capacitor will be lower than the other end. So if RA is, say, 500Ω and RB is 50Ω, C2 will be the larger value. If we redraw the pi network as in **Fig 1.22** we see how the components are connected.

The π network is a parallel resonant circuit formed by C1 and C2 in series and L1 in parallel. Hence, at resonance $X_L \approx X_{C1} + X_{C2}$.

T AND L MATCH NETWORKS

The T match is a popular network for antenna tuning units. It also consists of two capacitors and an inductor, but behaves as a high-pass network so it does not attenuate harmonics generated in the amplifier. It gets its name because it looks like the letter T.

The L match network is a partial π network with one capacitor omitted. Very often this missing capacitor is present in the device to be matched to, so it becomes a true π network.

The L match gets its name because it looks like the letter L. The higher impedance is at the end where the capacitor C1 is connected in shunt to ground. The L match is a popular network to match to a transistor or valve amplifier input or output with minimal losses. Some-

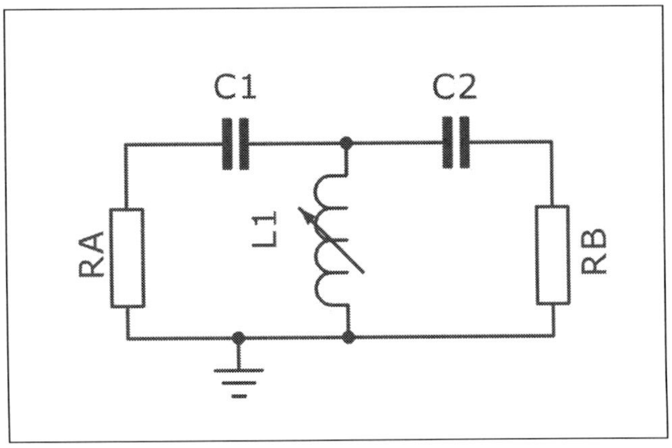

Fig 1.23. T match network

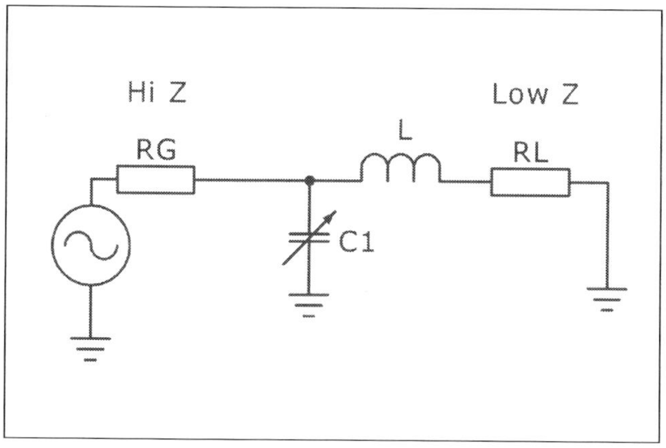

Fig 1.24. L match network

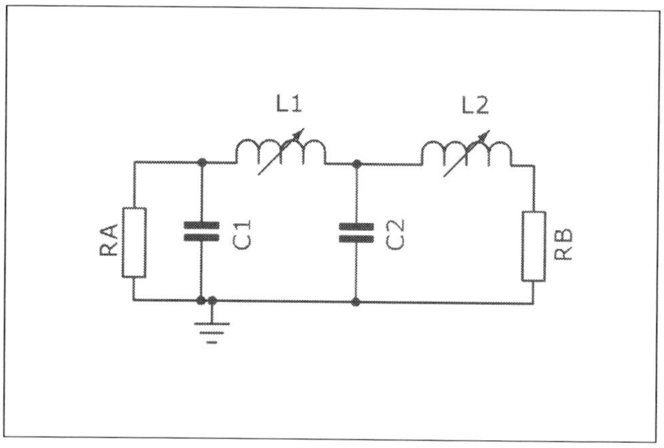

Fig 1.25. Pi-L network

times for DC blocking purposes an extra capacitor is connected in series with one end, normally the higher impedance one. This is normally a low reactance capacitor (high capacitance value) that does not affect the matching. The formulae to calculate the L-match components for a step-down transformation are:

$$X_L = \sqrt{(R_L \times R_G) - R_L^2} \text{ Ohms}$$

$$X_C = \frac{R_L \times R_G}{X_L} \text{ Ohms}$$

where:
R_L is the low impedance load resistor
R_G is the high impedance source resistor.

The values of reactance are converted into component values using the standard formulae at the frequency of interest.

If the L-match needs to be a step-up transformation the circuit is simply reversed, the rule is that the shunt C is always across the higher impedance end. The single section L match can match over an impedance ratio of about 15:1 maximum, beyond this the pi network is a better option.

PI-L NETWORK

The pi network is limited to the ratio of impedance it can match over satisfactorily, normally 100:1 is considered the practical upper limit. Hence, if the anode load is 5000Ω and the antenna impedance is 50Ω this sets the upper limit. When the transformation ratio is too high for the normal pi network, impractical values for the capacitors and inductor may be required. In cases like this we can add an extra L network on the end to complete the transformation. This looks like a double pi network but is really an L match tacked on the end. Capacitor C2 is shared between the pi and L network. The advantage of the Pi-L network for power amplifiers is that it offers more harmonic attenuation than a standard pi network. Up to 50dB of second harmonic attenuation is normal, whereas the pi network will only attain about 35dB.

UNLOADED Q AND LOADED Q

For power amplifier duty we need to consider the two Q values. The unloaded Q (Q_U) will always be higher than the loaded Q (Q_L). The unloaded Q is determined by the component Q. If the value of loaded Q is high, the amplitude of the circulating RF current in the network can reach high levels, which increase the power dissipation in the components. Higher RF voltages will also be present. For power amplifiers using valves loaded Q values of between 8 to 20 are accepted to be the limits for practical amplifiers that need to cover a moderate bandwidth. For VHF and UHF power amplifiers we can often accept higher values of loaded Q. If the components are inherently low loss types a loaded Q of 25 or higher is not unusual for some narrow band cavity amplifiers used at UHF and microwaves.

The formula to calculate Q_L is:

$$Q_L = \frac{X}{R_S + R_L}$$

where
X is the reactance of the inductor or capacitor
R_S is the total loss resistance of the network components
R_L is the load resistance

Tank circuit efficiency as a percentage is given by:

$$\text{Tank Effy} = \frac{R_L}{R_L + R_S} \times 100\%$$

And the formula can be rewritten to be:

$$\text{Tank Effy} = (1 - \frac{Q_L}{Q_U}) \times 100\%$$

Hence, for best efficiency (lowest circuit losses) the value of Q_L needs to be low, but for best harmonic suppression Q_L needs to be high. A compromise obviously must be made and often a value between 10 and 15 is chosen. In an HF amplifier using a pi network covering a relatively wide bandwidth, for example the 160m or 80m band, the value of Q_L will vary across the tuning range because the inductor is often a fixed value. Provided that the value of loaded Q remains within the accepted limits little trouble should be experienced.

A USEFUL TUNING TOOL

A useful tool when assessing if an air wound inductor in a tuned circuit requires more or less inductance is a 'magic wand'. This is a piece of insulating material with a piece of ferrite attached to one end and a piece of brass or aluminium attached to the other. The insulating material can be a paxolin or plastic rod; the ferrite and brass can be glued to the ends with two-part epoxy glue. If the end with the ferrite is placed near the inductor the apparent inductance increases by about 10%. Conversely if the end with the brass or aluminium is brought close to the inductor the apparent inductance falls by a similar amount. For air wound coils we can adjust the inductance by spreading apart the turns, this lowers the inductance by a small amount, or by squeezing them closer together this raises the inductance by a small amount.

The tuning aid is made from an 8mm-diameter paxolin rod about 180mm long. The left-hand end has a brass sleeve pressed on. The right-hand end has a ferrite slug glued into a hole drilled in the rod. This tool was part of the tool kit supplied with the RCA AR88 communications receiver made during WWII for aligning the RF tuned circuits. A selection of different types is useful when adjusting smaller or larger diameter inductors.

Fig 1.26. Commercial 'magic wand' tuning aid

MATCHING TO A GROUNDED GRID AMPLIFIER INPUT

Often when constructing a valve amplifier we are faced with certain mechanical problems and also conflicting electrical ones. A common example is connecting two valves together in parallel in a low inductance manner. If the valve pins are spaced a fair distance apart this makes it tricky. Also the reduction in the number of components is a good move to reduce circuit losses. In a grounded grid amplifier the two cathodes need to be driven with equal amplitude signals and in phase. A thick strap of copper is one way, but this can exert too much mechanical stress on the fragile pins.

Let us look at ways to simplify the construction and eliminate one or two components along the way. **Fig 1.27** shows a 50MHz amplifier using two Russian Gi7bt triodes under construction. The cathode of the Gi7bt at 50MHz looks like a resistive portion of about 300Ω in parallel with the grid-cathode capacitance of 10pF. Hence, we need a step-up transformation from the 50Ω driver. Because we have two valves and these are spaced a fair distance apart a strap is not a good option at 50MHz.

The filament chokes and filament connectors are not fitted at this stage in construction. They will run from the central insulating standoff between the valves. The chokes are ≈20 turns of 0.8mm diameter enamel copper wire wound on a 5mm diameter ferrite rod approx 20mm long and cemented to hold the turns in place. The cathode and filament clamps are made from 1mm thick brass bent around a former. A M3 screw and nut provide clamping pressure to obtain a low resistance contact.

To get around this problem we decided to employ a modified Pi-L match. From the air-variable tuning capacitor situated midway between the two valves two pieces of flexible copper strap or thick wire run from the capacitor stator pins, one to each cathode ring. (These could also be copper braid about 3mm wide and ≈1.5mm thick. Braid from coaxial cable such as URM-43 can be utilised if rolled flat.) The distance between the stator pins and the cathodes is approx 40mm if a straight piece of wire is used. A length of wire of 1mm diameter presents an inductance of about 1nH per mm of length, so the connecting strap presents an inductance of about 30 to 40nH between the variable capacitor and the cathode. Provided the two straps are equal lengths the cathodes will be driven in phase. In **Fig 1.27** the connection is made by a piece of U shaped 1.6mm tinned copper wire formed to connect the two cathode clamps and the air variable stator pin. The equivalent circuit is shown in **Fig 1.28**. The pi section uses VC1, L2/L3 and the cathode capacitance of the valves. L1 is the input L-match.

The 50Ω input has a low reactance DC blocking capacitor C1. This performs no part in the matching and is simply to prevent the cathode voltage being shorted to ground. The value of C1 is not too critical and any value between about 470pF and 10nF will be suitable at 50MHz. L1 and VC1 perform the matching, depending on the valves chosen L1 will need some optimisation to obtain the best input match. These transform the 50Ω input impedance up to some higher value. VC1 needs to have a low minimum capacity and a maximum of 25pF. At resonance the value is approx 12pF with two Gi7bt triodes.

Fig 1.27. 50MHz amplifier under construction

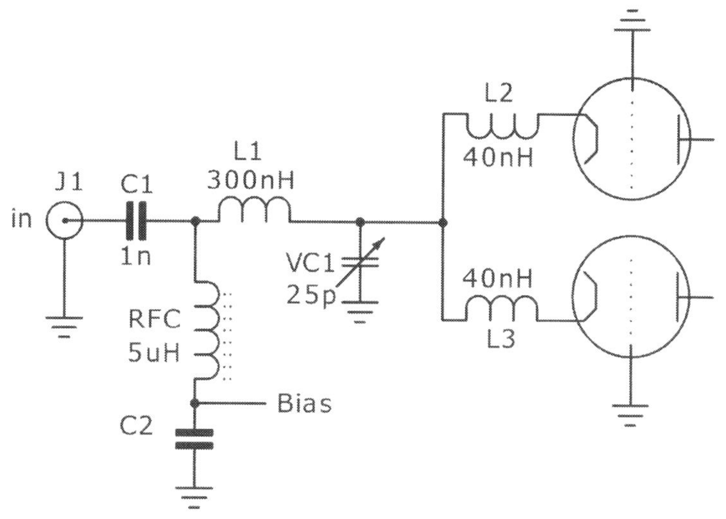

Fig 1.28. Schematic of the input network

The 40nH inductors (L2 and L3) from VC1 to each valve cathode complete the transformation. Because the inductors are floating above DC ground at the cathode voltage required for the correct biasing we can now use one RFC to connect to the bias generator circuitry. This saves one high current inductor. Also, as this is now connected across the 50Ω point and not across each cathode, we can use a lower value of RFC. (The minimum reactance of RFC needs to be 500Ω, which at 50MHz is 1.6μH.) Capacitor C2 decouples the RFC; its value is not critical and 1nF to 10nF will work fine. One side of the filament transformer secondary connects to here and the other side connects via RFCs to each filament pin. The transformer secondary must be floating above ground. The bias circuit connects to the cathode RFC.

SOFTWARE

As a companion to this book is a suite of software that performs various calculations. It is available on the supporting website at http://www.rsgb.org/books/extra/rfdesign.htm The software uses standard formulae but speeds up the design process considerably. The pi tank network calculations are included. All you need to do is the run the software and input the required parameters such as anode load resistance, loaded Q and frequency and the component reactance and values are displayed.

Chapter 2

Radio Frequency measurements

WHEN making measurements on RF circuits we often use several different ways to express the results. If the measurement is power flowing in a transmission line a common measuring instrument is a 'through-line wattmeter'.

IN CIRCUIT RF POWER MEASUREMENT

A through-line wattmeter consists of a directional coupler that can determine the magnitude of the power flowing from the source to the load. By reversing the meter connections we can also measure the power reflected from a mismatched load, the reflected power. Many instruments for precision measurement of power in a section of transmission line are dual directional couplers hence the forward and reflected power can be displayed simultaneously. The name given to the forward and reverse power is the *incident power* and the *reflected power*.

Fig 2.1 shows a precision directional coupler made by Narda. It covers 1 to 12.4GHz and has a coupling factor of -10dB.

> The instrument does not measure power, it measures the voltage existing on the transmission line and by using the standard formula $P = V^2/R$ converts this to a power measurement in watts. Power could also be found by measuring the current flowing, but this is much more difficult to achieve in practice.

The magnitude of the reflected power will depend on the degree of mismatch existing between the transmission line and the load. If the load impedance is the same numeric value as the transmission line impedance the mismatch will be zero, or zero reflected power occurring.

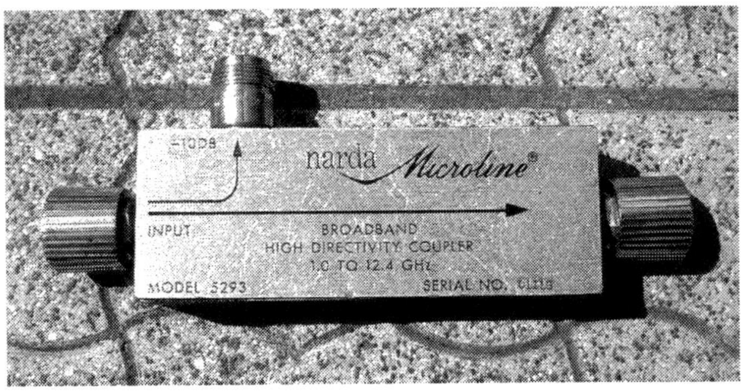

Fig 2.1. Precision directional coupler

Fig 2.2. Dual-directional couplers

In **Fig 2.2**, the top coupler is made by Telletra, Italy, and was used in a 1.7 to 2.3GHz microwave link system as an antenna VSWR and power meter. The lower version is made by Hewlett Packard and was used in a Vector Network Analyser covering 1 to 1000MHz. The coupling factor of the top item is -29dB and it can handle 1kW at 23cm. The two sampling ports have 14dB attenuators to calibrate the power measurement. Power detector assemblies with point-contact microwave diodes convert the RF voltage to DC. The lower item has a coupling factor of -30dB and a directivity of approx 50dB; it is rated at 1W maximum. It uses a printed circuit strip-line coupling method. The APC-7 connectors have adapters to N-type using UT-250 semi-rigid coax. The two silver 'thimbles' contain the 50Ω sample line termination resistors.

HOME-MADE DIRECTIONAL COUPLERS

For most measurements we do not need the supreme accuracy of a laboratory grade coupler. Often these appear on the surplus market and if one can be found for a reasonable price they are a good item to purchase. However, some are fitted with microwave connectors such as the APC-7, which require adapters to more normal types such an N-type. The coupler shown in Fig 2.1 has APC-7 sexless connectors for which adapters are very expensive, often more than the item cost to purchase from a surplus source.

> An important condition is the ability of the directional coupler to distinguish between the incident and reflected power. This is termed the 'directivity' and is normally expressed in decibels (dB). To be able to measure a very small reflected power level in the presence of a large forward power requires a high directivity. If the directivity is 20dB, the smallest reflected power it could accurately determine is 1% of the forward power. This is a VSWR of 1.22:1 and this sets the lowest VSWR it can measure accurately. Laboratory grade directional couplers have directivity figures of 30dB or more and can accurately determine reflected power levels down to 0.1% or lower, a VSWR of 1.066:1 or better.
>
> A second factor is the 'coupling factor'. This is the level of the power picked up by the directional coupler referred to the actual power flowing. For maximum sensitivity the coupling factor needs to be high, that is a low number when expressed in dB. If the coupling factor is -10dB, when 100W is flowing in the main line 10W will be coupled into the sampling line and presented to the measuring instrument for detection. With tight coupling the directivity suffers. For this reason the measurement is often restricted to very low levels of incident power, typically a few tens of mW in practice.

Fig 2.3. Home-made directional coupler

A simple way to imagine what is occurring is to visualise a large pipe with one end closed off. If we fire a tennis ball with a cannon from the open end of the pipe to the far end (without it touching the pipe walls) it will strike the far end and be returned back to the open end. This single 'pulse of energy' is occurring continuously in a transmission line. If we could use a very high-speed camera with lots of tennis balls being continuously fired down the pipe we would see just one tennis ball going down the pipe and being reflected back to the open end; this is very close to the real thing. The tennis balls travelling in opposite direction would tend to collide with one another and this gives rise to the standing waves.

An alternative is to construct a coupler using readily available materials. **Fig 2.3** shows a single coupler built in a die-cast box using a piece of copper tube as the main line. This coupler is designed for a characteristic impedance of 50Ω. The sampling line is a piece of wire connected between the two PTFE feed-throughs on the lid. Placing the sample line nearer or further away from the main line varies the coupling. The sample line termination is two 100Ω carbon-composition resistors in parallel and a detector diode to drive a meter.

The main line is made from a piece of 10mm diameter copper tube with the ends flattened to solder to the connector centre pins. This design works well from 50MHz to 432MHz with minimum insertion loss.

The formula to calculate this is quite complex, but the program to calculate it for different materials and frequencies is available on the website.

Another method is to etch a coupler on printed circuit board material. This consists of two parallel tracks, one being the main line and the other the sampling line. The degree of coupling is defined by the gap between the main line and the sample line. This line needs to be a certain minimum length for adequate coupling.

SPECIAL MEASURING INSTRUMENTATION

The laboratory dual directional coupler allows not only the amplitude of the incident and reflected voltage to be ascertained but also the phase angle of the two voltages to be measured. This measurement is often made using a *Vector Voltmeter* or *Vector Network Analyser*. Knowing the phase angle lets us establish the type of load mismatch. For example, if a section of transmission line of 50Ω impedance is terminated with a pure 50Ω resistor the reflected power will be zero and the phase angle between the incident and reflected voltage at the sending end will be 0°. However, the phase angle between the incident and reflected wave at the far end of the transmission line is always 180°, this is because of the direction of travel.

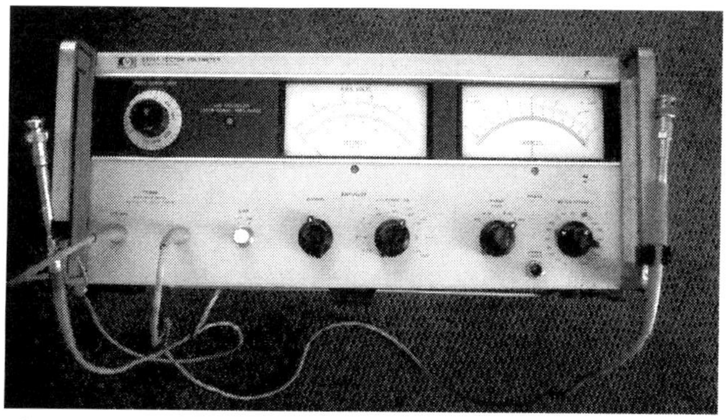

Fig 2.4. Hewlett Packard 8405A vector volt-meter

The Hewlett Packard 8405A vector voltmeter shown in **Fig 2.4** covers 1 to 1000MHz and can measure RF voltages from 1V down to 100μV and phase angles with an accuracy of ±0.25°.

If, however, the far end of the transmission line is terminated in a pure reactance the phase angle of the incident wave will be either -90° or +90°. The case when the line is terminated in a capacitive reactance will give a phase angle of -90° and if the reactance is inductive the phase angle will be +90°. In order to quantify the exact angle and sign of the mismatch present at the far end of the line (the load end), by looking at the sending end with a directional coupler, we need to know the electrical length of the transmission line. We generally explain this in the following way. The transmission line is assumed to have zero loss, where a loss occurs we need to measure this and calibrate the measuring system.

If the electrical length is λ/2 (180°) the angle existing between the incident and reflected voltage at the sending end of the line are the same. If the signal is a voltage with 0° phase angle at the sending end and it is completely reflected by the far end, the reflected voltage we know will reverse in phase by 180° when it reaches the far end. The incident wave from the sending end also experiences a phase shift of 180° by travelling along the λ/2 length of transmission line. Because the two voltages are now in phase and of equal amplitude, the voltage appearing across the open end is now twice the voltage

A line stretcher is a piece of transmission line whose electrical length can be varied over some number of degrees, sometimes as much as 90° a bit like an 'elastic coax'. This often takes the form of a 'trombone' of two air-spaced lines operated within a cavity, one of which can be telescoped inside the other to alter the physical length. A more convenient version used a shorting bar bridging two parallel microstriplines. By moving the shorting bar the electrical length was varied. The early versions of the modern Vector Network Analyser were known as a 'Transmission Reflection Coefficient Test Set' and were adjusted manually at each new measurement frequency to normalise the instrument. Today VNAs are microprocessor controlled and the normalisation is carried out in software with a swept frequency source giving continuous coverage of the range of frequencies under investigation. The early equipment used an external signal generator to supply the input signal and measurement was only possible at fixed frequencies. If many points were required this became a time consuming process to change the generator frequency and normalise the equipment for each new frequency.

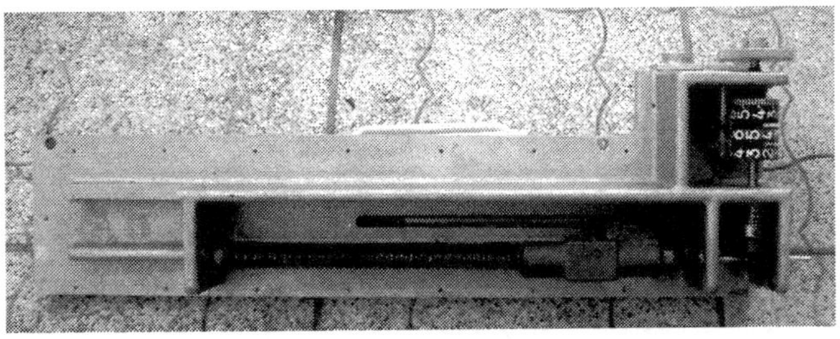

Fig 2.5. Line stretcher

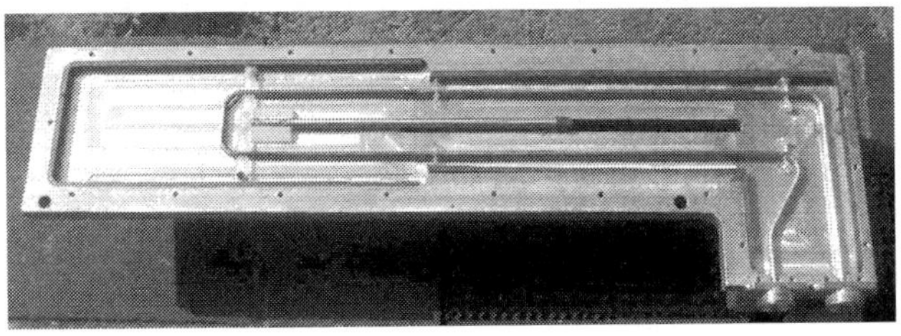

Fig 2.6. Internal details of the line stretcher

appearing at the sending end if the line was perfectly matched.

The electrical length being 180° means that there is also a phase shift of the same value as the reflected wave travels back to the sending end. Upon arriving back at the sending end the reflected wave voltage is now in phase with the sending voltage and the voltage also increases to twice the matched line condition. By using a dual directional coupler to separate out the two voltages we can determine the phase relationship.

Hence, the impedance seen at either end of the λ/2 line is the same magnitude and phase. If we arrange for the electrical length of the line to be always multiples of λ/2 (λ, 3λ/2, 2λ etc) then this simplifies the measurement. However, when we need to sweep across a range of frequencies this causes problems unless a special device known as a 'line stretcher' is employed to allow us to adjust the electrical length to λ/2 at each measuring frequency.

This item was used in an early HP Microwave Transmission Reflection Test Set, the forerunner of the modern VNA. This type uses the trombone section of line between two parallel lines to alter the electrical length. The position of the trombone is altered by the leadscrew and the repositioning is by the turns counter for accurate resetting.

The fixed line runs between the two APC-7 sexless connectors. The movable section is driven by the leadscrew.

> The instrument today most often used for impedance measurements on one-port or two-port networks is the Vector Network Analyser (VNA). This contains two dual directional couplers, one in series with the input and the second in series with the output. In addition it has a swept signal source and accurate amplitude and phase detectors so it is a self contained instrument. Hence all the parameters of a two-port network can be ascertained without having to reconnect the directional couplers during the testing.

QUARTER-WAVE LINE (λ/4)

Now consider the same measurement when the transmission line is λ/4. The line now possesses an inverting condition. If a reactive load is placed at the far end and we measure at the sending end the reactive sign is inverted. Suppose we connect an inductor with a reactance of +100Ω across the transmission line, the λ/4 line now makes the reactance to appear as a capacitive reactance of -100Ω. To obtain a true measurement we need to add a further λ/4 line in series to obtain a λ/2 line.

However, as long as we know the line is a λ/4 line we can simply change the sign of the reactance we measure at the sending end and get the correct result. This effectively is what the VNA does when we normalise it before making a measurement. It knows when the line is an open circuit and a short circuit during the calibration routine, so it can invert the sign of any signal it measures, which gives the incorrect reading during the calibration set up. Suppose we have a connecting cable between the VNA and the equipment we wish to measure. At 100MHz it just so happens it is λ/4 and so at 200MHz it will look like λ/2. If we sweep from 50MHz to 500MHz the cable will behave like multiples of λ/4 at some frequencies and multiples of λ/2 at others. During calibration the VNA finds these anomalies and corrects the data to compensate.

It is also able to work out the varying electrical length and the insertion loss of the cable as the signal is swept across the wanted frequency range. Once it has all this data stored as 'constants' it can then compute the correct reactance and amplitude at the far end of the line at any frequency over which it is calibrated for.

Most VNAs allow the operator to recall calibration data to use in a measurement. If for example last week we measured the impedance of an antenna using a long length of coax and then we find we need to repeat the measurement because we have made some changes to the antenna design. Normally this would be a slow process while someone goes up the tower and disconnects the coax and puts the open, short and load terminations on in the correct order while the VNA operator does the calibration routine. By simply retrieving the calibration data for the previous test we can be up and running in a few seconds using last week's data.

MEASURING IMPEDANCE AT THE END OF A LONG TRANSMISSION LINE

The way this problem is countered is by calibrating the VNA measuring system across a broad range of frequencies. The swept frequency range chosen is normally so that it covers about 25% more at the lower and upper frequencies of interest. For example, if we wanted to ascertain the impedance of a 23cm antenna we would sweep between about 1GHz and 1.6GHz.

Before a measurement can be made the VNA needs to be 'calibrated'. Firstly, the end of the transmission line that will connect between the VNA and the circuit under examination is open circuited and the instrument calibrated for this condition across the range of frequencies. Next the far end has a short placed across it and the instrument calibrated for the short circuit condition. Finally a load resistor is connected across the far end and the instrument calibrated for the matched load condition.

This process is known as *normalising* the test system. Once the instrument has been calibrated in this way it is possible to calculate any impedance at the far end by observing the voltage and phase of the incident and reflected waves at the sending end. When antenna impedance measurements are being made the coaxial cable can be very long in terms of wavelength. As long as the calibration is performed correctly at the end that will connect to the antenna, the instrument operator sitting as much as 100m from the antenna can see the exact condition occurring at the antenna feed point by measuring at the far end of the line. If the length of cable is changed the whole normalising routine needs to be repeated for the new length of cable.

This is for a one-port measurement where we do not measure insertion gain or loss but simply the input impedance measurement, the same as a through-line wattmeter does. When a true two-port measurement is made there are two cables between the VNA and the circuit under examination and both then need to be normalised before a measurement can be made accurately. Two-port measurement is typically used for filters, amplifiers and other networks where the insertion loss or gain and the input and output match need to be examined.

> One simple way to obtain crude VSWR information with just a simple RF voltmeter is firstly to open circuit the far end of the transmission line and measure the RF voltage occurring at the sending end. When the line is then perfectly matched at the far end the voltage at the sending end will fall to exactly half of the open circuit condition. This of course assumes the line has negligible loss. This is the same as the signal generator EMF and PD measurement. When the signal generator output is open circuit the voltage appearing is twice the matched condition.

For the case where the far end of the transmission line is either open circuit or short circuit the mismatch condition will be infinite. In the case of the open circuit line condition the voltage existing at the open end of the line will be twice the voltage when the line is correctly terminated. Similarly, when the far end is shorted the current at the end of the transmission line will be twice the value of a correctly matched line.

MEASUREMENT VALUES

The ratio between the incident and reflected signal amplitudes can be written in several ways. It can be simply expressed as a percentage of the reflected power to the forward power. It can also be expressed as a ratio, this we call the *Voltage Standing Wave Ratio* (VSWR) because the equipment measures the voltage and not the power. For the case where the line is perfectly matched to the load the reflected power will be zero and hence the magnitude of the VSWR is 1. Because this is a ratio it is written as 1:1. If the far end of the line is open or short-circuit the VSWR seen at the sending end will be infinite if the line has zero loss. In practice it is very difficult to obtain an infinite VSWR because even a short piece of cable has some small loss, but very high values are common. Any VSWR greater than ≈30:1 can be considered sufficiently high as to be approaching infinity.

To convert the incident and reflected power levels to *reflection coefficient* format we use the formula:

$$\rho = \sqrt{\frac{P_r}{P_f}}$$

where P_r and P_f are the reflected power and incident power in watts and ρ (Greek letter rho) is the numeric value of the reflection coefficient. The reflection coefficient is normally referred to as 'lin mag' format (linear magnitude).

For example: A through-line power meter measures 100W forward and 25W reflected. The value of ρ is therefore 0.5. To convert between the reflection coefficient value r and VSWR we use the formula:

$$VSWR = \frac{1 + \rho}{1 - \rho}$$

A reflection coefficient of 0.5 is the same as a VSWR of 3:1. To convert reflection coefficient to logarithmic format in dB (log dB) we use the following formula:

$$\text{Measurement in dB} = 10 \times \log_{10} (\rho)^2$$

For a VSWR of 3:1 the measurement in dB is -6.02dB.

Graphs for converting percentage-reflected power to VSWR are contained at the end of the chapter. Some useful reference points to remember are:

1%	=	1.22:1
2%	=	1.35:1
4%	=	1.5:1
11%	=	2:1
25%	=	3:1
33%	=	4.2:1
50%	=	6:1
67%	=	10:1

VSWR AND WHAT IT MEANS

VSWR is simply the ratio between two different impedances. If we are measuring in a 50Ω system impedance and the load impedance is different the VSWR will be greater than 1:1. It is illuminating to make a few measurements with some non-inductive resistors and a low power transmitter to appreciate the limitations of the common VSWR meter. If we have a number of 2W carbon composition resistors we can learn a lot about how the VSWR meter measures different impedances.

If we connect two 100Ω resistors in parallel the effective resistance value is 50Ω. Provided the lead lengths are short and we use a relatively low frequency the VSWR meter should indicate a perfect VSWR of 1:1. What we have effectively done is to make a low power dummy load.

If we now replace this with one 100Ω resistor, the VSWR meter should indicate 2:1. Now parallel four of the 100Ω resistors to make a 25Ω load and connect it across the output of the VSWR meter. The VSWR indicated should also be 2:1. By connecting resistors in series or parallel (with very short leads to keep the inductance low) we can make terminations of 200Ω, 300Ω, 400Ω etc and 25Ω, 12.5Ω, 6.25Ω etc. For each termination we should get different VSWR readings.

VSWR is simply the ratio between the system impedance and the unknown impedance. So if the termination is 200Ω the VSWR indicated is 200/50 = 4:1. Similarly for 300Ω it is 6:1 and for 400Ω it is 8:1 etc. For lower values of termination the same argument holds. With 25Ω the VSWR is 50/25 = 2:1, for 12.5Ω it is 4:1 and for 6.25Ω it is 8:1. So it should be clear that the average VSWR meter doesn't tell us anything of much use, except the value of VSWR. For an indicated VSWR of 3:1 this could be due to a 150Ω load or one of 16.6Ω, we simply have no way of telling which one it is.

> I can almost guarantee that if you perform this type of measurement on your VSWR meter you will find errors between the indicated and actual VSWR. Using just three resistor combinations can confirm if the meter indicates correctly. With a 50Ω, 100Ω and 25Ω load you can check the 1:1 and 2:1 VSWR points. It is very unlikely that the 2:1 VSWR indication will be correct for both loads when this is done.

ALTERNATIVE NOMENCLATURE

Another way to express reflection coefficient is to refer to it as the *return loss* in dB. This is what the 'log dB' figure is in VSWR parameter data. If an attenuator is used in a system the input reflection coefficient when the attenuator is perfectly matched is 0, a VSWR of 1:1. If the far end of the attenuator is open circuit or short circuit the *return loss* is a value of twice the attenuator value. So when we see, for example, a log dB value of, say, -2.2dB it is behaving the same as an attenuator of 1.1dB insertion loss placed between the sending signal and the input of the device, which is open or short circuit.

If a 3dB attenuator is used as a means of reducing the VSWR on the input of an amplifier the worse case (totally mismatched) return loss is -6dB. This is a VSWR of 3:1. In input reflection coefficient terms expressed as linear magnitude this is a value of 0.5, because the measurement is the amplitude of the voltage. A -6dB return loss means that the reflected power arriving back at the input is 25% of the sending power, because the incident power travels from the input through the attenuator suffering 3dB of loss. It is totally reflected back from the output through the attenuator to the input, suffering another 3dB loss. Hence, the return loss is twice the attenuator value. Power is V^2/R, hence to obtain 25% of the power means the voltage is 50% or 0.5, which is the input reflection coefficient value in magnitude terms for a 3:1 VSWR.

A 6dB attenuator has a return loss of -12dB; this is a VSWR of 1.67:1 (a reflection coefficient of 0.2512).

If a 30dB attenuator is used and the far end left open circuit the return loss is -60dB and this is a VSWR of 1.002:1 (a reflection coefficient of 0.001). Any value of return loss of greater than ≈-16dB (a reflection coefficient of less than 0.1585) can be considered a reasonably good VSWR match.

A long piece of lossy coaxial cable, such as RG-58/U, will behave like a dummy load, even if the far end is left open-circuit or is short-circuited, because the return loss is twice the cable attenuation. (RG-58/U at 432MHz has a loss of approx 30dB per 100m). All the power entering the coaxial cable will be dissipated as heat.

> **WATER-POWERED POWER METER!**
> This technique has been used at 23cm and higher frequencies to make an accurate power meter. Peter Blair, G3LTF, many years ago demonstrated his method of measuring power with a piece of RG-58/U or URM-43 coax and a styrene cooler box. The end of the coax had a low wattage 47Ω carbon resistor to terminate the coaxial cable. The cable was coiled up and placed in a styrene cooler box and the box filled with water. The temperature of the water was accurately measured and then power applied for a few minutes. The temperature rise of the water was measured. Knowing the volume of water, the temperature rise and the time the transmitter was operational allows the calculation of the average power with an accuracy of better than 10%.

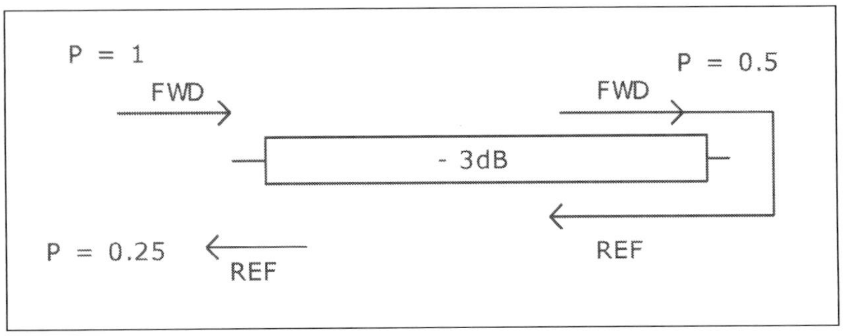

Fig 2.7. Attenuator return loss

POWER MEASUREMENT PARAMETERS

The basic unit of power is the watt, named after the Scottish scientist James Watt, who defined the horsepower in measurements (1HP = 746W). Larger powers use multiples of the watt, 1000 watts is conventionally written as 1kW. Smaller levels of power are often given in milliwatts (mW), this being one thousandth of a watt. For measuring instruments we often use a different notation. When expressing power as a ratio we can use the dB notation. This is power expressed logarithmically referred to some standard, in this case the decibel. The decibel is one tenth of a Bel, a measurement system devised by the Bell Telephone Corporation. Today the Bel is no longer used and the decibel is universally used when logarithmic power ratios are expressed.

For example, 1W can be written as 0dBW, meaning it is referred to 1W as the basis of the measurement. So a power of 400W can be expressed as +26dBW, meaning it is 26dB greater than 1W. Normally the + sign is not used except where confusion may arise.

For power levels smaller than 1W we can use the milliwatt (mW) as a reference. In this case we refer everything to a standard power of 1mW. 1mW is then written as 0dBm, where the m indicates 1mW. So a power of 2mW is 3dB greater than 1mW and the power would be written as +3dBm. It is important to use the + or - sign when dealing with mW measurements. A power level of one tenth of a mW would be written as -10dBm, the minus sign indicating it is below the reference level of 1mW. Lower power levels can also use the microwatt (μW) which indicates the power is one millionth lower than 1W. Hence 1μW can be expressed as 0.000001W, or -60dBW, or -30dBm or

0dBμW, although the last version is hardly ever used today. Even smaller powers can use the picowatt reference, 1pW is equal to one thousandth of a microwatt, or -90dBm.

Today most RF design engineers only use the "-dBm" notation when expressing small power levels and it is easy to juggle the numbers in the head to get the correct meaning. For example, a receiver sensitivity might be expressed as so many microwatts in 50Ω or as the voltage required in microvolts (μV) or as "-100dBm". Unless otherwise mentioned the standard system measurement impedance is accepted to be 50Ω, although for absolute correctness a measurement should clearly identify the system impedance where there can be doubt.

Some useful reference points to remember are:

1mW	=	0dBm
2mW	=	+3dBm
4mW	=	+6dBm
5mW	=	+7dBm
10mW	=	+10dBm
20mW	=	+13dBm
40mW	=	+16dBm
50mW	=	+17dBm
100mW	=	+20dBm
1W	=	+30dBm
10W	=	+40dBm
100W	=	+50dBm etc.

So 20W can be expressed as +43dBm and 1kW as +60dBm. To convert between dBm and dBW we subtract 30 from the dBm number and change the m to W. Hence 400W is +56dBm or +26dBW. Similarly to convert from dBW to dBm we add 30 to the dBW value.

To convert between power as a *ratio* to dB we use the following formula:

$$dB = 10 \times \log_{10}(\text{power ratio})$$

For a power *gain* of 100 the value is +20dB. Hence, if an antenna has a forward gain of 13dB, this as a power ratio is 20 times. Gain is denoted by the + sign and loss by the - sign. Where we state *gain* we do not normally need to include the + sign in the result, this is implied by the statement.

ATTENUATOR POWER RATING

Very often when measuring spectral purity of a transmitter we need to reduce the transmitter power to some very low value that the measuring instrument can handle without endangering the sensitive input networks. A typical spectrum analyser might have a safe maximum input level of 10mW (+10dBm). Assume we have a nominal 100W transmitter and the spectrum analyser for optimum measurement of harmonic levels cannot tolerate more than 1mW (0dBm). If the spectrum analyser input is over driven the mixer will generate harmonics of the input signal and this can cause a false reading to be made. So we need to reduce the ≈100W down to 1mW or less. 100W is +50dBm so we need a minimum of 50dB of attenuation between the transmitter and the spectrum analyser.

If we used a 50dB attenuator this would need to be rated at ≈150W to handle the anticipated power safely. Hence, the power dissipated in the attenuator would be 100W less the 1mW sent to the analyser, a value of 99.999W, which in round numbers is 100W. In practice it is preferable to split the attenuation into several smaller pieces to allow a better power dissipation. We could choose to use five 10dB attenuators connected in series. The first attenuator needs to be able to safely dissipate 90W, because the power leaving the attenuator is 10% of the input power, +40dBm or 10W. This still requires about a 150W attenuator so is not very good. So we choose to use a lower value attenuator to reduce the power dissipation. Suppose we choose a 6dB attenuator, the power leaving the attenuator is now 25% of the input power, so 75W is dissipated in the attenuator, not much better than the other two options. If we choose a 3dB attenuator the power leaving it will be 50W and hence 50W are dissipated in the attenuator. This means we can use a 100W rated attenuator, which gives us a 100% safety margin should something go wrong. The nominal 100W could vary by as much as ±1dB, which is 80W to 120W in round numbers.

As we still have a power of 50W to dissipate our second attenuator could be a 100W type with 47dB attenuation. 47dB is not a standard item but we can use lower power attenuators connected in series to get to where we need to be. Let us select a 6dB attenuator for the second stage. With 50W entering it will pass 12.5W onto the next attenuator. This means this attenuator needs to dissipate 37.5W and a 50W type would be suitable. The next attenuator is presented with 12.5W, if this is a 10dB attenuator the power leaving it will be 1.25W and 11.25W will be dissipated in it. A 15W or 20W attenuator will be suitable. We still have a further 31dB of attenuation to connect, so a 30dB attenuator rated at 2W would be suitable. The final 1dB attenuator has 1.25mW presented to it so a 100mW attenuator would be suitable.

A better arrangement would be to place the 1dB first in line and it would then dissipate only ≈20W when 100W is applied, so a 25W rating would be suitable. Hence our attenuator line up would consist of the following attenuators: 1dB, 3dB, 6dB, 10dB and 30dB.

A low cost replacement for laboratory grade attenuators it to use lossy coax as the first element. If a long enough piece of RG-58/U or URM-43 coax is used the attenuation can be quite high and it can safely dissipate the power because of the large surface area.

VSWR MEASUREMENT OF ANTENNAS

When we use a through-line directional power meter to measure the VSWR of an antenna it is important to place the VSWR measuring instrument in the correct place. If the feeder cable from the transmitter to the antenna has loss the VSWR measured will be incorrect if the VSWR meter is connected between the transmitter and the coaxial cable input.

Assume we have a 100W transmitter and the coaxial cable feeding the antenna has a loss of 3dB. If the VSWR meter is placed at the transmitter end of the cable the result will be a VSWR measurement lower than actually occurring at the antenna feedpoint. In this case the power arriving at the antenna will be half of the 100W generated by the transmitter, because of the 3dB coax loss. Assume the antenna impedance is very mismatched to the transmission line; let us suppose it is a total open or short circuit due to some fault. Since only 50W is available at the antenna end and all of this will be reflected back to the transmitter it will also suffer the same 3dB attenuation. Hence, when it arrives back at the transmitter the VSWR meter will measure a reflected power of 25W. This is a reflected power ratio of 25% of the forward power, or a VSWR of 3:1. But the VSWR at the antenna is infinite. By placing the VSWR meter at the antenna instead of the transmitter end of the coaxial cable we obtain a true reading. The only case where it is permissible to place the VSWR meter at the transmitter end of the coax cable is where the cable has virtually zero loss. 1dB of coaxial cable loss will measure a VSWR of ≈9:1 when the antenna is in reality an infinite VSWR.

> There are directional power meters available with a remote-measuring head (the part containing the directional couplers) and these can be used with a meter in the shack. They just need a 3-core cable between the measuring head and the meter to give accurate VSWR readings. The measuring head is mounted close to the antenna and the 3-core lead is run down to the meter display unit.

DUMMY LOAD POWER METERS

An accurate method of measuring power is to use a type of power meter known as a 'dummy load power meter'. These have a terminating load resistor and a detector head built in. They range in power from about 25W up to many kilowatts. The higher power types are normally filled with oil for better dissipation of the heat. The very high power types, as used in broadcast transmitter testing, are normally water-cooled to dissipate the power. These contain a relatively small load resistor through which cooling water is pumped to extract the heat. The load resistor has a flow meter fitted and thermometers to measure the input water temperature and the output temperature. Knowing these allows the user to look up on a graph the actual RF power reading. This method is normally within 5%.

An alternative method for modest power levels is to utilise a type of power meter with a *bolometer* head. These normally are limited to quite low power, normally not more than 100mW. The bolometer head consists of a load resistor to suit the system impedance, usually 50Ω, that operates as a dummy load and a temperature measuring element thermally coupled to the resistor. By measuring the temperature change the power can be measured very accurately over a wide frequency range. Most of these bolometer power meters are designed for microwave application but will work at lower frequencies. The most common type is the Hewlett Packard HP-432A. This covers from 10MHz to 10GHz with an accuracy of approx ±2% worst case, and typically ±1% across the majority of the frequency range. The maximum power rating is only 10mW (+10dBm) so we need to connect attenuators before it to dissipate the bulk of the power. (The measuring heads are quite fragile and above ≈1W input are known to burn out!)

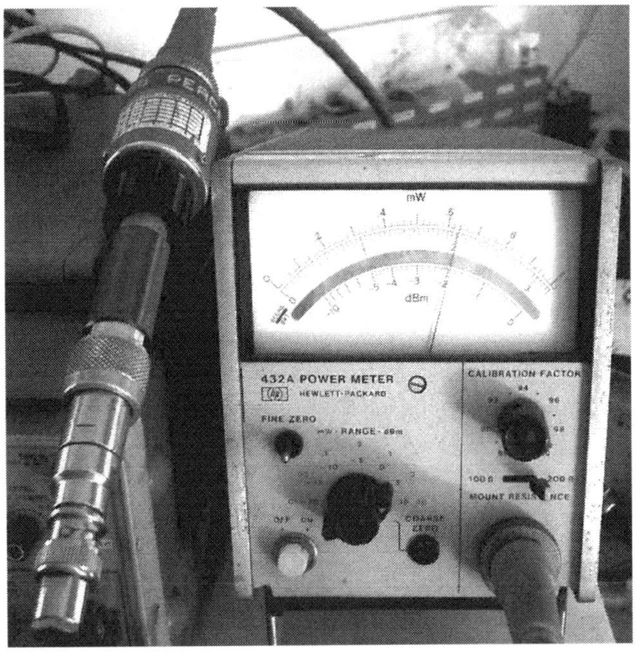

Fig 2.8. HP-432A power meter with attenuator being used to measure a low power transmitter

The measuring head of the HP-432A in **Fig 2.8** has a 20dB attenuator connected in series to the output of an oscillator multiplier chain for a 2m transmitter. **Fig 2.9** shows a close up of the meter scale.

The HP-432A power meter has several switched ranges from +10dBm down to -20dBm in 5dB steps. The range switch is on the +5dBm scale and the attenuator being used is a 20dB. The needle is reading 2dB below full scale. Because full scale on this range is +5dBm the actual power is (5 - 2) = +3dBm or 2mW as read on the top scale. The true power into the attenuator is therefore (20 + 3) = +23dBm or 200mW. The actual power is slightly higher because we have used a fairly long piece of RG-58/U coax to connect to the meter. By moving the measuring head to the output of the multiplier chain we would expect to see a little more power by eliminating the loss in the coax. It is important to know the attenuator value with some precision to obtain the best accuracy. Normally laboratory grade attenuators and power meters are calibrated on a regular basis to ensure the accuracy does not suffer.

Another common technique is to use the HP-432A connected to the sample port of a directional coupler to measure power in circuit. Again the coupling factor and any attenuators need to be taken into account to obtain an exact result.

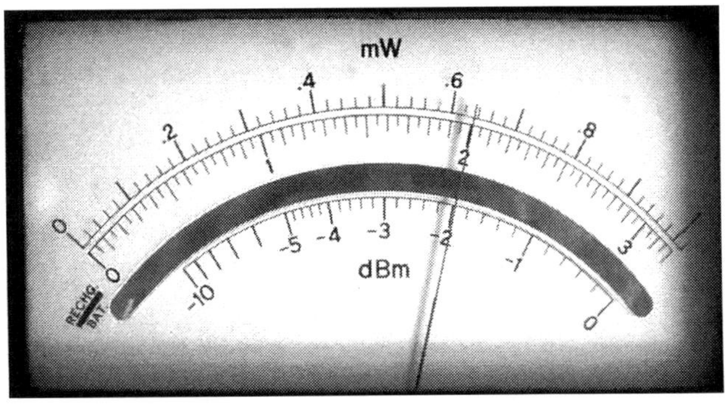

Fig 2.9. Close-up view of the meter scale

MEASURING SPECTRAL PURITY

The most common method to ascertain how 'clean' a transmitter is uses an item of test equipment called a spectrum analyser. This is basically a superhet receiver with a swept local oscillator and a fixed IF. The signal to be examined is converted via a mixer to the IF and the IF amplifier consists of an accurate power measuring detector. This is normally performed by a 'logarithmic amplifier' that provides a voltage or current proportional to the logarithm of the input power. Hence, the display can be scaled directly in dB.

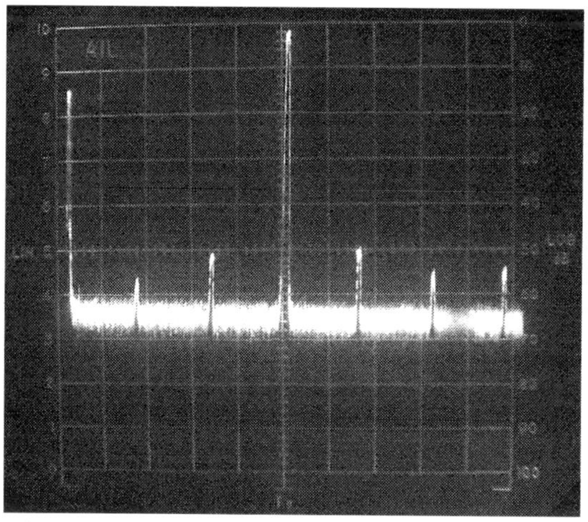

Fig 2.10. Spectrum analyser display for transmitter spurious products

Spectrum analysers can cover very wide frequency ranges in one sweep, for example from 10kHz to as much as 2GHz. For the higher frequency bands the sweep is often split into ranges, for example 2 to 4GHz, 4 to 8GHz etc for convenience.

By applying an input signal the spectrum analyser displays the carrier and any spurious products such as harmonics, IMD etc. **Fig 2.10** shows a typical display screen for the spurious products of a low power 2m exciter.

The vertical scale is graduated in 10dB steps and the carrier is just below the top line, about 2dB down. The products adjacent to the carrier are formed by the multiplier stages. The worst products are approx -48dBc (dB below the carrier); the further away ones are about -55dBc. The frequency span is 0 to 300MHz with the 144MHz carrier centered. The analyser noise floor is about -60dBc and this limits the extent to which we can see any spurious products. The 60dB measurement range is known as the *on-screen dynamic range*.

EXTENDING THE RANGE OF MEASURING EQUIPMENT

A common problem with some spectrum analysers is the limited dynamic range when trying to assess the levels of harmonics in the presence of a strong carrier. The overall limitation is the dynamic range of the mixer and if this is over-driven it will generate harmonics not present in the transmitter output. Most simple spectrum analysers only have about 60dB dynamic range. If we need to measure to greater values we have a problem.

One simple way to overcome this limitation is to reduce the amplitude of the carrier fed to the analyser. If we use additional attenuation this would lower all the signals by the same amount, so we are no better off than before. But if we arrange a notch filter tuned to the carrier frequency that reduces it by, say, 20dB and leaves all the other signals at the previous level we get an extension of the range of the same order. When this technique is used we need to calibrate the system to allow for any insertion loss at the frequencies other than the carrier. Usually a signal generator is used at the second and third harmonic frequency and the measured amplitude compared to the signal generator output. In most cases the insertion loss of the notch filter at the second and third harmonic frequency is quite small, perhaps only a dB or so. In such a case the absolute level of the harmonic signals are measured and then converted back to the 'below carrier' level in dB, the dBc figure.

Often when this technique is used we find we can then remove some of the attenuation before the analyser to gain extra sensitivity, allowing us to see further down than the 20dB we achieved. Using this method with a deep notch (30dB or more is not difficult to achieve) can allow examination to as much as -100dBc. The danger with this measurement technique is the notch filter being accidentally de-tuned when high power is used. This could damage the spectrum analyser mixer. The safe way to adjust the notch filter is using a signal generator at the carrier frequency to null the carrier before the transmitter is attached. A good substitute for a tunable notch filter is a short-circuited $\lambda/2$ coaxial stub connected in parallel with the spectrum analyser input and cut to the carrier frequency.

Another commonly-used method is to place a steep cut-off high-pass filter between the output of the final attenuator and the spectrum analyser. This filter needs to roll up from maximum attenuation to minimum

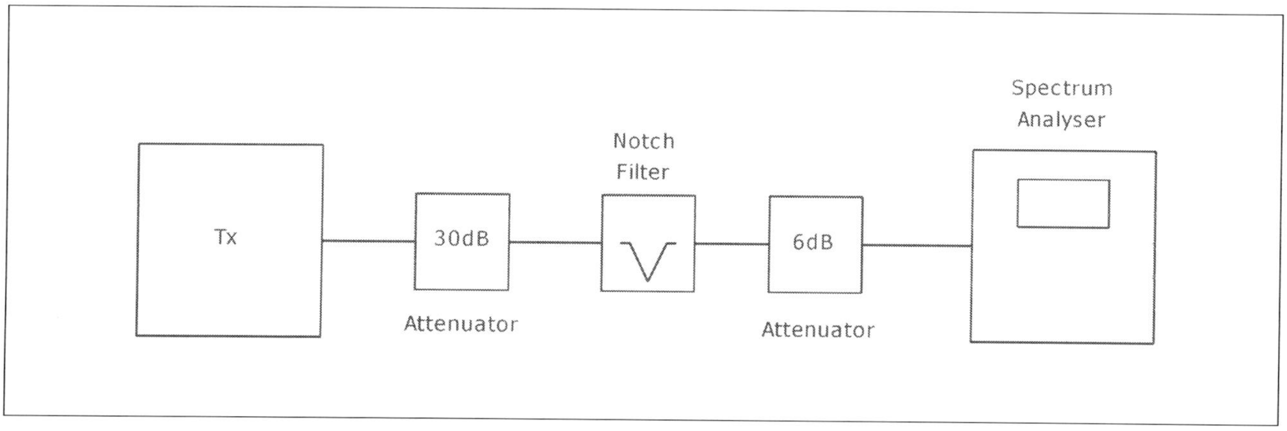

Fig 2.11. Harmonic measurement with notch filter

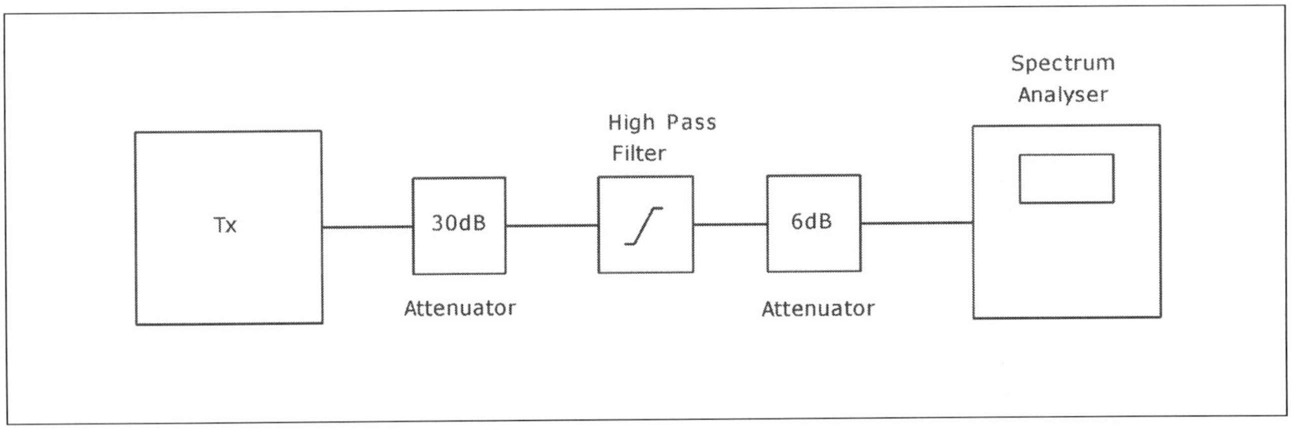

Fig 2.12. Harmonic measurement with HPF

attenuation at some frequency lower than the second harmonic. Again the system needs to be calibrated with a signal generator to establish the calibration factors. If the parameters of the high-pass filter are known, by sweeping the response with a VNA for example, this step can be omitted, but it is recommended to confirm the original calibration. Of the two methods the latter is the one least likely to cause damage to the spectrum analyser because the high pass filter is normally a fixed tuned piece of equipment.

If the response of the high-pass filter is not accurately known then by injecting a signal of, say, 0dBm into the high-pass filter at the carrier frequency and at the second and other harmonic frequencies the calibration can be accurately determined to within ±1dB uncertainty. Using this method I was able to measure the harmonic rejection of a UHF transmitter to better than -90dBc although the spectrum analyser only had a dynamic range of 50dB. This used a nine-pole HPF with a cut-off of ≈600MHz and rejection of the carrier frequency of 45dB.

MEASURING BROADBAND NOISE FLOOR OF TRANSMITTERS

Many amateur transmitters have a poor broadband noise performance: you can often hear the 'hash' on a receiver when they transmit by listening about 250kHz from the transmit frequency. Unfortunately this is not an accurate way of measuring the noise output. (It could also be caused by *reciprocal mixing* in our receiver because of the noise performance of the local oscillator.) In order to ascertain the noise performance we need to hook up some additional equipment to a spectrum analyser to allow us to see this noise level in the presence of the strong carrier. Even the very best spectrum analyser cannot measure the noise floor without some help.

This measurement is similar to the harmonic level measurement above but we need to dig much further down below the harmonic levels to be able to measure the noise. Many of the modern amateur transceivers are very poor compared with equipment manufactured 20 or more years ago. This is because today most amateur transceivers use frequency synthesisers to generate the carrier. These simply do not have the same low broadband noise performance of the older crystal-controlled equipment.

> The Kenwood TS-700G 2m transceiver of 1970s vintage is still regarded by many as the benchmark against which modern transceivers should be measured. This used a mixer VFO scheme to generate the receiver and transmitter LO with low noise crystal oscillators. As well as this it used a narrow-band tracking-filter in the transmitter low-power amplifier stages to suppress broadband noise by a significant amount. The designers of this piece of equipment really knew what they were doing.

Whilst most modern amateur transceivers are probably acceptable when run on their own, when a high-power linear amplifier is added to increase the output power this always raises the level of the broadband noise level close to the carrier. Solid-state linear amplifiers are the worst offenders and valve linear amplifiers are quite a bit better. If a 10W transceiver is driving a linear amplifier to attain 400W output this is a gain of 40 in ratio terms and 16dB in power gain terms. Hence, even if the linear does not add any additional noise, which is very unlikely because a high power amplifier has a very poor noise figure, the broadband noise level will be increased by the same amount. This may now make the signal audible on a quiet band at a distance of 10km or more, especially if high gain antennas

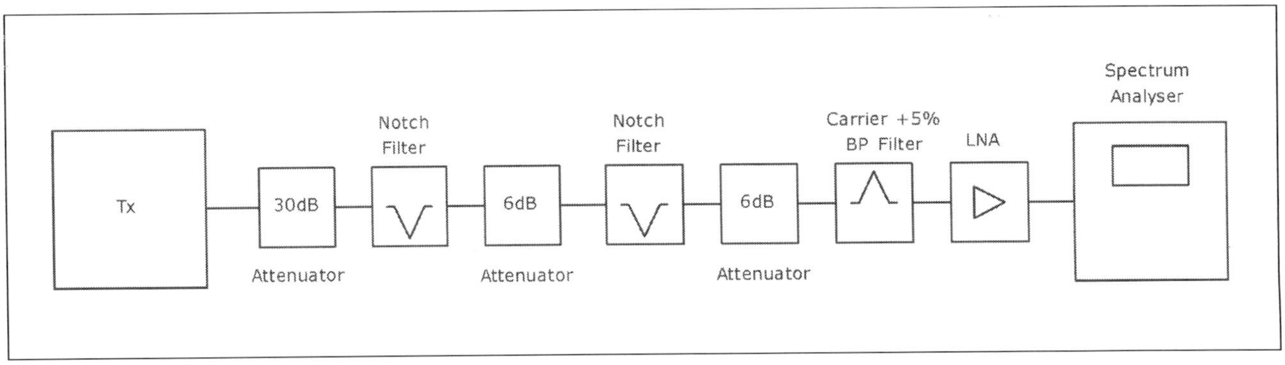

Fig 2.13. Broad-band noise measurement

and masthead LNAs are also used. The basic system to measure broadband noise is shown below.

Because we need to null the carrier as far as possible we need two notch filters connected in series. To ensure the filters are correctly terminated means we need to insert attenuators on the input and output ports to give a good VSWR termination. With good notch filters 90dB of carrier rejection is attainable. Following the second filter is a narrow band-pass filter to increase the suppression of the carrier and eliminate any harmonic signals, which the notch filters do not attenuate to any great extent. This band-pass filter needs to be centred on a portion of spectrum about 5% higher or lower than the carrier frequency. This is the portion of spectrum where we will be looking for the broadband noise energy. Because the noise level is likely to be very low and the spectrum analyser is not a very sensitive receiver (it typically has a noise figure of 10dB or more) this LNA needs to have a fair gain and low noise figure. In most cases a noise figure of ≈2dB and a gain of 20dB or more is normally required.

Before the transmitter is connected we use a signal generator to adjust the notch filters to maximum attenuation at the carrier frequency, with the spectrum analyser centre frequency at the carrier frequency. Next we change the spectrum analyser frequency to the carrier +5% frequency and inject a signal at this frequency into the 30dB attenuator. This gives an absolute power level to measure the noise level against. The spectrum analyser IF bandwidth and sweep range is adjusted to 1kHz BW and 5kHz sweep or similar. If the analyser has a very narrow band IF filter, such as 100Hz or lower, this increases the system sensitivity considerably; in this case we only need to sweep over about 1kHz. With a signal into the 30dB attenuator of -120dBm the analyser output should be easily observed. Sometimes the value of the 30dB attenuator needs to be reduced, in some cases as low as 10dB to get sufficient sensitivity. In this case it is imperative to check the adjustment of the notch filters for absolute maximum rejection of the carrier to prevent damage to the LNA or spectrum analyser.

Having set everything up, we can now connect the transmitter and transmit. You should observe an increase in the noise floor of several dB if the sensitivity is sufficient. Record the highest deflection of the trace and switch off. Without changing any settings reconnect the signal generator and adjust its output to achieve the same deflection of the trace. The level required is the absolute value of the noise power generated by the transmitter at that frequency.

Now we only need to establish the final piece of the puzzle to calculate the noise floor. This is the transmitter power output. This can be found be inserting a through-line wattmeter between the transmitter and the 30dB attenuator.

Suppose the signal generator level we measured was -100dBm to achieve the same deflection as the transmitter noise. Assume this is a 10W transmitter, +40dBm. Then the noise floor is 100 + 40 = -140dBc at a 5% offset. This, however, is still a noise signal of -100dBm being radiated into the antenna and will be heard over a considerable distance by stations with sensitive receivers. If we now attach the 16dB gain linear amplifier the

> In professional RF design laboratories the band-pass filter is normally a narrow band crystal filter with very steep skirts to eliminate adjacent channel spurious signals. Using a narrow band crystal filter allows us to get really close to the carrier - often within a few kHz - without overloading the spectrum analyser, and this allows us to dig about 40dB further down into the noise. The way this method works is to mix the carrier frequency down to some convenient IF, such as 10.7MHz, filter the unwanted signals out and then look at the 10.7MHz signal on the spectrum analyser. This, however, needs a very low noise signal generator to drive the mixer LO port to see far enough into the noise floor. Calibrating such a system is not a five-minute task!

level of noise power being radiated increases to at least -84dBm, even with no noise contribution by the amplifier, which is very high and will be heard over a wide area even by stations with average receivers.

Broadband noise falls off in a more or less linear manner as we move away from the carrier frequency. We were only able to measure at 5% off the channel because of the primitive set up we used, but we can see that as we get closer to the carrier the noise level will rise. At an offset of 100kHz the noise is likely to 10dB or higher than at 5% away.

As an indication of what can be achieved, I worked for a military contractor that manufactured VHF transceivers for the South African Defence Force. These were rated at 50W carrier level and used very low noise frequency synthesisers and narrow band tracking filters in the transmit chain to suppress the broadband noise spectrum. The power amplifier used low noise MosFets. The receiver front end used very complex high dynamic range amplifiers, tracking filters and mixers and it had a sensitivity of -110dBm, ≈0.5µV. This radio was a frequency hopping FM transceiver. The design specification required that two of these transceivers could be fitted into a vehicle with the antennas situated two metres apart. When one was receiving on one channel the second transceiver at 10% away in frequency would transmit at full power with no discernible increase in noise floor on the receiving radio when receiving a signal at receiver sensitivity level. Of course these cost quite a bit more than an amateur transceiver! The RF power coming down the receiver coax when the other radio was transmitting was about 2W because of the small antenna separation. The theoretical minimum noise floor is -174dBm/Hz; we attained almost -160dBm/Hz.

Measuring the transmitter broadband noise floor was very difficult at 50W and we had to construct special equipment to attain the measurement limits. Even so we had about another 6dB in hand to dig even lower if we needed to. A later development was an aircraft transceiver with even tighter specifications and the test set up was pressed into service with minor modifications.

It is not only broadband noise that limits the sensitivity but also any spurious coherent products from the mixing and synthesisers; these need to be comparable to the noise floor to attain the required performance. The receiver front end used 10W RF power transistors operating in a push-pull amplifier to attain the required dynamic range when a 2W interfering carrier only 10% off frequency is present.

DUMMY LOADS

Every amateur station needs a dummy load and in some countries it is a licence requirement to have one in the station. They can be purchased from amateur equipment suppliers or ancillary equipment manufacturers or bought on the surplus market for a reasonable cost. They are also simple to make at home using some low inductance resistors connected in parallel to obtain the required impedance. A typical surplus dummy load is shown in **Fig 2.14**. This is a 500W rated dummy load for continuous carrier power or 5kW-peak pulse for radar applications or SSB.

This is not a resistive dummy load in the normal sense. The main line connects from the centre pin of the N-type connector to the far end, where it is grounded. The line is a rectangular copper bar silver-plated. On either side of the line is a piece of microwave absorbing ferrite, which is thermally bonded to the heatsink. Although it works well between 2GHz and 12.4GHz at 432MHz it appears as a high VSWR mismatch. If an ohmmeter is connected across the connector it reads 0Ω.

In the old days some amateurs used a light bulb as a dummy load, often contained in a biscuit tin to reduce radiation. These are far from being accurate because the filament temperature varies over a wide range with applied power and hence the resistance rises with increased filament temperature. They are also quite inductive and only suitable for lower frequencies, perhaps up to 10MHz at best.

Fig 2.14. Narda Microline 50Ω dummy load for 2 to 12.4GHz

Fig 2.15. Internal details of a microwave dummy load. The ferrite blocks and the central main line are shown

Fig 2.16. HP 3406A broadband sampling voltmeter

MEASURING RF VOLTAGE

If a suitable cathode ray oscilloscope (CRO) with sufficient bandwidth is available, we can use this to measure RF voltage. Most low cost CROs have limited bandwidth and so are not suitable for very high frequencies. In professional laboratories an instrument known as a *broadband sampling voltmeter* (BBSV) is often used. These have a wide bandwidth, typically several GHz for the best types. A typical example is shown in **Fig 2.16**. This is a Hewlett Packard BBSV that can often be found on the surplus test equipment market for a reasonable cost.

This item covers from 1mV rms fsd to 3V rms fsd up to 1GHz. The meter scale is graduated in volts and dB for easier measurements.

SIMPLE TEST EQUIPMENT TO CONSTRUCT AT HOME

Every amateur will need certain items of test equipment when making measurement. Many of these can be constructed with very little effort. Some of the items useful to have are dummy loads or termination loads. For a simple dummy load we can parallel some low inductance resistors to obtain the required resistance value and power rating. A typical homemade dummy load is shown in **Fig 2.17**. This uses twenty 1kΩ resistors of 2W rating. Hence, the dummy load can dissipate 40W approx. By immersing this dummy load in water the power rating doubles.

For low power terminations, a number of low wattage resistors connected across an RF

Fig 2.17. Home made dummy load using 2W resistors

Fig 2.18. SMA 50Ω termination

connector will suffice. **Fig 2.18** shows a 50Ω termination using two 100Ω 0.25W resistors soldered on to a SMA connector. This performs well up to 2GHz. For the best performance the resistors can be chip (SMD) types. 1206 chip resistors behave as resistors up to at least 3GHz.

Another item it is handy to have is a selection of attenuators. The low power varieties can be bought quite cheaply or can be constructed using normal resistors if we can tolerate slight inaccuracies. Some varieties are shown in **Fig 2.19**.

Fig 2.19. Top: two home made attenuators; Middle: JFW Industries 20dB attenuator rated at 250W when bolted to a heatsink; Bottom: Mini-Circuits Labs 10dB and 20dB N-type attenuators

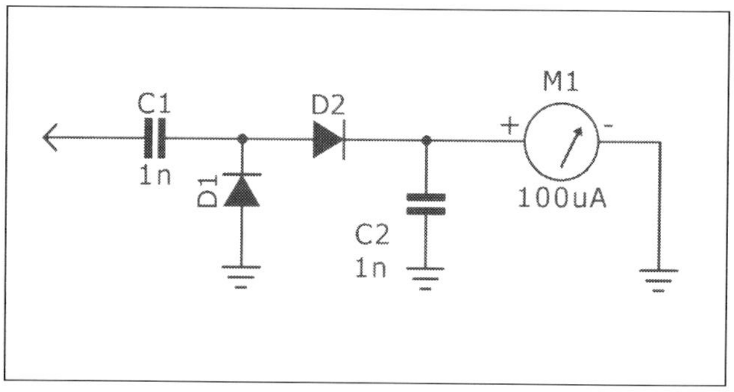

Fig 2.20. Diode RF probe

For measurement of small RF voltages a diode probe is the most common type used. This can be made in an evening and used with a normal analog micro-ammeter to indicate the amplitude of RF voltage. See **Fig 2.20**. It consists of two Schottky or normal silicon diodes (1N914 or 1N4148) and a few other components. It will measure RF voltages as low as 50mV. Although not absolutely calibrated it can be used as an indication when a circuit is being aligned.

Another useful item that costs almost nothing to make is an 'RF sniffer' (**Fig 2.21**). This consists of a few turns of wire on the end of either a piece of coax or soldered to a RF connector. Using this we can obtain a small sample of RF energy from a tuned circuit and feed this to the RF probe for detection. These two examples use enamelled copper wire to form the coupling loops.

Fig 2.21. Two RF sniffer probes on BNC connectors

USEFUL VSWR GRAPHS

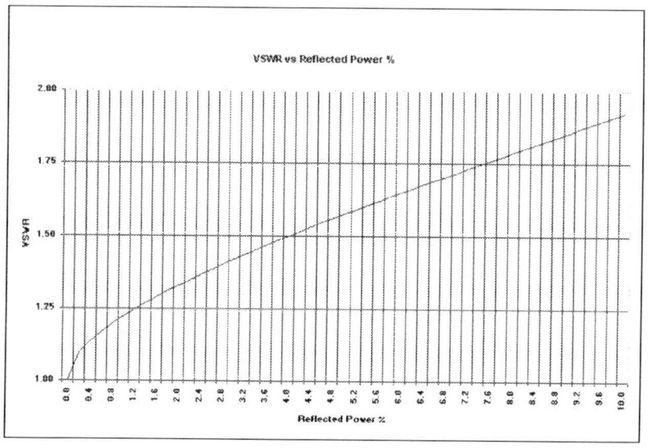

Fig 2.22. VSWR versus reflected power in percentage from 0 to 10%

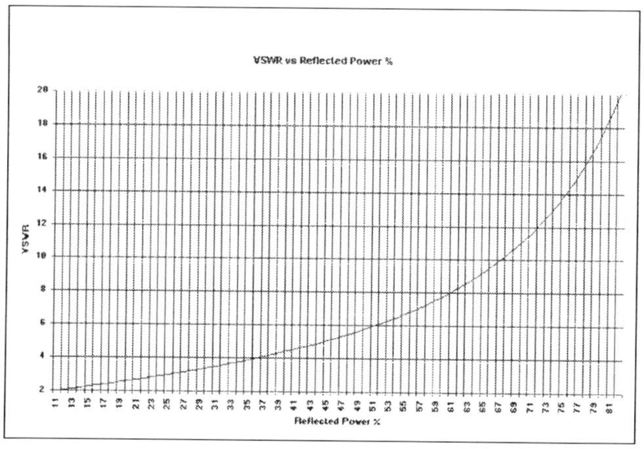

Fig 2.23. VSWR versus reflected power in percentage from 10% to 80%

Chapter 3

Understanding 'S' Parameters

SMALL signal 'Scattering parameters' (S parameters) are used to describe the impedance at the terminals of a multi-port device such as an RF transistor. Using S parameters a designer has an accurate measurement of how the transistor will behave when operated in a practical circuit. The input and output of the transistors are referred to as *ports*, hence in a transistor there are two ports; the common ground connection is not referred to as it is assumed to be zero impedance. An attenuator or filter can also be treated as a two-port network. A load resistor or an antenna is treated as a one-port network as it only has an input and no output. A diplexer, which has two inputs and one output, is treated as a three-port network.

S parameter data of transistors are obtained with specialised instruments that measure the input and output impedance while the transistor is biased up in a normal running condition. For small signal devices this means that the input drive level needs to be less than that which would cause gain compression to occur. For power transistors the drive level needs to be much higher to generate the rated output power of the device.

The magnitude of the reflected power will depend on the degree of mismatch existing between the transmission line and the load. If the load impedance is the same numeric value as the transmission line impedance the mismatch will be zero, or zero reflected power occurring.

SEMICONDUCTOR DATA

S parameters are commonly used for RF semiconductors to express the mismatch between the input and output terminals when measured in a specific impedance system. By default, the universal impedance used today is nearly always 50Ω. However, some devices intended for television tuners are sometimes expressed in a system impedance of 60Ω or 75Ω. If the data does not specifically state the impedance we can normally assume it is 50Ω.

Because the device being characterised is treated as a two-port network there are four different sets of data. These are S_{11}, S_{21}, S_{12} and S_{22} and these are input reflection coefficient, (input VSWR) forward gain, reverse isolation and output match (output VSWR) expressed as the magnitude of the reflection coefficients. The value of S_{11} and S_{21} are determined with a signal generator connected to the input with characteristic impedance of R_g and the output terminated in a load R_L. The values of S_{12} and S_{22} are determined by terminating the input with a load R_L and the signal generator is moved to the output port. R_g and R_L are normally 50Ω.

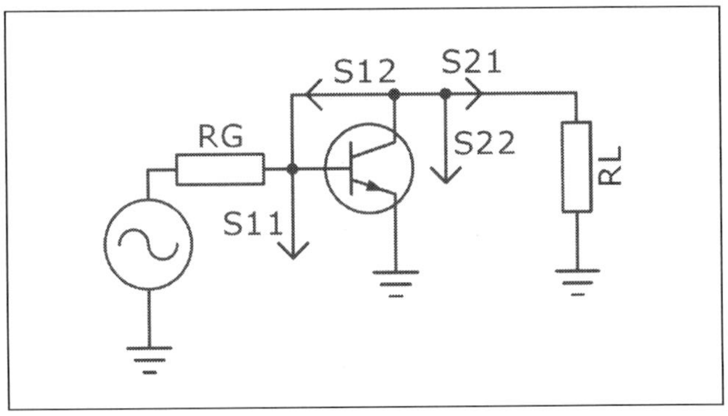

Fig 3.1. S parameter derivation

See **Fig 3.1**. The derivation of the four S parameters is as follows:

S_{11} is the complex impedance seen looking into the input terminal of the device when the input is driven by a source of impedance R_g. It is the same as the measurement made with a through-line power meter where we determine the VSWR the input presents to the source signal. The output is terminated in a load R_L.

S_{21} is the forward gain when the input is driven by a source of impedance R_g used for the measurement, in most cases this is 50Ω. The output is terminated in a load R_L.

S_{12} is the magnitude of the output signal, after amplification, which travels back through the device to the input. The smaller the value the more stable the device will be. The input is terminated in a load R_L.

S_{22} is the measure of how well the device output matches to the load termination. The lower the value the lower the VSWR seen by the device output terminal. The input is terminated in a load R_L.

> Note that the correct way of pronouncing these four sets of data: S_{11} is pronounced as "S one-one" and not as "S eleven". Similarly the others are "S two-one", "S one-two" and "S two-two".

The order in which they are presented in a data sheet should always be as listed above: S_{11}, S_{21}, S_{12} and S_{22}. The magnitude and phase angle is presented for each of the four data types. These are shown for a variety of different frequencies running from some low frequency to as high as the manufacturer's upper frequency limit. Hence, a microwave RF transistor intended for use at up to 2GHz could start as low as 50MHz in some cases and the upper frequency for the S parameter data could be 4GHz or higher.

To convert the reflection coefficient values for each data type requires some mathematical juggling which we will discuss a little later. There is, however, a second way of displaying the data which is quite common. This expresses the reflection coefficient in decibels (dB). Sometimes the manufacturer will give the data using either or both magnitude and logarithmic terms and sometimes only the value for forward gain (S_{21}) is shown in dB, whereas all the others are shown in linear magnitude format.

The value of gain (S_{21}) when expressed in dB is *not* the gain to be expected when the device is used in a circuit, because the impedances required to obtain maximum gain are often higher or lower than the 50Ω system impedance used to measure the values. The gain when measured in 50Ω impedance is normally much lower than it will achieve in a correctly matched circuit. Usually the data sheet also gives the typical gain to be expected at some frequency in a test circuit as an indication of what device is capable of.

As we saw in the chapter on measurements, to convert the incident and reflected power levels to *reflection coefficient* format we use the formula:

$$\rho = \sqrt{\frac{P_r}{P_f}}$$

where: P_r and P_f are the reflected power and incident power in watts. ρ (Greek letter rho) is the numeric value of the reflection coefficient. The reflection coefficient is normally referred to as 'lin mag' (linear magnitude) format.

$$\text{VSWR} = \frac{1+\rho}{1-\rho}$$

To convert the reflection coefficient to logarithmic format in dB we use the following formula:

$$\text{Measurement in dB} = 10 \times \log_{10} (S_{xx})^2$$

where S_{xx} is the S parameter.

The reason we square the S parameter is that this is a voltage measurement and to obtain power we need to square the voltage. To obtain the value in dB we take the logarithm of the S_{xx} parameter squared and multiply by 10.

For example, a device has a forward gain of 2.5 when expressed in linear magnitude terms. This is a logarithmic gain of 7.95dB in a 50Ω system. (When correctly matched in a circuit the forward gain could be as high as five times this figure, typically 30dB or more is quite common at lower frequencies.)

A bipolar junction transistor has a S_{11} linear magnitude value of 0.85. This is an input reflection coefficient of -2.82dB or input VSWR of 12.3:1 when measured in a 50Ω system. If the device is a small signal device we would normally expect the input impedance to be high, very roughly equivalent to 12.3 x 50Ω or about 600Ω. The input matching network needs to transform up from the 50Ω source to the base impedance of ≈600Ω if maximum gain is required. If, however, the device is a power transistor the impedance is more likely to be very low, in the case of the 12.3:1 VSWR this would be about 4Ω and is simply the ratio between the measuring impedance of 50Ω and the device input impedance.

Device Z = VSWR x System Z, or
Device Z = System Z / VSWR

ALTERNATIVE NOMENCLATURE

Another way to express reflection coefficient is to refer to it as the *return loss* in dB. This is what the log dB figure is in the S parameter data. If an attenuator is used in a system the input reflection coefficient when the attenuator is perfectly matched is 0, a VSWR of 1:1. If the far end of the attenuator is open circuit or short circuit the *return loss* is a value of twice the attenuator value. So when we see, for example, an S_{11} log dB value of, say, -2.2dB it is behaving the same as an attenuator of 1.1dB insertion loss placed between the sending signal and the input of the device, which is open or short circuit.

INTERPRETING SEMICONDUCTOR S PARAMETERS

Common Source, Z_o = 50Ω, T_A = 25°C, V_{DS} = 2V, I_{DS} = 25mA.

Freq	S_{11}		S_{21}			S_{12}			S_{22}	
GHz	Mag	Ang	dB	Mag	Ang	dB	Mag	Ang	Mag	Ang
0.5	.98	-18	14.5	5.32	163	-34.0	.020	78	.35	-9
1.0	.93	-33	14.3	5.19	147	-28.4	.038	67	.36	-19
2.0	.79	-66	13.3	4.64	113	-22.6	.074	99	.30	-31
3.0	.61	-94	12.2	4.07	87	-19.2	.110	44	.27	-42
4.0	.54	-120	11.1	3.60	61	-17.3	.137	31	.22	-49
5.0	.47	-155	10.1	3.20	37	-15.5	.167	13	.16	-54
6.0	.45	162	9.2	2.88	13	-14.3	.193	-2	.08	-17
7.0	.50	120	8.0	2.51	-10	-13.9	.203	-19	.16	45
8.0	.60	87	6.4	2.09	-32	-13.6	.210	-36	.32	18
9.0	.68	61	4.9	1.75	-51	-13.6	.209	-46	.44	38
10.0	.73	42	3.6	1.52	-66	-13.7	.207	-58	.51	34
11.0	.77	26	2.0	1.26	-82	-13.8	.205	-73	.54	27
12.0	.80	14	1.0	1.12	-97	-14.0	.200	-82	.54	15

S parameter data for the ATF-10136 GaAsFet

Above are the S parameter data for a typical low noise GaAsFet, the Agilent / Avago Semiconductors ATF-10136. In this data the system measuring impedance (Z_o) is confirmed as being 50Ω, the device is measured in the common source topology and the data is for a drain-source voltage of 2V and a drain current of 25mA. The drain-source voltage and drain current shown are the optimum values for lowest noise performance, as determined by the manufacturer.

The values for S_{21} and S_{12} are shown in linear magnitude format as well as in logarithmic format (dB). Inspection of the values for S_{11} magnitude and phase angle shows that the device at lower frequencies is predominantly a capacitive input with a high mismatch ratio between the gate and the 50Ω measuring system. At 0.5GHz the input match to 50Ω of 0.98 is equivalent to a VSWR of approx 99:1, which is roughly 5kΩ impedance for a conjugate power match. Hence, the input network at 0.5GHz needs to take the form of a step up transformer with an impedance ratio of 99:1 to match to 50Ω. This, however, is not the condition for lowest noise matching.

STABILITY CRITERIA

Of great importance with any amplifier is how good the reverse isolation (S_{12}) is. If this value is poor, some of the power exiting the device can find its way back to the input via the parasitic internal capacitance between drain and gate, and cause the device to oscillate. Generally, for maximum stability, the value for S_{12} needs to several dB greater than the anticipated forward gain value of S_{21}. In the data at 0.5GHz this is a value of -34dB when the forward gain is 14.5dB, hence there is an excess reverse isolation over the forward gain of about 20dB which means the device should be very stable. Perusing the other frequency values shows that S_{12} is always several dB greater than S_{21}, so the device should be stable.

However, this is when the input and output are terminated in 50Ω, which is how the S parameters are obtained. In a practical circuit this is often not the case and as we saw earlier the input impedance needs to be very high to achieve maximum gain. As the input and output impedance are raised in value the internal feedback capacitance, although a very low value, becomes more significant and the device no longer has an excess of S_{12} over S_{21}. Using the S parameters we can establish the stability criteria for any device when terminated in either 50Ω or any other impedance. This is known as the Rollet stability factor (k). If the Rollet Stability factor is less than 1.00 the device is potentially unstable for some conditions of input or output mismatch. When the value of k is equal to or greater than 1.00 the device is totally stable: no matter what the input and output termination mismatch are it will not oscillate.

> The fact that the manufacturer was able to measure the device S parameters in a 50Ω measuring system shows that the device must have been stable. Sometimes this is not the case, especially for power transistors. In this case a different measurement technique is used, which we will discuss later.

MAXIMUM AVAILABLE GAIN AND MAXIMUM STABLE GAIN

Using the S parameters we can establish two other important factors, these are the *maximum available gain* (MAG) and the *maximum stable gain* (MSG) criteria for the device. Although these can be obtained from the standard formulae, today it is far easier to use the analysis software to derive them. (To derive these from the S parameters using a scientific calculator takes about 30 minutes, using the software it takes less than 1s). This not only greatly reduces the time required, but the software can also interpolate between the points for any frequency of interest. Also the software can with a great confidence interpolate for higher and lower frequencies than the data table frequency points. Hence, a GaAsFet may only be characterised down to 100MHz in the S parameters but with the analysis software we can determine the MAG and MSG values for lower frequencies not covered by the table. From this we can see if the device is suitable for, say, a 50MHz amplifier. We can also establish the Rollet Stability factor for any frequency of interest, this factor is very important.

MAG by definition is the 'available output power versus input power when the device is conjugately matched'. In other words, when the input and output ports exhibit the lowest VSWR in a 50Ω environment. From this the value of MAG can be established in dB of forward gain. This assumes the device is unconditionally stable, with a Rollet Stability factor of 1.00 or more.

For the more common situation where the device may not be unconditionally stable, the value we need to use is MSG. This by definition is 'the available power output versus power input when the device is not conjugately matched to the source and / or load and the Rollet Stability factor is potentially less than 1.00 at some frequency'. For most amplifier designs this is the likely situation. If the value of gain exceeds the value of MSG the device will oscillate.

Where the device is potentially unstable with some source or load termination impedance it is not possible to design for the MAG condition. This is especially the case for noise matched amplifiers where we need to mismatch the input deliberately to achieve the noise performance. Hence, it is very common to have to reduce the gain to the MSG value, or below, to ensure the device is totally stable. In practice this means we have to sacrifice several dB of gain (in some cases as much as 20dB!) to achieve a stable design.

INPUT MATCHING

Of interest is the way S_{11} changes with frequency. Most small signal RF semiconductors have a small capacitance to ground when the input terminal is measured and for bipolar microwave transistors this is of the order of 1 to 4pF. An FET type will also be similar values. By comparison a microwave RF power transistor input capacity could be 10pF or more. (Because this is a difficult measurement at high frequency it is often performed at a low frequency of 1MHz in a shielded enclosure to eliminate stray capacity.) If the S parameter data for S_{11} of the ATF-10136 is examined this shows the general trend of most small signal transistors.

At frequencies below ≈5GHz the input appears as a capacitive reactance, shown by the minus sign of the phase angle for S_{11}. Above ≈5GHz the internal bond wire inductance and package gate lead inductance cancels the gate capacitive reactance and the input becomes net inductive. At some frequency between 5GHz and 6GHz the input swings through 0° phase angle and behaves as a pure resistance with a value of approx 135Ω, and hence is very easy to match to. The ATF-10136 is intended for use at 4GHz in satellite television low noise front-end amplifiers.

To match to the gate at lower frequencies means we need an equal and opposite reactance to cancel out the predominantly capacitive reactance. This makes the gate look like a pure resistance. This means we need either a shunt or series inductor, or a section of transmission line that appears to be inductive. Above ≈6GHz the situation reverses and as the gate now looks inductive we need an equal and opposite capacitive reactance to cancel out the inductive reactance. This is the condition for a conjugate power match when maximum gain, and hence lowest input VSWR, is required.

However, for optimum noise performance this is not the method used. In order to achieve the lowest noise performance the input of the device needs to be termi-

Generally, GaAsFet and PHEMT low noise transistors have a very poor input VSWR at lower frequencies (≈100MHz and below) because the resistive portion of the complex impedance is very high, in some devices as much as 1MΩ. To achieve the ultimate noise performance means the device input needs to be severely mismatched for lower frequency applications. At higher frequencies the situation is not as severe and low VSWR matching is usually possible above about 3GHz which allows printed microstrip circuits to be used. Bipolar transistors have lower resistive portions and so are easier to match to at lower frequencies.

nated in the impedance that equals the value for lowest noise, Γ_{opt}. (Gamma opt) For devices such as GaAsFets and PHEMTs at lower frequencies this in most cases is a severe mismatch, possibly a VSWR as high as 50:1 to a 50Ω system.

This can cause a problem when a low noise amplifier is connected to an antenna that does not present 50Ω non-reactive matching to the input. When the low noise amplifier is adjusted it is normally performed using a noise figure measuring system with a very low VSWR 50Ω noise source. When the low noise amplifier is then connected to a typical antenna, which is not well matched to 50Ω across a broad range of frequencies, the amplifier may oscillate because the mismatch is now favoring the unstable operating region. A noise figure measuring system terminates the amplifier input with a low VSWR 50Ω termination over a wide range of frequencies; antennas in general do not.

Consider the diagram shown in **Fig 3.2**. This is the VSWR plot for a microwave feed horn for a hydrogen line radio telescope at 1420MHz. The feed horn behaves as a 'waveguide beyond cutoff' below about 1200MHz and the VSWR presented to the LNA is very high. At 1050MHz it exceeds a VSWR of 10:1. This is typical of most microwave feed horns.

For this reason it is imperative firstly to stabilise the device with external components to ensure the Rollet Stability factor is above 1.00. Very few antennas offer a low VSWR across a broad bandwidth. Amateur antennas are often very narrow band, typically not more than ≈2% for a 3:1 VSWR, and at frequencies as little as ≈5% from the band centre frequency the VSWR can become excessive, causing instability if the device is not correctly stabilised. Many of the published designs in amateur journals are only marginally stable even in a 50Ω environment and can be expected to give poor noise figures because the device may oscillate at some frequency. Most amateur low noise amplifiers require retuning when the antenna is connected to bring the noise figure to optimum.

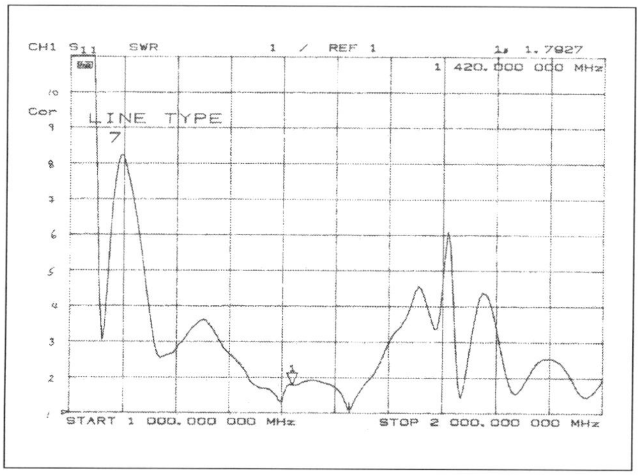

Fig 3.2. VSWR plot of feed horn for 1420MHz

NOISE DATA

V_{DS} = 2V, I_{DS} = 25mA

Freq. GHz	NF dB	Γ_{opt} Mag.	Ang.	R_N / 50
0.5	0.35	0.93	12	0.80
1.0	0.4	0.85	24	0.70
2.0	0.4	0.70	47	0.46
4.0	0.5	0.39	126	0.36
6.0	0.8	0.36	-170	0.12
8.0	1.1	0.45	-100	0.38

ATF-10136 noise data

The noise data for the ATF-10136 is shown above. This shows that the optimum noise matching on the input (Γ_{opt}) occurs at a different linear magnitude and phase angle than that for maximum gain. In virtually every low noise device when the optimum noise match is achieved the forward gain (S_{21}) falls a little, perhaps 1dB or more. The value given for NF_o is the ultimate noise figure in dB that can be obtained when perfectly noise matched. The input network between the source (antenna) and the device input needs to establish the values of magnitude and angle listed in the table.

These values of magnitude and angle are what we need to present to the device for optimum noise matching. In other words it is the opposite of the S_{11} parameters derived when we effectively 'look into' the device from the 50Ω measuring instrument. The way the manufacturer establishes these values is to construct an amplifier and adjust the input-matching network for the lowest noise figure when the noise source is 50Ω. Having adjusted the network for lowest noise figure the device is then removed and the S parameter measuring system is connected to the point where the input terminal of the device was originally connected and a 50Ω termination is placed across the normal input connector. The measuring instrument 'looks into' the connection point to see what the device is terminated with for lowest noise figure. This is a labour intensive process and so this is why only a few frequency points are normally listed. At each measurement frequency the input network needs to be adjusted for lowest noise figure and then the device removed. However, with as little as three spot frequencies a computer analysis package can find the *slope of the curve* connecting the spot frequencies and calculate the optimum S parameter for other frequencies with a high degree of accuracy. The more points available the better the analysis software can ascertain the correct noise match network. If one of the spot frequencies is very close to the wanted operating frequency the analysis is even more accurate.

An alternative method is instead of connecting the

measuring instrument to the input terminal connection point; the device is also removed. Next a short circuit, an open circuit and 50Ω load are connected at this point while the S parameter measuring system is still connected to the normal input matching network. This is the same technique as when the VNA is normalised. This is a better practical method as it can be very difficult to make a test fixture with good high frequency performance, especially in the higher microwave spectrum. Many manufacturers probe at the wafer level in a special test fixture to ascertain the noise match and then 'back calibrate' by including the package parasitic factors. This is because the same die is often used in several similar devices with different packages.

This is also the way that power transistors are characterised, because many are not stable when presented with input and output terminations of 50Ω. Again a test amplifier is made and the device matching networks are empirically adjusted for optimum performance at several different frequencies. At each frequency the device is removed and the instrumentation looks back into the network to extract the S parameter data.

Because a power amplifier is running at watts of input and output power this makes it very difficult when the typical VNA only has an output power of a few mW. To get to the required drive level needs additional power amplifiers to raise the level. This requires additional external directional couplers, rated for the input and output power expected. Calibrating a system such as this is fraught with accuracy problems, especially if the design needs to be wide band. Also, because the VNA is a small signal device, reducing the output power to levels the VNA can handle requires very small coupling factors in the external directional couplers. In general small coupling factors do not give the best directivity and so the accuracy suffers. It can take several hours to calibrate such a system at each measurement frequency and hence is avoided if at all possible.

Examining the values for optimum noise matching confirms what we have already concluded. At lower frequencies we need an inductor to cancel the capacitive reactance of the gate to source terminal. The required optimum reflection coefficient at 0.5GHz is 0.93 with a phase angle of +12°, i.e. an inductor. (For comparison purposes for a conjugate power match the inductor needed is equivalent to +18°, so the two matching schemes although similar are different. If the input tuning has sufficient range the input can be adjusted for a conjugate match or the lowest noise figure condition by retuning.) At higher frequencies, 6GHz and above, the angle required changes sign to denote the requirement now being for a capacitor instead on an inductor to resonate with the gate-source terminal.

One thing we can establish from this examination is the best circuit topology to 'noise-match' a device. If the input terminal is already capacitive, which most microwave RF transistors are below about 5GHz, we clearly need to cancel this with an inductor, either in shunt or series. Hence, placing a shunt capacitor across the device input to ground will degrade the noise match; what we require is an inductor not a capacitor. Ideally for the lowest loss input matching network we would use the gate or base capacity to resonate an inductor to obtain the required value of Γ_{opt}. At high frequencies this is a practical method, but at lower frequencies the value of inductance required becomes very large and obtaining sufficient Q for the inductor is a great problem.

> A common mistake made is to assume an infinite Q value for the inductor. Using analysis software we can explore the minimum inductor Q required to achieve the value of F_{min}. Because the Q of an inductor at low frequencies is limited to about 300 often we cannot attain the ultimate noise figure the device is capable of.

Hence, we normally have no option but to use a shunt capacitor across the gate to lower the inductance value required, this is why it is not possible to attain the potential noise figure at lower frequencies. It is also the reason that microwave transistors like GaAsFets, HEMTs and PHEMTs are not a good choice for bands such as 144MHz and lower. Even 432MHz presents a difficult matching problem with a microwave transistor intended for operation on frequencies of 4GHz and above. The potential gain a GaAsFet or a similar device can attain at frequencies below 1GHz means that the device gain needs to be severely limited to obtain a good stability factor and often 10dB or more gain reduction needs to be used to prevent oscillation.

The factor Rn/50 needs some explanation. This is the noise resistance of the impedance we need to present to the device input for optimum noise performance when measured in a 50Ω system. However, this value is not measured during the S parameter evaluation but it is calculated from the minimum noise figure and the noise figure obtained when the frequency is varied about the nominal frequency. It is only used when a computer analysis package is used to ascertain the optimum component values for a noise match.

If we now examine the S parameters for a typical bipolar RF transistor we see a different input match criteria. Below are the S parameters for the AT-41435 microwave transistor made by Agilent / Avago Semiconductors.

Common Emitter, Z_o = 50Ω, T_A = 25°C, V_{ce} = 8V, I_c = 10mA.

Freq	S_{11}		S_{21}			S_{12}			S_{22}	
GHz	Mag	Ang	dB	Mag	Ang	dB	Mag	Ang	Mag	Ang
0.1	.80	-32	28.0	24.99	157	-39.2	.011	82	.93	-12
0.5	.50	-110	21.8	12.30	108	-29.6	.033	52	.61	-28
1.0	.40	-152	16.6	6.73	85	-26.2	.049	56	.51	-30
1.5	.38	-176	13.3	4.63	71	-24.0	.063	59	.48	-32
2.0	.39	166	11.0	3.54	60	-21.9	.080	58	.46	-37
2.5	.41	156	9.3	2.91	53	-20.4	.095	61	.44	-40
3.0	.44	145	7.9	2.47	43	-18.8	.115	61	.43	-48
3.5	.46	137	6.7	2.15	33	-17.5	.133	58	.43	-58
4.0	.46	127	5.6	1.91	23	-16.0	.153	53	.45	-68
4.5	.47	116	4.7	1.72	13	-15.0	.178	50	.48	-75
5.0	.49	104	4.0	1.58	3	-13.9	.201	47	.48	-82
5.5	.52	91	3.3	1.45	-7	-13.0	.224	40	.47	-89
6.0	.59	81	2.5	1.34	-17	-12.1	.247	36	.43	-101

AT-41435 S parameters

By examining the magnitude of S_{11} across the frequency range we can see that the input is much easier to match to a 50Ω source, the worst match is at 100MHz where the value is 0.8. This is a VSWR of 9:1, impedance of ≈450Ω. Again the input is capacitive up to about 2GHz and then it behaves as an inductive reactance. At 1.5GHz the value of magnitude is 0.38 which is equivalent to a VSWR of 2.22:1 or about 111Ω. This is very different to the ATF-10136 GaAsFet and far easier to match to.

We can also see that the value of forward gain in a 50Ω system is higher, being 21.8dB at 0.5GHz. This shows that the collector impedance needed is a lot lower than the comparable GaAsFet device. Because the value of S_{21} is nearer to being a good match the value of S_{12} is also more accurate and we can see that the device is very stable. The collector impedance required is shown by the value for S_{22} of 0.48 at 1.5GHz, this is approx 142Ω. At a pinch we could use a 150Ω resistor as the collector load to obtain a good output match.

By examining the noise parameter data we can also get a good idea of what is required to match the AT-41435 for optimum noise performance. The noise figure possible is not as low as the ATF-10136 because of the device limitations, but it is suitable for a simple broad bandwidth LNA design.

V_{ce} = 8V, I_c = 10mA

Freq GHz	NF dB	Γ_{opt} Mag.	Ang.	R_N / 50
0.1	1.20	0.12	3	0.17
0.5	1.2	0.10	14	0.17
1.0	1.3	0.05	28	0.17
2.0	1.7	0.30	-154	0.16
4.0	3.0	0.54	-118	0.35

AT-41435 Noise data

The value of magnitude required is low; hence it is close to 50Ω. At 1GHz it is almost a perfect match to 50Ω. In fact if the device were simply connected directly to a 50Ω source and load, with suitable DC blocking capacitors to prevent shorting the collector and base bias, the device will offer reasonable gain and noise figure across a wide bandwidth without any matching. The AT-41435 and similar devices are a good choice for a second stage amplifier in a high gain LNA or where the ultimate noise figure is not necessary. At 23cm it can be expected to offer a noise figure of approx 1.4dB with about 12dB of gain. At 144MHz the potential noise figure is 1.2dB with a gain of about 20dB, so it would suit an LNA for 144MHz EME applications.

Other popular bipolar transistors are also suitable, the BFW92A data does not give the ultimate figure but simply the noise figure when used in a 50Ω system. I have used the BFW92A in several wide band LNA designs. The manufacturer's data state the noise figure at 800MHz is typically 2.5dB when a 50Ω source and load is used. When correctly noise matched a noise figure of ≈1.5dB and gain of ≈13dB were obtained at 350MHz to 450MHz in a wide band circuit; this gives some indication of its potential.

NARROW BAND AMPLIFIERS

A narrow band amplifier is normally considered one where the bandwidth covered is less than ≈10% of the centre frequency. For the amateur bands we can normally employ narrow band amplifiers, which simplifies the design and affords more gain than a wide bandwidth type. During the design the input and output are simply conjugately matched for maximum gain, if the noise figure is not an important criterion. Where the noise performance is important the input is deliberately mismatched to achieve the required noise figure and this causes a small loss in gain to occur. The output matching normally has little effect on the noise performance and so it is usual to design this for a conjugate match to maximise the gain.

WIDE BAND AMPLIFIERS

A wide band amplifier is generally considered to be a fixed tuned amplifier where the bandwidth is considerably greater than 10%; there is, however, no hard and fast definition as to what constitutes a wide band amplifier. For one covering 50MHz bandwidth at 800MHz some designers would treat it as a narrow band design and others as a wide band design. Generally, when the bandwidth exceeds 40 to 50% it is essential to design as a wide band amplifier and not a narrow band one. An example might be an amplifier to insert in front of a spectrum analyser covering 10MHz to 2GHz with a flat gain response to increase the sensitivity to small signals whilst maintaining the calibration accuracy for widely spaced signals. Here the bandwidth is obviously 1.99GHz out of 2GHz, approx 99%, and this would be classified as a wide band design. However, the device we would need to use for this would need to have useful gain up to as high as 6GHz, so the bandwidth is more like 33% of the device's potential gain.

Sometimes we required an amplifier to span a large portion of spectrum. An example could be a television masthead amplifier for VHF and UHF. If required to cover this total spectrum an amplifier with a flat gain between about 200MHz to about 1GHz is required. Because the gain of any semiconductor falls off with increasing frequency this does not suit our application. The way we address this problem is to deliberately mismatch the input or output, or both, at the lower frequencies and as the frequency increases progressively provide a better match. For the mismatched case the gain achieved is lower than would be obtained in a narrow band amplifier. At the higher frequencies, where the device gain would in any case be lower, the input or output matching is closer to the maximum gain narrow band condition. Hence, the input and output matching networks behave like high-pass filters.

For the case of the television masthead amplifier we also need to take into account the coaxial cable loss between the amplifier and the television receiver. As the cable loss increases with increasing frequency, for a given length of cable, we need to tilt the *gain-bandwidth* response more in favour of the higher frequency end of the spectrum. Another factor could be the sensitivity of the television receiver. At VHF it is not difficult to achieve sufficient sensitivity, but at UHF the higher noise figure and lower gain means we need more signal voltage to obtain a good picture quality. When the attenuation between the transmitter and the receiver is taken into account, the free space path loss will be at least twice as high at UHF as at VHF and hence the signal from the antenna (assuming they are the same gain for VHF and UHF) will be less than for VHF. This can cause the gain slope to be severely tilted towards the UHF end of the bandwidth to compensate. Because all semiconductor devices exhibit a 6dB/octave loss of gain as the frequency rises, if we need a flat gain versus frequency the matching networks need to have a +6dB/octave slope to compensate. For the other cases we may need to modify this to about +10dB/octave to counter the higher losses at UHF.

However, the gain that can be achieved with this type of amplifier is often much less than a narrow band amplifier. For example, an amplifier for a fixed frequency of 200MHz using one transistor in a narrow band design may be able to develop as much as 20dB of gain. But an amplifier covering 200MHz to 1GHz with a fairly flat frequency response (within ±1dB) may only achieve an average of 10dB across this wide range. If a gain of 20dB is required this often means we need to employ two devices in series to obtain the required gain.

This technique is not used for television tuners. These often are designed as narrow band amplifiers and the centre frequency is varied to cover the band by varicap diodes or normal tuning capacitors to maximise the sensitivity and to reject adjacent channel signals.

A better option for the television masthead amplifier is to use two different amplifiers, one optimised to cover the VHF portion and one for the UHF portion. In many cases the VHF and UHF antennas will be separate items and can be tailored to suit the anticipated path loss for each band. The amplifiers are now narrower bandwidth hence they can be tuned for more gain and a lower noise figure. The two amplifier outputs can be combined with a zero degree power combiner or a common wide band buffer amplifier stage with high dynamic range to feed a common coaxial down lead. This also eases the rejection of spurious signals from adjacent channel transmitters.

Chapter 4

Impedance matching

THIS is a very complex subject and whole textbooks have been written on just this subject, so we can only attempt to cover the basics here. Impedance matching in essence is the *transformation of one impedance to another one*, either higher or lower in value.

The definitive book on the subject for amateurs is Walter Maxwell's, W2DU, book *Reflections* published by the ARRL and this is recommended reading for serious amateurs.

Assume we need to match between the antenna impedance of 50Ω to the input of an amplifier device such as a triode valve. I know that today valves are old fashioned, but the explanation is the first part of the story. It could equally apply to a modern device such as a junction FET.

The grid of a triode valve (or gate of a JFET) is essentially a very high resistance (RG) and a small capacitance shunting it to ground (CG). Typical small signal valves such as the EF91 pentode have an input capacitance of about 3pF and a resistance of about 20kΩ. If we used a transformer to make the transformation, the impedance ratio is 400:1. In a transformer the turns-ratio required is the square root of the impedance ratio. So our transformer requires 20:1 between the grid and the 50Ω antenna. This might use one turn at the primary (antenna) and 20 turns across the grid to ground, or two turns on the primary and 40 turns on the secondary etc. However, making such a transformer is not easy because the stray capacity of the winding can make the windings resonate at some unwanted frequency and we cannot easily adjust the turns ratio without starting again from scratch. So let us look at a simpler method that can allow some adjustment.

It should be obvious that we require a high value of RD to get to 20kΩ and so the L needs to be quite large. Suppose we are operating at 7MHz, we can therefore calculate the required value with a practical inductor Q. Suppose we choose to use a ferrite toroid which can achieve a Q of 200 at 7MHz. Now we need to select a suitable value of tuning capacitor, but we need to be aware that there are already 3pF effectively in shunt with the grid to ground inherent in the EF91 grid-cathode electrodes. Hence, we need to subtract 3pF from the calculated result.

We will start by trying an inductor value of 10µH. This at 7MHz has a reactance of 440Ω and hence requires a capacitor of 51pF to resonate. Subtracting the ≈3pF grid value means we need a 48pF capacitor and a standard value 47pF will be suitable. The loss resistance RS of the inductor is made up from the copper loss, skin effect and core losses. With a reactance of 440Ω and a Q of 200 RS is 440 / 200 ≈2.2Ω. Hence, since RD = X^2/RS, this is a value of 88kΩ. This is much more than required so we need either to shunt the tuned circuit with a resistor to bring it to 20kΩ or choose a lower L-C ratio.

With an inductor of 5µH the capacitor value required is ≈102pF. (Note that since we have halved the inductor value we need to double the capacitor value to maintain the same resonant frequency). This gives a reactance of 220Ω, with the inductor Q of 200 and a RD value of ≈44kΩ, which will require shunting with a resistor of approx 39kΩ. This should be suitable. (We could reduce the value of L to 2.5µH and with the same Q this would give a value for RD of approx 22kΩ which would require a capacitor of ≈204pF to resonate.) However, we haven't yet devised a way to connect the 50Ω input to the network yet. There are several options; we could tap the inductor near the ground end and feed the antenna signal in this way, the auto transformer method. Finding the correct tapping point is difficult.

Another method is to use separate primary and secondary windings but again finding the exact turns ratio

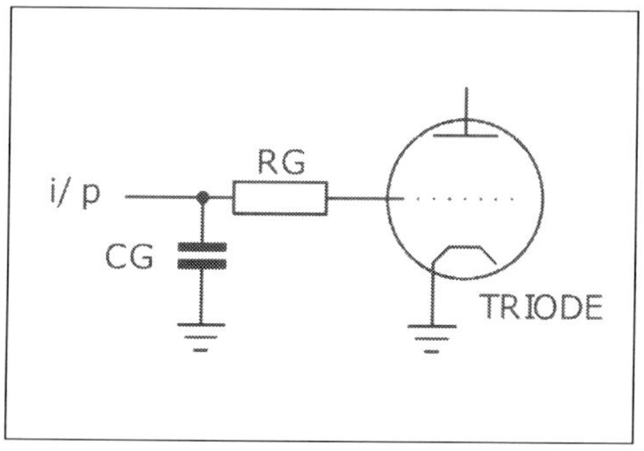

Fig 4.1. Equivalent grid circuit

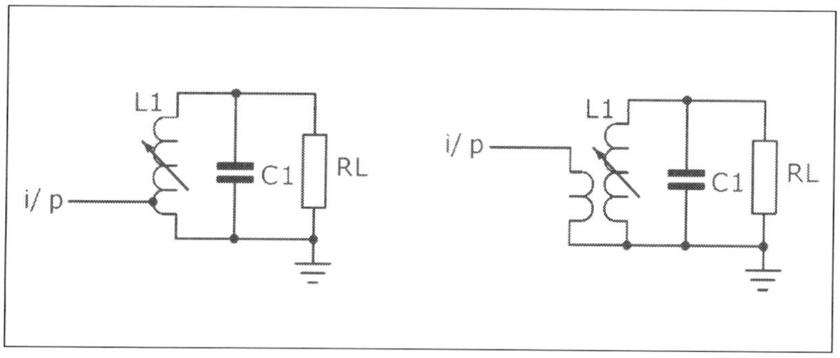

Fig 4.2. Tapped inductor and transformer coupling

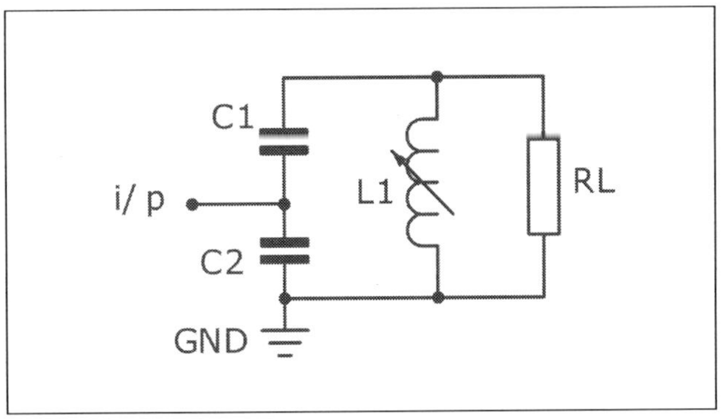

Fig 4.3. Tapped capacitor impedance matching

can be tedious and insufficient coupling between the two windings often results in extra loss. An alternative method is a capacitor-tap scheme. This is shown in **Fig 4.3**.

In this method the capacitor resonating the inductor is split into two series-connected ones and the low impedance input is connected in shunt with the bottom capacitor C2. Because we now have two capacitors in series we can vary the ratio of the capacitor values whilst achieving the same total capacitance to achieve resonance and to correctly match to both the 50Ω and the effective grid load of 20kΩ.

If we made each capacitor twice the value required to resonate with the inductor, the impedance presented to the antenna is half the impedance appearing across the grid, or about 10kΩ. By making the bottom capacitor (C2) much larger in value than the top one we have a variable tapping point, a variable potential divider. C1 could then be made up with a variable capacitor of approx 30pF maximum and a parallel fixed capacitor to bring the network to resonance. Alternatively the inductor could be made variable over a small range and then the capacitors could both be fixed values. The values required are:

C1 = 120pF - total
C2 = 2000pF – fixed (2 x 1nF in parallel)
L = 5μH ±10%
R = 39kΩ

The formulae to calculate these values are quite involved and interact with the other components. For example, if we know the value of L and the total tuning capacity effectively across the inductor we can find either C1 or C2 but not both at the same time, so to make life simple I wrote Basic programs allowing me to solve for any unknown component. These are contained with the software on the website associated with this book.

TRANSFORMATION 'GAIN'

One thing that isn't obvious from this is the apparent 'gain' a step-up transformation allows. (In fact an impedance transformation cannot have gain because it is a passive device and is more likely to have loss, but the voltage step up is quite significant, especially if high resistance devices such as valves or JFETs are used.) Because valves and FETs are voltage driven devices, and have very high input resistance, if we transform up from some low impedance, such as 50Ω, to a few tens of kΩ we get an increase of voltage. Power is V^2/R and assuming the transformation is made with minimal loss then the power is substantially the same.

The anode current variation of a valve is a function of the voltage swing on the grid, generally expressed in mA/V, called the 'conductance' and known as g_m. A small signal valve might have a g_m of 5mA/V. That is for a 1V variation of the grid voltage the anode current will vary by 5mA. If the anode load is 50kΩ and the g_m is 5mA/V a signal at the grid of 1V rms this will cause the anode current to change by 5mA rms across 50kΩ. This is an equivalent anode voltage swing of 250V. If the anode load resistance is 10kΩ the anode voltage swing is 50V rms. Hence, very high voltage amplification is possible with valves.

To obtain a grid voltage swing of 1V across 20kΩ when the transformation ratio is 400 is equivalent to 1V/400 at the 50Ω input. This is a voltage of 2.5mV across 50Ω. An anode voltage swing across 10kΩ of 50V is a power of V^2/R = 50 x 50 / 10000 = 250mW. The input power is $(2.5mV)^2$ / 50 = 125 x 10^{-9}W a power gain of 2 million times or 63dB. In practice it is not normally possible to attain such high power gains due to instability but 50dB per stage is not uncommon at small signal levels for low frequencies.

MATCHING TO A LOW IMPEDANCE

Very often when RF power transistors are used we need to match from the 50Ω impedance of the driver to the base of the power amplifier. Perusing the manufacturer's data on the device will give us typical impedance values at several frequencies. Normally we will pick the values closest to the frequency of operation as a good starting point. Suppose we need to match from 50Ω into the base of a medium-power transistor, the impedance will be of the order of a few ohms and the base has a fairly high capacity. This requires a 'step-down' transformation. Let us assume the base impedance is 3Ω and the base capacity is 30pF. The 3Ω is the *resistive portion* or as it is known the *real part* of the impedance. Remembering that we can only dissipate power in a pure resistance and not a reactance means that we need to make the 30pF disappear somehow.

To cancel the 30pF base capacity means we need an equal and opposite reactance, an inductance. However, we need to calculate the reactance of the 30pF before we know what that inductance is. If the frequency is 30MHz then the reactance is 177Ω. The equivalent inductance is therefore 940nH, or a little less than 1μH . If we connect a ≈1μH RF inductor between the base and ground this takes care of the reactive component. Now the base 'looks like' a pure resistance of 3Ω. To get from 50Ω to 3Ω is an impedance ratio of 50/3 = 16.66:1 and so a 4:1 turns-ratio transformer would do the job. The 1μH inductor can be decoupled to ground with a suitable capacitor and used to feed in the base bias current so saving an extra component.

Also note that this type of transformation causes a step-down in voltage across the base but a step-up in current flowing into the base. Assuming the transformation is loss-less then the same power will be at the 50Ω input as the base input. If the input current into the base is 1A with a 10:1 step-down in impedance than the current flowing at the 50Ω point is 0.1A. Hence, when high levels are used, say 10W at 50Ω driving a base impedance of, say 1Ω or less, the current flowing in the matching network can reach levels of tens of amps and the RS and hence Q of the components become critical. Very low-loss capacitors are required for this type of application. A useful publication to read is the *RF Capacitor Handbook* published by the American company ATC (American Technical Ceramics). The vice president of engineering of ATC, Vince Perna Jr, gave some details of how the RS losses in a typical RF ceramic capacitor cause problems. In one amplifier for 450MHz the loss resistance of the series input capacitor made by a competitor was so high that it melted the solder and forced extra heat energy into the transistor, causing it to exceed its safe junction temperature and fail.

PI NETWORK MATCHING

The pi network is one of the cornerstones of amateur designs and its basis was discussed in the chapter on tuned circuits. The pi network is often used to match the anode of an HF transmitter power amplifier valve to the antenna. It offers low loss and flexible matching, so is suitable for matching an amplifier to an antenna if the impedance varies over the band, such as is the case with a simple dipole. Another benefit is that it is inherently a low-pass network and so attenuates the harmonic content.

In practice the two capacitors shown in **Fig 4.4** are normally made variable and the inductor is a fixed value. Where several bands need to be covered the inductor is provided with taps and a multi-pole switch used to select the appropriate tapping point. At resonance the anode of the valve appears to behave as a pure resistance.

THE L-MATCH

Another common matching technique is the 'L-match', so called because it looks like the letter L. In practice it is really a modified pi-match. You will remember that the matching to the base of a power transistor required us to eliminate the base capacity with a shunt inductor. The L-match performs the same function but uses a series inductor to eliminate the base capacity. But the major difference is that in the L-match the base capacity becomes the second capacitor in the pi-match (C2 in **Fig 4.4**).

The components CB and RB are the base capacity and the base resistance (real part). The L-match only requires two components, L1 and C1 to transform from the high impedance to the low impedance at the base. Normally to prevent upsetting the DC bias on the base a second capacitor is often needed in series with the input. This capacitor is simply a coupling capacitor with a low reactance, i.e. a high value chosen to suit the operating frequency; it performs no part in the matching.

An important thing to note about the L match is the position of the shunt capacitor C1. This is always at the end of the inductor that terminates in the higher impedance. By reversing the network we can transform up or down in impedance. In **Fig 4.7** and **Fig 4.8** the two different topologies are shown. Note that C2 is not part of the matching network, it simply serves as a DC block and it is a low reactance chosen to suit the frequency of operation. It can be omitted when no DC bias is present.

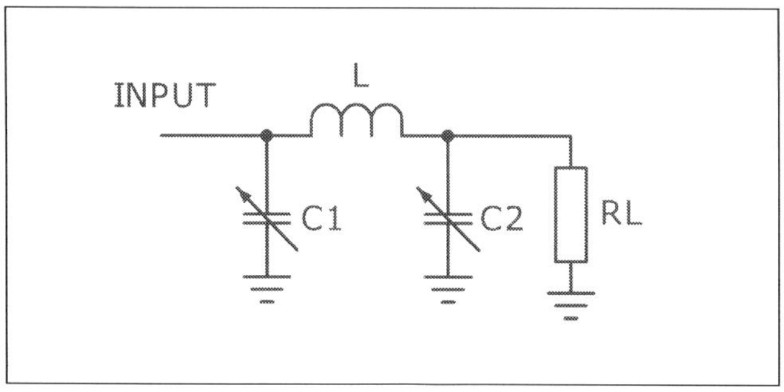

Fig 4.4. Pi network

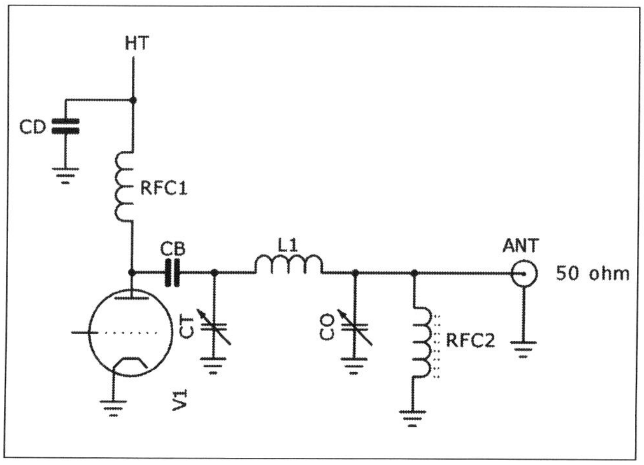

Fig 4.5. Normal pi tank network

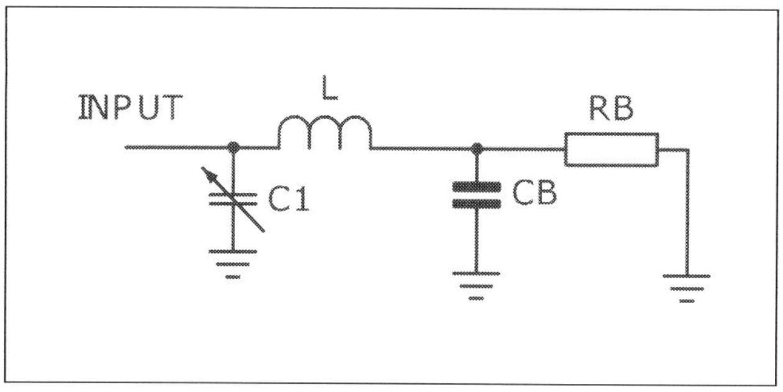

Fig 4.6. L-match network

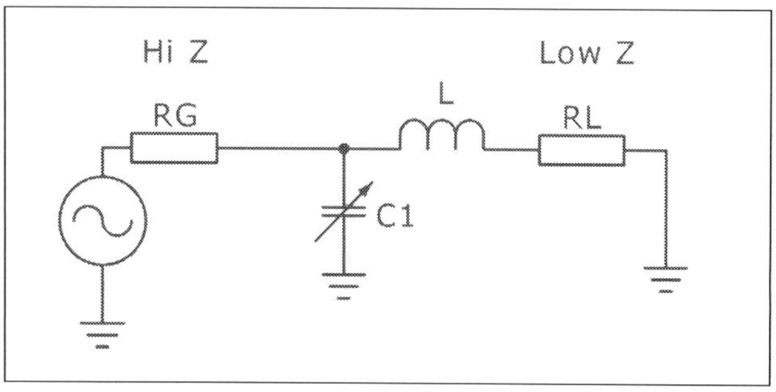

Fig 4.7. Transforming from high Z to low Z

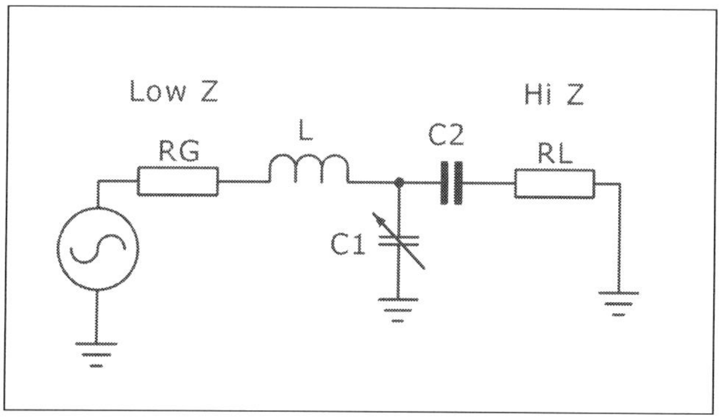

Fig 4.8. Transforming from low Z to high Z

IMPEDANCE MATCHING TO AN ANTENNA

This is one of the items that causes the greatest problems to amateurs. Although we might have a good VSWR meter it actually isn't of much use when the antenna matching is a long way from perfect. All the VSWR meter is able to tell us is the *mismatch ratio*. If the VSWR meter is designed for a system impedance of 50Ω it is only suitable for this impedance, so everything has be done using 50Ω.

As we saw in the chapter on RF measurements, VSWR is simply the ratio between the system impedance and the unknown impedance. So if the termination is 200Ω resistive the VSWR indicated is 200/50 = 4:1. For lower values of termination the same argument holds: with 25Ω the VSWR is 50/25 = 2:1. So it should be clear that the average VSWR meter doesn't tell us anything of much use, except the value of VSWR. For an indicated VSWR of 3:1 this could be due to a 150Ω load or one of 16.6Ω, we simply have no way of telling which one it is. However, armed with this insight we can use a technique to establish at least whether the antenna impedance is too high or too low, so causing the VSWR to be greater than 1:1.

Suppose we have an antenna that indicates 2:1 VSWR: this could be either 100Ω or 25Ω. By placing a 100Ω resistor across the antenna feed point one of two things could happen. The first is that the VSWR now falls to 1:1, which tells us the impedance was too high. The second thing that could happen is that the VSWR only changes by a very small amount or not at all. This tells us the impedance is too low and towards 25Ω. To correct this condition we could place a 25Ω resistor in series with the centre of the coax and it should then indicate 1:1. However, this is not a practical solution for either case as the resistor will absorb 50% of the transmitter power.

The average VSWR meter only tells us the 'percentage of error' but not the exact terminating load value. For this we need more complex test equipment or a little common sense as we have just demonstrated.

An impedance of 50Ω can be made up from myriad combinations of resistance and reactance, either series or parallel connected. When an antenna designed for 50Ω is measured using an ohmmeter we do not expect to measure a resistance of 50Ω, it could be an open circuit or a short circuit to DC.

MATCHING WITH QUARTER-WAVE LINE

A common technique where two impedances need to be matched is to utilise a section of coaxial cable cut to be an electrical quarter-wave length. The formula to calculate the required coaxial cable impedance Z_C is:

$$Z_C = \sqrt{Z1 \times Z2}$$

where Z1 and Z2 are the two impedances to be matched.

For example, an antenna of 100Ω impedance is matched to a 50Ω feeder cable with a quarter wavelength of coax of 70.7Ω. In practice a standard 72Ω coaxial cable will normally suffice with little error. The calculation of the physical length needs to take into account the velocity factor of the coaxial cable used. The effective matching

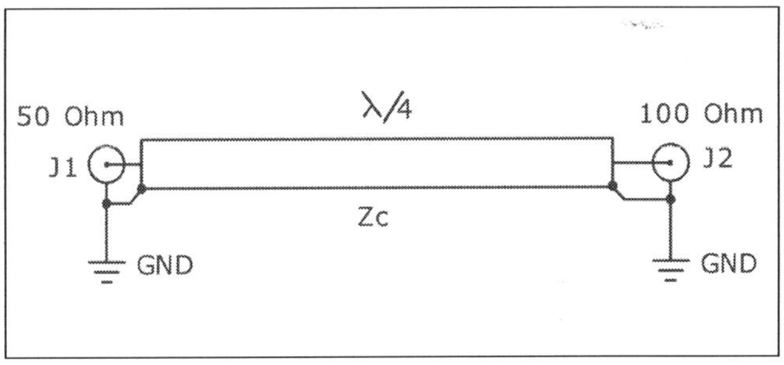

Fig 4.9. Impedance matching with a quarter-wave line

bandwidth will be approx 20% before a severe error occurs, so this method is useful at VHF and UHF where the bands are quite narrow in percentage terms.

To find the physical length of a λ/4 section we use the formula:

$$\text{Length} = \frac{300}{f\,(\text{MHz})} \times (\text{Vel factor} \times 0.25) \text{ metres}$$

Hence, a λ/4 coaxial line for 144MHz using RG-58/U or URM-43 is approx 343mm long as the velocity factor of the cable is 0.66.

The principle of matching with a section of λ/4 line can be expanded to use lumped components to simulate a section of coaxial line. If the lumped network is designed to have a 90° phase shift from one end of the line to the other and the component values are selected correctly we can replace a piece of coaxial cable with a pi network. This is shown in **Fig 4.10**. An advantage of this method is that at lower frequencies the lumped network occupies a far smaller volume. (At 80m a quarter wavelength of RG-213/U coax is approx 13.2m in length.) The second advantage is that any characteristic impedance 'line' can be obtained by selecting the component values correctly. With coaxial cable we are somewhat limited by the available impedances manufactured but using a pi network we can match over a larger ratio of impedances.

The formulae to calculate the component values are:

$$L = \frac{Z_o}{2\pi f} \text{ and } C1, C2 = \frac{1}{2\pi f Z_o}$$

where Z_o is the impedance required and *f* is the operating frequency in Hertz.

The response of a typical pi network used to make

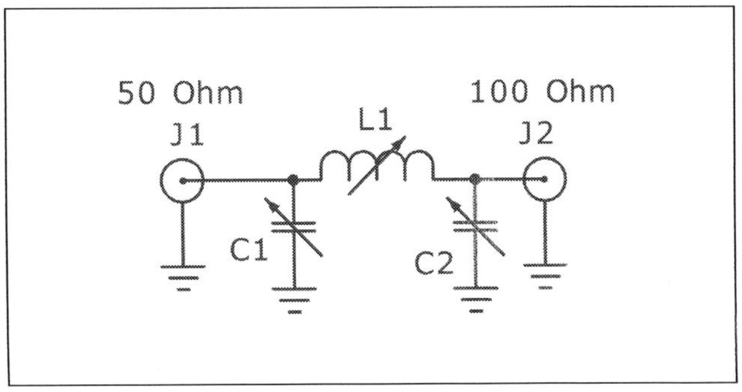

Fig 4.10. Simulating a λ/4 line with lumped components

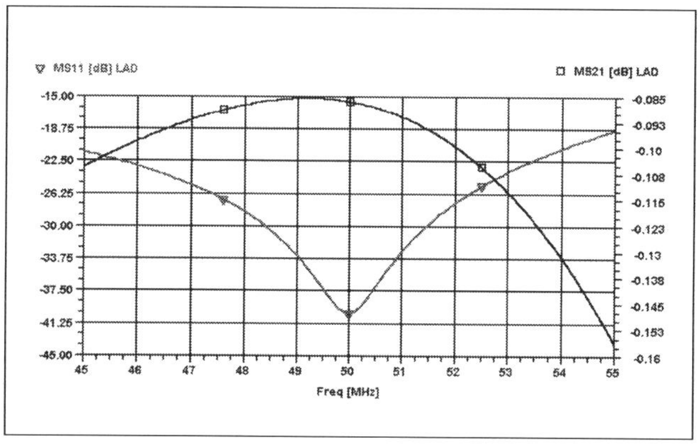

Fig 4.11. Response plot of pi network used to simulate a λ/4 line

a simulated λ/4 line is shown in **Fig 4.11**. The top trace is the insertion loss and the lower trace is the input match in dB format. Although the network looks like a low-pass filter it behaves more like a band-pass filter.

REACTANCE INVERSION OF λ/4 LINE

What isn't so obvious about the different electrical lengths of transmission line is how reactances are inverted. If the line is λ/2 the impedance looking into one end is exactly the same as occurring at the other end. In other words, if we place a reactance of +100Ω at the far end of the line and 'look into' the other end we see the same +100Ω reactance. But if the line is λ/4 the reactance is now inverted and if we look into the line the reactance is -100Ω.

This is a very powerful tool for impedance matching. If for example we need to 'place' a reactance of inductive nature across a transmission line but cannot tolerate a DC short circuit condition, a λ/4 line with a capacitor across the far end can do the job. The impedance of the coaxial cable or open wire lines can be any suitable value, it does not need to be same as the main line.

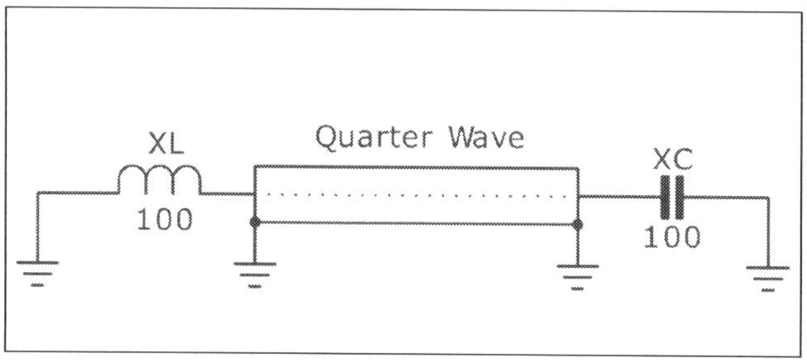

Fig 4.12. Reactance inverting property of λ/4 line

CONJUGATE MATCHING

Most impedance matching is known as 'conjugate matching'. This means we use an equal value but opposite sign reactance to cancel the mismatch. Assume the load is 50 +j25Ω, this means the resistive part is 50Ω and the reactive portion is 25Ω with a positive sign, hence it is inductive. To match correctly to this we need to generate an impedance of 50 -j25Ω, that is we need to generate a capacitive mismatch. When the two impedances are conjugately matched we obtain maximum power transfer (neglecting any loss caused by the matching network components due to non-infinite Qs. Generally all practical matching networks have some small loss, as long as the loss is less than ≈5% of the total power we can normally ignore this fact.)

THE ANTENNA TUNING UNIT (ATU)

When the antenna being used is inherently narrow band, such as a λ/2 dipole for 80m, the transmitter may have difficulty loading into it if the frequency selected is a long way from the resonant point. For modern solid-state transmitters this is a problem. Because the matching networks in the power amplifier are fixed tuned, they are often unable to develop full power into a reactive load and the VSWR detector in the transmitter backs off the power to prevent damage to the transistors. With older valve transmitters this is not normally a problem because the pi tank network can be adjusted over a wide range, often as much as a VSWR of 5:1, and so attaining full power output is not a problem. Many modern solid-state transmitters now include a built-in ATU to counter these problems. But due to the small volume available for the ATU the insertion loss is quite high because the inductor Q able to be achieved is less than optimum. This is like placing a low value attenuator in series with the transmitter output to absorb some of the mismatched power.

Popular types of ATUs are the pi network and the T network. The pi network offers better suppression to transmitter harmonics because it is inherently a low-pass network. The T network with series capacitors and a shunt inductor on the other hand is a high-pass network and so does not offer much attenuation to harmonics, in some cases it can accentuate them and this can be a serious problem. The alternative T network consists of two inductors in series with a shunt capacitor connected at the mid-point to ground. This offers better harmonic suppression but limits the range of reactance it can match over. Many commercial ATUs can be configured as pi or T match networks by switching.

Connecting an ATU between the transmitter and the antenna feed line corrects the problem because the transmitter now 'sees' a low reactive termination close to the design impedance. With a sufficiently adaptive ATU we can load into almost any impedance. The party trick of an amateur I knew in the UK was to load up a metal bed frame on 160m and then conduct a QSO with the locals! For random length 'long-wire' antennas an ATU is essential (in fact a 'long-wire' antenna by definition is an antenna which is several wavelengths long, so many of the random-length wire antennas used by amateurs are not in fact long wires but 'short wires'.)

Walter Maxwell, W2DU, in his book *Reflections* explodes more than a few myths about antenna matching and misconceptions by some amateurs about VSWR and its effects on radiation efficiency. To quote: "Probably the most serious and widespread misconception concerning SWR prevailing throughout the amateur fraternity is the erroneous notion that there is a direct one-for-one relationship between reduction in reflected power and the resulting increase in radiated power. In other words, many believe that for every decreased watt of reflected power there is an additional watt of increased radiated power. Not so, but the number of amateurs who have been misled to believe this invalid and unscientific premise is unbelievable.

"Another related concept, popular, but also erroneous, is that when terminated in a mismatch, the coaxial feed line becomes part of the radiator, causing radiation from the feed line because of the standing wave. This is untrue because the line voltages and currents, and the standing wave resulting from the mismatch, are entirely contained in and between the outer and inner conductors, inside the coax. No standing wave develops on the outside of the coax because of impedance mismatch. However, feed line radiation may result from standing waves on the outside, if a balanced dipole is fed without using a balun. There is also no radiation from tuned open line feeders, because the current is flowing in opposite directions; hence the fields developed by the currents in each wire are cancelled."

Many amateurs believe they understand the simple impedance matching technique of a $\lambda/4$ section of coaxial cable to transform from one impedance to another. The pi and T networks are simply different ways of achieving the same effect. The pi network is popular for this application and it can be made to simulate any wavelength between about 0.1λ up to about 0.375λ by simply varying the network values. Effectively we have a variable characteristic impedance and variable line length by tuning the air variables and selecting a different tapping point on the inductor.

But what most amateurs do not appreciate is that although the transmitter is made to 'see' a low VSWR looking into the ATU input terminals, the impedance the ATU 'sees' at its output terminals is still the same as before. In other words, we haven't changed the antenna feed line impedance - to do this we would have to change the physical dimensions of the coaxial cable - we have simply fooled the transmitter into thinking it now 'sees' a dummy load! So the radiation efficiency due to the antenna VSWR hasn't changed at all. All that has happened is that the transmitter is now able to deliver more power into the ATU input terminals.

The ATU will have some inherent through loss because the inductor and capacitor Qs are not infinite and hence some power will be dissipated in the ATU. The important factor is whether or not the extra power we can develop from the transmitter is more than the insertion loss of the ATU. If the ATU through loss is more than we achieved as extra power then we are actually worse off than without the ATU in circuit. For example, with a moderately high VSWR the transmitter may only develop 80% of the rated power without an ATU. With the ATU in circuit and adjusted for lowest VSWR the transmitter now develops 100% of its rated power into the ATU, but the ATU may have an insertion loss of 1dB. 1dB in round numbers is $\approx$20%, so only 80% of the power generated by the transmitter is now applied to the input of the feed line. Hence, we haven't gained anything at all. Modern solid-state transmitters with very compact ATUs shoehorned into a limited space have relatively high losses and this serves to degrade the potential power output.

The equivalent circuit that the antenna presents to the feeder is shown in **Fig 4.13**. When the antenna is resonant XC and XL cancel and only R is left. This gives a VSWR of 1:1 if the feed line is the same value as R. At frequencies either side of resonance either XC or XL will be larger and so the antenna looks like a complex impedance containing resistance and reactance. As has already been mentioned, if the antenna resistive part is different to the feed line impedance the VSWR will be greater than 1:1. For example, a half-wave dipole at resonance looks like an impedance of 73.1 +j0Ω. A suitable coaxial cable impedance would be either 72 or 75Ω, this would cause a VSWR of about 1.02:1. If a 50Ω coaxial cable is chosen the VSWR would be about 1.5:1. Therefore if your dipole is a true 73.1 +j0Ω at resonance and you are using 50Ω coaxial cable and a VSWR meter designed for 50Ω, if the VSWR meter shows 1.5:1 you are bang on resonance. Unfortunately most amateurs think this is wrong and unnecessarily fuss to get the VSWR to 1:1. If you really must see 1:1, then changing the coax to 72Ω and the VSWR meter to a 72Ω type will indicate 1.02:1 which is so close to 1:1 it is academic. However, it is very unlikely your dipole will be exactly 73.1 +j0Ω at resonance and certainly won't be if you move a few tens of kHz off resonance.

Any VSWR below 2:1 is quite satisfactory and striving to lower the VSWR beyond this does not increase the signal strength observed by a distant station by any significant degree. If the extra loss caused by a high VSWR on the transmission line is 1dB the station at the far end will not be able to detect any degradation because the typical S-meter is calibrated in 6dB per S-point. A 1dB loss is hence about one sixth of an S-point and no commercially-manufactured amateur transceiver has an S-meter calibrated that accurately!

A further fallacy is that a high VSWR causes higher harmonics to be radiated and this may cause television or radio interference. There is absolutely no truth in this belief and it can be proven that a transmitter's harmonic attenuation is not affected by a high VSWR on the output. What is a problem with solid-state transmitters is that the power amplifier stage is very sensitive to mismatch, especially reactive mismatch. This causes the intermodulation performance to suffer. This is why the manufacturer's fit 'reflected power back off' to hold the IMD products to something reasonable. Valve power amplifiers do not generally suffer from this problem and can happily work into a load with a VSWR of as high as 10:1 with no ill effects. In fact many of the older valve broadcast transmitters used very high feed line and antenna VSWR. The BBC Radio 4 198kHz long-wave transmitter at Droitwich used open wire feeders to carry the transmitter power to the antenna with a 'designed for VSWR' of 30:1 and did so for more than 50 years with no problem.

What is not so obvious from the various impedance matching methods is that all they work by generat-

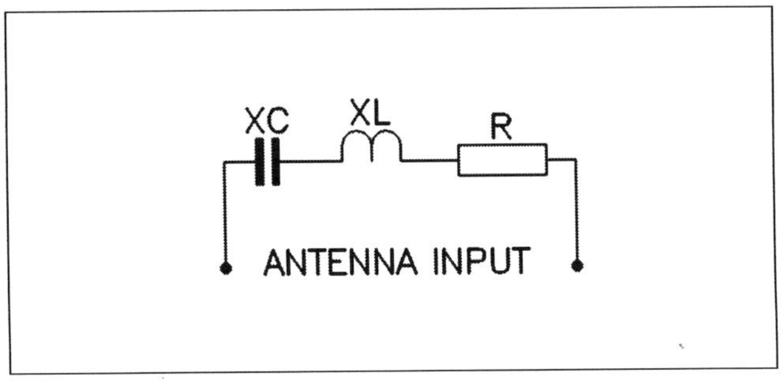

Fig 4.13. Equivalent circuit of antenna off resonance

ing a high VSWR somewhere in the network. This *is* a case of 'two wrongs making a right', in some cases the VSWR can be as much as 100:1 and is absolutely no problem as long as the components in the networks have sufficient voltage or current rating. Under an infinite VSWR condition with an open circuit line the line voltage only rises to a value of twice the matched line condition. Similarly, the current under a short circuit condition only rises to a value of twice the matched line condition, the voltage of course being zero. To dissipate power we need both voltage and current, in neither case is any power 'lost'. In the case of the open circuit line condition, although the line voltage at the end of the line is twice the normal, there is no current flowing. In the shorted line condition although the current is twice the normal there is no voltage present and again no power can be 'lost'. Again we need to remember that no power can be dissipated in a pure reactance, one without any associated loss resistance. For power to be dissipated the voltage and current need to be in phase and in a pure reactance the current and voltage are 90° out of phase.

These facts are well understood by professional engineers but a lot of the published articles by amateurs have failed to see the true facts and clouded the issue with assumptions, some of which defy the laws of physics. Unfortunately amateurs read these publications and believe them to be gospel but do not delve into the subject to see if the quoted facts hold water.

In fact there is nothing preventing us from using a vastly different transmission line impedance than the antenna or transmitter. In some cases where low loss transmission is essential we can convert from the 50Ω transmitter to as high as 600Ω balanced wire lines to reduce the line attenuation over long runs. At the far end we simply convert to the antenna impedance, whatever it happens to be. So a 50Ω transmitter could drive an antenna of, say 100Ω feed point via a long balanced wire transmission line with suitable conversion networks. As long as the conversion networks have low insertion loss the total line attenuation will be far lower than the commonly used coaxial cable. The higher the balanced line impedance the lower the capacitive losses between the two parallel lines so little power is lost due to this. Also, the wider the line spacing the higher the impedance and also the higher the breakdown voltage. This requires the spacers between the parallel lines to be made from a low loss material with minimal high voltage leakage characteristics. For the power levels used by amateurs often wooden dowels boiled in wax can be utilised or plastic rods made of synthetic material such as nylon.

> Line impedance can be visualised as two different diameter water pipes. For high impedance lines, the pipe is a small diameter and hence needs a higher pressure to cause water to flow to deliver a certain volume in a given time. The electrical version of 'pressure' is voltage and hence a high impedance line for a given power flowing (the volume of water) requires a higher RF voltage. A low impedance line is the analogy of a large diameter pipe and now we need much less pressure to make a certain volume to flow in a given time. Hence, a low impedance line has a lower 'pressure' of RF voltage.

DOES MY ATU *REALLY* TUNE MY ANTENNA?

One of the firm beliefs of some amateurs is that an ATU does *not* actually 'tune' the antenna. This can be considered from two viewpoints. If the feeder cable is disconnected from the antenna feed point and a Vector Network Analyser connected instead, we can measure the resistive and reactive portions presented by the antenna. Under this condition the VSWR still exists. However, if we connect the feeder and observe before the ATU the VSWR is 1:1. Moving the Vector Network Analyser to the point after the ATU (disconnecting the ATU) allow us to see the impedance presented at the input to the transmission line.

If the feeder is an exact multiple of half wavelengths the impedance we should see is exactly the same as at the antenna feed point. But this is a very big *if*. What is more likely is that the feed line is not an exact multiple of half waves and certainly won't be as we move about the band. Also the antenna will only present a low VSWR at one spot frequency. So the ATU 'impedance' (actually its reactive portion) is transformed back up the feed line incurring inversions of the reactance every 90° of cable electrical length (every $\lambda/4$) until it reaches the antenna feed point. Under this condition the 'impedance' the antenna presents is now modified by the ATU on the far end of the feed line.

So yes, the ATU *does* in fact 'tune' the antenna. A more correct analysis is that the ATU 'tunes' the antenna *and* the feed line connected in series, and we cannot divorce the two because one cannot work without the other. If an extra $\lambda/4$ section of coaxial cable is connected we have to flip from, say, a capacitive reactance generated by the ATU to an inductive one of the same value because the $\lambda/4$ inverts the reactive portion. This will bring the VSWR back to 1:1. If we add an extra $\lambda/2$ section we don't need to do anything because we have a 'double inversion' which brings the reactance back to its original value. Therefore with one length of feed line we need a certain ATU setting to obtain a VSWR of 1:1 but change the feed line length or frequency and we have a whole new set of variables to deal with. The only case where the feed line length is immaterial is when we use a dummy load instead of the antenna. Then we don't need the ATU because the dummy load presents a flat match over a very wide bandwidth, assuming the dummy load is the same resistance as the feed line impedance. Unfortunately no known antenna behaves like a dummy load across a wide bandwidth. Those that appear to do so have high loss resistances, which dissipate the power but do not radiate a significant portion.

So striving for a feed line which is an exact multiple of a half wavelength is another fallacy we can easily blow out of the water. If the electrical length of the feed line happens to be 5.7619λ this is completely irrelevant because an ATU will be able to 'tune' the antenna from the transmitter end as long as it has sufficient tuning range. If the electrical length is analysed the important part is the decimal fraction, in this case 0.7619 and since this is made up of $\lambda/4$ sections we subtract $\lambda/2$ to find the 'effective electrical length'. This is 0.2619λ and so is close to a real $\lambda/4$. If we substituted another cable with an electrical length of 0.2619λ it will behave in exactly the same way as the original cable.

WHAT HAPPENS TO THE REFLECTED POWER?

Another popular myth or urban legend is what happens to the reflected power when it arrives back at the output terminal of the transmitter. Some published articles tell us it is absorbed by the power amplifier stage and cause untold mayhem. Some articles are adamant this causes excessive dissipation in the output device, which can lead to damage. This is absolute nonsense and can be easily proven to be incorrect.

I presume this assumption is made is because the output stage is thought of as being similar to a signal generator. A signal generator normally has a fairly high value attenuator between the oscillator or output stage so that the impedance it presents to the load is constant. Hence, any power reflected by the load is absorbed and dissipated as heat. A power amplifier does not work like this. Remember the 'conjugate match' we discussed earlier: the power amplifier is a conjugately matched device and hence does not look like a resistive source or a load. It can be thought of simplistically as a 'one-way valve' that only allows power to pass from the amplifier to the load. If any reflected power is sent back to the power amplifier it closes the valve and will not allows it to pass back into the amplifier stage.

Conjugate matching is the opposite reactance to that which the load presents. Hence, no significant power is passed back into the amplifier stage. Instead the reflected power is 'repelled' by the amplifier and forced to travel back to the antenna. What this means is that if, say the VSWR mismatch causes the antenna to reflect 25% of the power (a 3:1 VSWR), and the transmission line has no loss, the amplifier only needs to provide 75% of the maximum power to the load. The reflected 25% when it is turned around in direction adds to the transmitter power to make 100% of the required power. Hence, under a high VSWR condition the amplifier doesn't need to work as hard to generate all the power. Hence, it is not as heavily loaded. By adjusting the amplifier loading controls we can in fact deliver more than 100% of the rated power into the transmission line. This is called 'reflection gain'. In this case the forward power meter would read 125%. This can be proven in a laboratory with accurate directional couplers with high directivity to reject the reflected power portion. Unfortunately the grade of instruments amateurs are able to acquire (because of economic factors) have insufficient directivity to see this and the reflected power appears to be subtracted from the forward power because of this measuring inaccuracy.

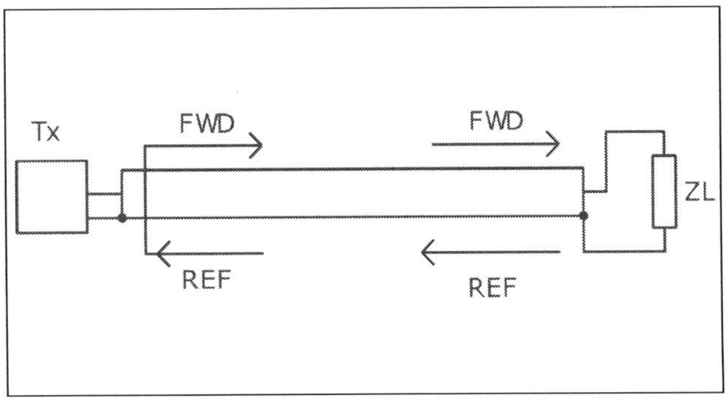

Fig 4.14. RF power flow under mismatched line condition

Depending on how you analyse the power amplifier it can be thought of as a current source or a voltage source. A current source has zero output impedance and a voltage source has infinite output impedance. Since neither is anywhere close to the feed line impedance it presents an infinite VSWR to the reflected power and it turns around and travels back to the antenna again, just like a high VSWR antenna. Most amateurs understand the reflection occurring at the antenna but cannot visualise the same thing happening at the transmitter, but in fact there is no difference.

The power travelling from the transmitter to the mismatched load will suffer some attenuation in a practical feed line. The same attenuation occurs to the reflected power travelling back from the antenna to the transmitter. If the feed line has 3dB of loss, the power arriving at the antenna is one-half of the power input to the feed line by the transmitter. Similarly, the power arriving back at the transmitter is also only half the power turned back by the antenna due to the mismatch. If the VSWR meter is placed at the input to the feed line (transmitter end) it will give an erroneous reading. If the VSWR meter is placed at the antenna end it will give the correct reading. With the VSWR meter at the transmitter end we need to know the feed line attenuation to calculate the antenna VSWR correctly.

> If the transmitter is delivering 100W into the feed line and the feed line has a loss of 3dB, only 50W arrives at the antenna. If the antenna is totally mismatched, either an open circuit or a short circuit, all of the 50W will be turned around and travel back to the transmitter. Because this power also suffers the same feed line attenuation when it arrives back at the transmitter it is now only 25W. The VSWR meter uses the 100W and 25W values to calculate the VSWR and shows 3:1. But the VSWR at the antenna is infinite.
>
> If the feed line has 6dB of loss, the VSWR appears to improve, it is now 100W forward and 6.25W reflected and now reads approx 1.65:1, which most amateurs consider acceptable.

What is a problem, as mentioned earlier, for a solid-state power amplifier is the power back off used by commercial amateur transceiver manufacturers. This is actually a crazy idea. The manufacturers of the transistor test them with very high VSWR loads to ensure they are able to withstand this sort of abuse. Often a VSWR of 30:1 is used, simply because it is difficult to generate anything much higher. Under testing the devices are driven to full load power and with a high line operating DC supply voltages and the VSWR is swung through the whole 360° of the Smith chart to see if any damage or instability occurs. This is a very severe test and the manufacturers can guarantee safe operation under these conditions. This testing is very much more severe than a device would normally encounter during use in a practical circuit.

A more serious problem is the IMD degradation under high VSWR conditions. This is because the transistor is no longer matched for optimum power transfer and the IMD products can become excessive under this loading. If the matching networks could be adjusted for this mismatch the amplifier would return the same low IMD products it does when terminated in a 50Ω load. Because the matching is 'fixed tuned' this is not possible. This is a penalty we pay for convenience and 'no-tune' operation.

Chapter 5

Receivers

BY definition, a radio receiver is 'a device which accepts the electrical signal from an antenna and by a process of amplification, filtering and detection outputs an intelligible signal'.

THE SUPERHETERODYNE RECEIVER

The superheterodyne (superhet) is the most commonly used type of receiver today, the reason being that the bulk of the amplification and filtering can be performed at a low frequency, the *intermediate frequency* or IF. The superhet uses a process of frequency mixing to bring the input signal to a lower frequency. At this lower frequency the amplification and signal filtering to reject out of band noise and other interfering signals are more easily performed. Hence, it is possible to achieve the very high signal amplification and good selectivity required without the difficulty of instability that could occur at the higher input frequency.

LIMITATIONS TO SENSITIVITY

In receiving systems the concept of *signal to noise ratio* (SNR) is used. The *signal* is the wanted output and the *noise* by definition is either internally generated in the receiver or externally generated by an interfering source. Noise, therefore by definition, is any signal other than the desired one.

The limitation to the ultimate sensitivity of the receiving system is the noise performance of the receiver first amplifier stages, the *front end* or *Low Noise Amplifier* (LNA). For maximum sensitivity, the front-end amplifier stages need to have the lowest possible noise figure with adequate gain.

As the first stages in a receiver effectively determine the overall receiver sensitivity, it is important to strive for the lowest noise figure in the early RF amplifier stage(s). Contributing factors to the overall receiver noise figure include the transmission line and connectors that connect the antenna terminals to the first amplifier stage. This can be a large contributor to the overall noise figure if lossy coaxial cable or connectors are used. In an optimum system the LNA would be connected directly to the antenna feed point, reducing the insertion loss of any cable to essentially that of the connector losses. The VSWR mismatch between the antenna and the LNA can add another fraction of a dB to the noise figure. In many cases the LNA will present a severe mismatch to the antenna feed point in order to obtain the optimum noise figure. VSWR mismatch of 10:1 or more is not uncommon for certain types of LNA.

IMAGE NOISE CONTRIBUTION

An important factor in a superhet receiver design is the noise power contained at the image frequency. Due to the mixing of the input signal to a lower IF, two possible input signal frequencies can produce the IF for a given local oscillator signal. One is the *upper sideband* and the other is the *lower sideband*. In a receiver *down-converter* using low-side LO injection, the upper sideband is the wanted signal and the image occurs at the lower sideband.

If the image frequency is not sufficiently suppressed by either filtering in the RF amplifier stages or by some other technique, the image noise, if it is an equal signal level to the wanted (the case for pure noise caused by resistive means) will cause a 3dB (50%) degradation of the signal-to-noise ratio. In many cases, due to interfering sources, the image noise power can be much greater than the wanted signal. There is however a trade-off to be made between the sensitivity gained by image noise filtering and the noise figure degradation caused by the loss in the first filter section before the amplifier stage.

It is possible to select too narrow a filter in an effort to suppress image noise; hence it will have appreciable loss. It is as well to remember that any loss before the first amplifier device will add directly to the noise figure of the amplifier. Whereas the receiver signal-to-noise ratio may be improved by reducing the image noise component, it may well be that this is more than negated by the noise figure degradation caused by the image filtering loss. Low noise amplifiers have little or no filtering before the active amplifier device in order to minimise the noise figure.

It also is prudent to select the local oscillator frequency to avoid potential interfering signals at the image frequency. For each input signal and IF combination there are two possible local oscillator frequencies that may be used. Often for ease of construction we would wish to choose the low-side injection in preference to high-side injection. Choosing low-side injection means that the multiplication for the local oscillator chain is less than for high-side injection and the LO frequency is consequently lower. But it may be that the image frequency potentially contains strong interfering carriers. For example, if 144MHz were the input frequency and the IF 21.4MHz, the choices of LO are (144 - 21.4) = 122.6MHz or (144 + 21.4) = 165.4MHz. The image frequencies would be 101.2MHz for low-side injection and 186.8MHz for high-side injection. The choice of low-side injection places the image in the middle of the FM broadcast band where strong interfering signals will be present. The choice of high-side injection places the image within the television broadcast Band III, where strong interfering signals at the image frequency may also be present. For these two cases the image rejection would need to be in excess of 100dB to prevent interference. By selecting a different IF, and hence LO, the image can be moved into a quieter portion of the spectrum.

IMAGE-REJECTION MIXERS

An alternative to image signal rejection by filtering is the use of an *image-rejection mixer*, which is a common technique used at microwave frequencies where filtering is often difficult. An image-rejection mixer (IRM) consists of two double balanced mixers, two 90° phase shifters and a 0° power combiner. By using this technique the unwanted image band of frequencies is cancelled out. A block diagram of a typical IRM is shown in **Fig 5.1**. Today for microwave applications the mixers and phase shifters are often constructed in chip-form directly on an IC.

The conversion loss of an image-rejection mixer is usually only a fraction of a dB higher than a normal type, in some cases it is a little less than a conventional mixer. However, one important fact to be aware of is that although image-rejection mixers reject coherent carriers at the image frequency they do not reduce the image noise, which is what we are more concerned with. So although we can utilise an image-rejection mixer we will also still need some image filtering to improve the system noise figure.

In practice, if the image signal noise power is reduced by approx 10dB the image noise contribution becomes insignificant. It is possible to achieve 25dB of image suppression to coherent carriers in an image-rejection mixer without introducing significant losses, and by also using wide-band low-loss filtering in the RF input a suppression of 40dB or more is achievable. Only in the case of the image frequency containing a strong coherent interfering signal would any greater suppression be required. Here it is preferable to utilise a notch or trap filter tuned to the image frequency that introduces minimal loss at the signal frequency. High performance HF communications receivers typically have image rejection figures of 70dB to 100dB due to input filtering alone.

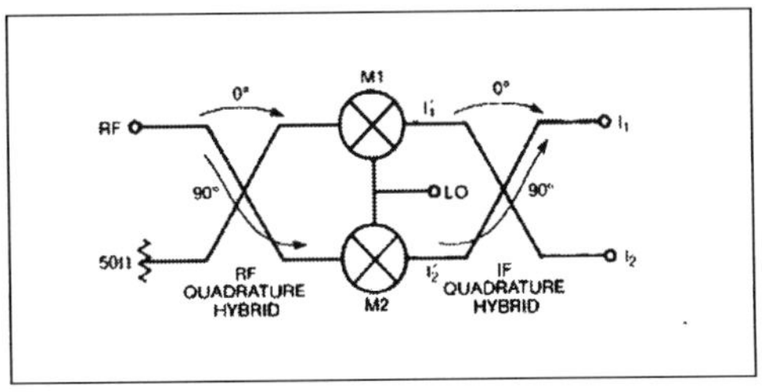

Fig 5.1. Image-rejection mixer block diagram

UP CONVERTING MIXER METHOD

Another way around the lossy image filter problem is to use a very high 1st IF, as the image frequency is situated at a frequency of twice the IF away from the wanted frequency. This for a low frequency input signal (in the HF or VHF range) will entail firstly mixing the signal up to a higher frequency of typically 1GHz, a second narrow-band down-converter is then used to bring the signal to a normal IF. The input amplifier band-pass filter can then be replaced with a low-pass filter rolling off in response a little above the receiver highest frequency. Low-pass filters generally have lower insertion loss than a narrow band symmetrical band-pass filter. The filter losses after the mixer at 1GHz are relatively small and can be tolerated in such a system. This is the technique used in a spectrum analyser where it is necessary to cover a wide frequency range in a single sweep. A sharp cut-off low-pass filter precedes the input to the mixer. The 1st LO then mixes the band up to a high 1st IF, perhaps as high as 2.2GHz for an analyser covering DC to 2GHz. After band-pass filtering at 2.2GHz the signal is mixed down to a lower 2nd IF and filtered for further signal processing and detection in a logarithmic amplifier. When a narrow range of frequencies needs to be scanned the 1st LO is phase-locked to an internally generated signal and the 2nd LO is swept over a narrow band.

Equipment for measuring the noise figure will, if the receive mixer is not preceded by an image filter, give an optimistic reading due to the image noise adding to the wanted signal. If the receiver consists of a wideband LNA (no image filtering) and a wideband double balanced mixer, the noise figure measured will be 3dB less than the actual figure. Noise figure measurements need to be done using a *single sideband* technique and not a *double sideband* type, which an unfiltered mixer with image noise contribution inherently gives.

LOCAL OSCILLATOR NOISE CONTRIBUTION

The assumption for optimum noise performance is that the local oscillator signal is a perfect sinusoidal carrier of zero bandwidth, but in the real world this is not the case. If the local oscillator is noisy, either in amplitude or phase and contains significant noise power at the image or other spurious response, the noise will be mixed on to the wanted IF signal. This is known as *reciprocal mixing* and causes degradation of the signal. Receiver local oscillators need to have extremely low noise performance in order to maximise the sensitivity. A crystal controlled os-

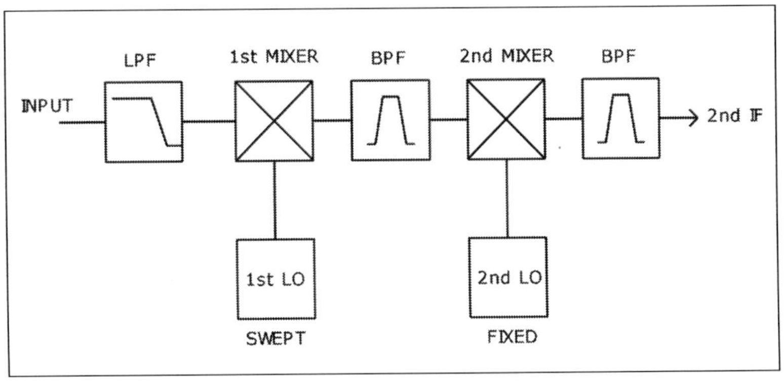

Fig 5.2. Block diagram of an up-converting spectrum analyser

cillator is usually far superior to a synthesised oscillator.

BROAD-BAND NOISE AND NOISE POWER

As the noise power contained in a 1Hz bandwidth for a particular condition is constant then the same follows for other bandwidths. Suppose we measure the absolute level of power in the 1Hz bandwidth, let us say it is 1μW in a 50Ω impedance for convenience. If we now use a measuring bandwidth of 1kHz the level of power we expect to measure is 1000 times as high, being 1mW. (You can imagine this is 1000 windows stacked side by side, each one is 1Hz wide, each of which lets 1μW pass. The total amount of power is therefore the sum of all the windows.) This is an increase of 30dB. If we use a 1MHz measuring bandwidth, the power level will be 1000 times higher still, or 1W. Therefore it is much easier to measure the power in a wide bandwidth than it is in a very narrow bandwidth. As the level of noise power in a 1Hz bandwidth is likely to be very small, it is easier to use a wider measuring bandwidth and then calculate the effective 1Hz bandwidth figure. In fact we can choose any suitable bandwidth and then work back to find the *1Hz-bandwidth* value. Everything can be scaled to a new bandwidth.

> The noise power in a resistor at an ambient temperature of +17°C (+290K) is a noise power of 4 x 10^{-21} W/Hz = -174dBm/Hz. Lowering the ambient temperature to close to absolute zero reduces the noise power considerably.

Now imagine a bandwidth 1MHz wide in the centre of which we place a carrier that occupies a bandwidth of 1Hz. If the noise is measured in a broadband power detector the carrier will need to be more than 1 million times greater in amplitude (60dB) to appear above the noise floor. If we now reduce the measuring bandwidth to 100kHz the signal only needs to be 100,000 times greater (50dB) to appear above the noise floor. Taking this concept further, if we further reduce the measuring bandwidth to, say, 1kHz the signal only needs to be 1000 times greater (30dB) than the noise to appear above the noise floor. When the measuring bandwidth is reduced to 1Hz only the signal will be left and all the other 1Hz segments containing noise will be excluded. Hence, by reducing the measuring bandwidth the signal appears to rise out of the noise floor. This causes the S/N ratio to improve as the system bandwidth is reduced.

SKY NOISE

The noise we measure when an antenna is pointed to different parts of the sky is determined by various factors. One of these is the operating frequency. At low frequencies the sky noise is high, reducing as we increase the frequency, hence there is no advantage to be gained by using a low noise figure for HF receivers because the sky noise is very high. The second factor is the elevation angle of the antenna and where the main antenna beam is pointed. At low elevation angles the antenna sees a far greater portion of the warm lower atmosphere than when pointed vertically. Excluding any radio stars that might be in the antenna beamwidth will give a uniform noise temperature for a particular operating frequency and antenna elevation angle.

There are, however, exceptions to this general rule. These are due to additional noise sources such as the oxygen frequency and others. These occur in the higher

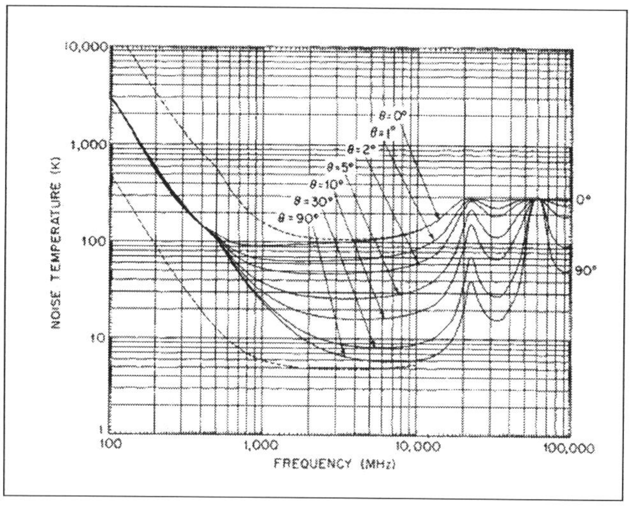

Fig 5.3. Sky noise versus frequency

(microwave bands) frequencies.

The graph in **Fig 5.3** shows the general level of noise to be expected for different frequencies and antenna elevation angles. The lowest frequencies have the highest noise temperatures and the higher frequency generally the lowest noise temperatures. There is, however, an ambient *background microwave radiation* (BMR) noise level of approx 2.7K at the higher frequencies.

From this graph can be determined the minimum and maximum sky temperatures for the various frequencies. The lower dashed curve is the lowest temperature to be expected when all other sources of noise are excluded. If you examine the lower dashed curve you will see that at 144MHz the lowest sky temperature is about 200K. This sets the lower limit to the noise figure required. Reducing the receiver noise figure below this point does not bring any significant improvement in sensitivity and is likely to cause intermodulation products to be generated by strong in-band signals if excessive gain is used.

Of interest are the two bumps in the noise curve at 22GHz and 60GHz, these are due to water absorption and the oxygen resonance effects respectively.

For an HF receiver we do not need very low noise figures. The lowest noise figure required is on 10m and even here 10dB is perfectly adequate. Lowering the noise figure at the expense of degrading the intermodulation performance is a common fault found in many amateur designs. On 160 and 80m the lowest noise figure practical is about 20dB because of the high level of static noise when an efficient antenna is used.

DIGITAL SIGNAL PROCESSING

With the advent of digital signal processing (DSP) receiver design has changed considerably over the last 10 or 15 years. Whereas in the early days of DSP it was difficult to manufacturer fast analog to digital (A-D) converters with acceptable noise performance and dynamic range, today it is possible to make very fast converters that work in real time and have clock rates up to 2GHz. Today the receiver does not consist of much hardware, the IF amplifiers have been replaced in some receivers with a simple wide band amplifier and then this is fed into the A-D converter.

In earlier days, because of the limitation in the available ICs, the A-D converter worked at a very low IF and the signal was down converted and then applied to the A-D converter. With very fast A-D converters becoming affordable the DSP now directly converts the noise spectrum at the input frequency into a digital bit stream for later processing. With computers becoming more powerful it is now possible to do the processing in real-time. Modern DSP receivers are zero IF types and the mixing, filtering and detection are performed directly in software using the raw digital bit stream. The advantage of the DSP approach is that it is possible to make continuously variable IF filters with only software and one piece of DSP hardware. This makes the DSP receiver approach much more cost effective as once the initial high development cost has been overcome the receiver depends entirely on the software used.

RECEIVER NOISE PERFORMANCE CALCULATIONS

The formula for calculating the system noise performance is listed below:

Noise Factor (F).

$$F = \frac{\text{Available S/N at input}}{\text{Available S/N at output}}$$

or

$$F = \frac{P_{si}}{P_{ni}} \times \frac{P_{no}}{P_{so}} = \frac{P_{no}}{G_a P_{in}}$$

where:

P_{si} = input signal noise power
P_{so} = output signal noise power
P_{no} = available noise power at the receiver output
G_a = available power gain
P_{ni} = available input noise power (Watts) due to kT_oB_n

Note: Noise Factor and gains are expressed as power ratios. Noise factor (F) is a unit-less measurement. Noise Figure (NF) is the logarithm of Noise Factor and is expressed in dB.

$$NF = 10 \log_{10} F$$

Hence: Noise Factor of 1 is equal to a Noise Figure of 0dB, Noise Factor of 2 is equal to a Noise Figure of 3dB, Noise Factor of 4 is equal to a Noise Figure of 6dB, etc.

The available input noise power, P_{ni}, is the noise power resulting from a matched source resistor at temperature T_o.

The signal-to-noise ratio at the output of an amplifier will always be less than at the input, as any practical amplifier will add some noise. Noise Factor (F) is a measure of the amount of noise added. F will always be greater than unity (or NF > 0dB).

When several stages are cascaded to form a system, the following formula is used (Friiss' formula):

$$F_{total} = F_1 + \frac{F_2 - 1}{G_1} + \frac{F_3 - 1}{G_1 \times G_2} + \frac{F_4 - 1}{G_1 \times G_2 \times G_3} \ldots$$

Note: Gains and noise factors are expressed as power ratios.

From this it can be seen that the overall noise performance is largely determined by the first stage (F1) and that the addition of later stages degrades the overall noise figure to a lesser extent, defined by the gain of the first stage (G1), providing the first stage is not a lossy line with G<1.

The noise characteristic of this stage can alternatively be represented by assuming that it has an equivalent noise temperature such that the noise power produced at the output of the stage due to internally generated noise is given by:

$$P_n = kT_eBG_a$$

$$\text{Thus } kT_eBG_a = \frac{(F-1)}{kT_oBG_a} \text{ or } T_e = \frac{(F-1)}{T_o}$$

The cascaded noise factor equation can be written in noise temperature format (modified version of Friis' equation):

$$T_e = T_e1 + \frac{T_e2}{G_a1} + \frac{T_e3}{G_a1 \times G_a2} \text{ etc} \ldots$$

As an example:
Noise Figure of a 200K device is:

$$NF = 10 \log_{10} 1 + \frac{200}{290} = 2.3\text{dB}$$

ANTENNA NOISE TEMPERATURE

The concept of equivalent noise temperature may be extended. If an arbitrary source of noise (thermal or non-thermal) is random, it can be modelled as an equivalent thermal noise source and characterised with an equivalent noise temperature. Thus, components and systems can be characterised by saying that they have an *equivalent noise temperature*.

For example a non-thermal device, such as an antenna, may be associated with an effective noise temperature, T. If the noise power received by the antenna is P_n in a system bandwidth, B, then T_{ant} may be expressed as:

$$T_{ant} = \frac{P_n}{kB}$$

where:
k is Boltzmann's constant = 1.38×10^{-23} joule/K
B is the system bandwidth in Hertz

The term T_{ant} includes all noise contributions up to the *reference plane*, which is normally taken as the input to the first amplifier stage (commonly referred to as the *Low Noise Amplifier*, LNA).

There are basically two major contributors to antenna noise temperature, namely sky noise and noise from ohmic losses in the antenna or coaxial cable. Sky noise comprises all external noise that enters the antenna. In satellite and radio astronomy systems, the sky noise is due to extra-terrestrial noise, thermal radiation from the atmosphere and man-made noise reflected from the earth.

The atmosphere affects external noise in two ways. It attenuates signals passing through it from outer space, and also generates noise because of the energy of its constituents. Sky noise generally varies with frequency, elevation angle of the antenna, (at low elevations the antenna 'sees' a greater portion of the atmosphere and some of the warm earth), and surface water vapour concentration.

EFFECTS OF SKY NOISE

As the sky noise reduces so dramatically as the operating frequency rises it is often more productive to choose a higher than a lower frequency for modes such as EME, even though the expected path loss is greater. One can safely assume that for frequencies up to about 144MHz half the system noise temperature is due to antenna and sky noise and the remainder is due to the LNA. Therefore there is no significant benefit to be gained by reducing the LNA noise figure much below half of the total system temperature.

For example, at 144MHz with the antenna pointed at the horizon in an urban environment the typical sky noise will be of the order of 1000K. With the antenna elevated to 45° it will be reduced to approx 200K. This is an equivalent LNA noise figure of (2.3 / 2) = 1.15dB. With the antenna pointed at the horizon with a 1000K sky noise the lowest noise figure required is about (6.5 / 2) = 3.25dB. In a rural environment the ambient noise with the antenna pointed at the horizon will be lower, perhaps as low as 300K, and then a lower noise figure can be used to advantage. Even so it is limited to about 3dB.

At 432MHz the sky is less noisy and the typical figures for these two cases will be approx 150K and 20K. This is an equivalent minimum LNA noise figure of 0.3dB.

At 1296MHz the sky noise has dropped to about 5K for an antenna pointed vertically upwards, assuming no interfering noise source is within the beam. This is an equivalent LNA noise figure of 0.07dB. It is extremely difficult to get anywhere near this with conventional microwave transistors at a modest cost. To achieve this very low noise temperature requires cryogenic cooling.

Turning this fact around explains why there is no great improvement to be gained by using an extremely low noise figure on the lower bands. At 144MHz for all modes except satellite or EME there is no advantage in striving for less than about 2dB NF. For 144MHz EME going down to as low as 1dB is all that is required. On 432MHz and above the lowest noise figure that can be achieved is of benefit as the sky noise is so low at the higher frequencies.

An equation for determining the antenna noise temperature up to the reference plane is:

$$T_{ant} = \frac{(l_a - 1)290 + T_{sky}}{l_a}$$

where l_a is the numeric value of the system ohmic losses up to the reference plane.

For an antenna-receiver system, the total effective system noise temperature, T_{sys}, can be defined:

$$T_{sys} = T_{ant} + T_{rx}$$

where T_{rx} is the equivalent noise temperature of the receiver.

At lower frequencies the required noise temperature is not that low due to the atmosphere generating considerable noise at these frequencies plus the sky noise. At HF there is no need for a low noise figure, at 21MHz a noise figure of 15dB is quite adequate and the receiver is rather designed for strong signal handling than low noise performance because of the amplitude of the signals expected. At frequencies in the UHF and SHF regions, the atmospheric sky noise drops rapidly and the receiver can benefit from a lower system temperature. However, at increasing frequencies it becomes increasingly more difficult to achieve low noise performance from active devices.

MULTIPLE CONVERSION FILTERING

In a conventional superhet receiver we might start with a first IF of 10.7MHz and then convert to 455kHz. The 10.7MHz filter does not need to be very narrow; approx 100kHz is adequate and this can easily be achieved with LC or ceramic filters. This filter is often referred to as a *roofing filter*. The main selectivity is obtained at 455kHz after the final mixing process. As long as the receiver amplifier stages preceding the roofing-filter have sufficient signal handling performance to strong signals within the roofing filter bandwidth this will produce acceptable results. At this point in the superhet receiver the total gain from the antenna to the roofing filter is normally low and so the amplitude of the signals is also low.

SYSTEM CALCULATIONS

In order to establish if the proposed receiver is able to fulfil the required parameters we need to do some calculations using the various building blocks. Whereas it is quite simple to do these with a scientific calculator the availability of analysis software makes the task less onerous and allows many different combinations to be examined in a short time. I use the program *SysCalc™*.

Fig 5.4 shows a typical analysis for a proposed Hydrogen Line receiver at 1420MHz. This receiver requires a very low noise figure and an IF bandwidth of approx 50kHz.

The various blocks are constructed with the gain or loss, noise figure and intermodulation parameters as-

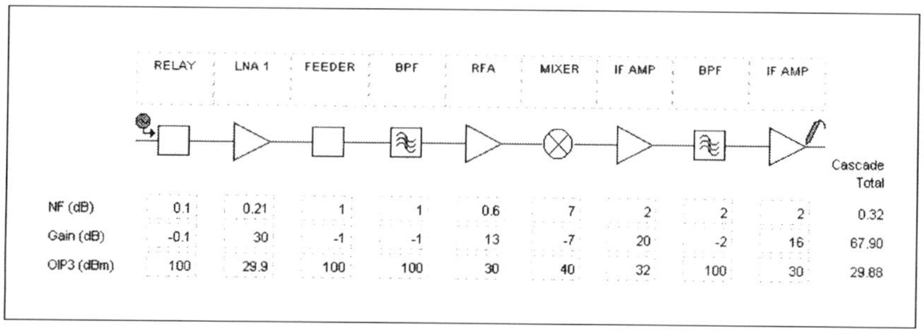

Fig 5.4. Analysis software output screen

signed. The system will consist of a two-stage PHEMT LNA mounted at the antenna and the main receiver is situated remotely with a feeder cable connecting the two units. The relay connecting the LNA to the antenna is used as part of a switching system to allow a load resistor or the antenna to be alternatively connected to the LNA; this was measured to have a loss of 0.1dB. This is a microwave change over relay obtained from a scrap Hewlett-Packard piece of test equipment that covers 1 to 18GHz. As this precedes the LNA this loss adds directly to the overall noise figure. The predicted noise figure of the LNA first stage is 0.18dB with a gain of 13dB; this design is using Agilent ATF-34143 PHEMT devices in a strip-line circuit. The first stage is optimised for the best noise figure and the second is optimised for best gain of 17dB consistent with obtaining a noise figure of 0.3dB. The overall LNA noise figure is 0.21dB and the gain is 30dB.

The input signal level assumed was -140dBm and so the power levels shown throughout the analysis results show how this level varies with the gains and losses of the various stages. At the output of the main receiver, prior to the detector, the level has risen to -72dBm because the overall gain is approx 68dB. The detector will be the AD606 log amplifier, which has an intercept point sensitivity of -88dBm and a useful input sensitivity of approx -80dBm. Therefore the receiver system will be able to resolve signals much below -140dBm, by altering the input power level we can see that the minimum discernible signal is -148dBm with the 50kHz bandwidth. If we choose to use a narrower IF bandwidth, say 10kHz, then the MSD level falls accordingly. However, a very narrow IF band width would require the LO to be variable in frequency to correct for the expected Doppler shift from distant objects.

The main receiver consists of a band-pass image rejection filter at 1420MHz and then an LNA prior to the mixer stage. Connecting the LNA and the receiver is a length of low loss feeder with a predicted loss of 1dB; this will be a piece of foam-filled microwave coax. It would be preferable to eliminate this extra loss by siting the receiver down converter at the antenna, but environmental problems concerning weather proofing and ambient temperature changes make this a difficult challenge. By changing the analysis file we can see what advantage is likely to occur if we choose this route. In fact there is no significant improvement, the overall noise figure remains the same, the only change is 1dB more output signal at the input to the log amplifier, so there is no need to go to all the trouble and inconvenience of mounting the receiver at the antenna. In fact with 6dB of feeder loss the overall noise figure only rises to 0.34dB, so the loss in this piece of cable is relatively insignificant.

The noise figure of the main receiver is approx 2.7dB, being the band-pass filter loss and the noise figure of the receiver LNA stage and the following stages.

The other thing the SysCalc software is able to do is to analyse how large a signal is needed to paralyse the receiver. If the input signal is above a certain level the later receiver stages will be driven into a non-linear region and will generate IMD products that will mask the wanted signal. With an input level of -140dBm the spurious free dynamic range (SFDR) is 59dB. So as long as the input signal does not exceed -140 -59 = -81dBm the receiver will perform satisfactorily. If the input level is anticipated as being greater than -81dBm it will be necessary to provide some form of manual gain control in the receiver, perhaps in the IF amplifiers following the mixer.

RECEIVER BAND-PASS FILTER DESIGN CONSIDERATIONS

The successful construction of a receiver front end requires various parameters to be considered and appropriate values assigned. If this groundwork is omitted or done poorly the likelihood is that the receiver performance will be inadequate for the task in hand.

Before we can ascertain the critical areas we need to specify the parameters. Once this is done we can then decide which parameter is the most important and list the various parameters in descending order of importance for our intended receiver design.

The various parameters (in no particular order) are:

1. Required sensitivity
2. Image rejection
3. Bandwidth
4. Large signal handling characteristics

The ultimate sensitivity depends on the noise performance of the early stages. If the amplifiers used for these have a high noise figure or the filter stage before the amplifier has a large insertion loss, the overall sensitivity will be degraded. The image rejection is related to the choice of IF and the local oscillator injection. The bandwidth is dependent on the percentage width of the amateur band we desire to cover. The amplifier device we choose largely determines the signal handling parameters and potential noise figure. Let us examine each parameter in detail.

SENSITIVITY

The potential sensitivity it is possible to achieve with a given noise figure is determined by the IF bandwidth required. For narrow band modulation schemes such as CW and SSB we can choose an IF bandwidth just wide enough to pass the wanted signal. This maximises the signal-to-noise ratio (SNR). In such a case we need not strive for such a low noise figure. If, however, we also desire to have good sensitivity for wideband modulation schemes such as FM, then we need to pay more attention to the front-end noise figure.

The overall system noise figure is determined by the first few stages in a receiver. If the stage preceding an amplifier consists of a loss, the system noise figure is degraded. In practical systems this will always be the case; the antenna feeder will exhibit some attenuation before the first amplifier stage. Also if the amplifier stage consists of a band-pass filter prior to the amplifier device the insertion loss of the filter and the feeder loss add to increase the net loss.

Assume we have a feeder loss of 3dB between the antenna and the first low noise amplifier. We will assume the low noise amplifier has a noise figure of 2dB. Therefore, the overall system noise figure cannot be less than the sum of the feeder loss and the amplifier noise figure; it will be greater than 5dB in this example. The higher the gain in the first amplifier stage the less this adds to the overall system nose figure.

However, it is very unwise to use more gain than necessary. We need to remember that at this point in the receiver the band-pass filtering before the amplifier is quite wide-band and will not reject strong signals several megahertz away from the channel frequency too well. Further down the receiver chain the bandwidth is far narrower and hence the unwanted signals will be reduced to insignificant levels. But at the first amplifier device it has very little frequency discrimination. Because of the noise performance aspects we cannot tolerate too much insertion loss in the first band-pass filter, this degrades the overall noise figure, so the first filter stage is often quite wide-band to reduce the loss as much as possible. In the case of mast-head mounted preamps this is even more of a concern. Low noise amplification and strong signal handling are generally not compatible with most commonly available devices.

The Q of the inductors and capacitors used for the filter determine the insertion loss of the filter. Normally the inductor Q is the dominant portion, for practical filters using good quality RF capacitors the capacitor Q is normally at least 10 times that of the inductor Q. As there is a practical limit to how high the inductor Q can be this limits the insertion loss from being zero.

STRONG SIGNAL PERFORMANCE

If there are strong signals from other transmitters close to the wanted frequency, the amplifier stage may begin to overload and cause intermodulation products to be generated. So the best option is to use very little gain in the first stage, only sufficient to overcome the filtering loss before and after the amplifier, consistent with maintaining an acceptable noise figure. We can analyse the performance of a front-end stage using computer software to assess the noise and signal handling performance. **Fig 5.5** shows a typical scenario for a 70cm receiver using commercial helical filters and a MMIC amplifier device.

The losses up to the first active stage consist of the feeder cable and the first band-pass filter, these add up to 3dB. The noise figure of the active device is 3dB and due to the high gain of the amplifier the system noise figure amounts to 6.01dB. In an ideal case the losses before the amplifier would be zero, if this were the case the system noise figure would be 3dB, being the noise figure of the amplifier. However, we need to look at what these equate to in system sensitivity. Taking our example receiver design with the feeder and filter loss we achieve the following:

Modulation	Bandwidth Required	Sensitivity	MDS
FM	15kHz	-114.2dBm	-126.2dBm
AM	6kHz	-118.2dBm	-130.2dBm
SSB	3kHz	-121.2dBm	-133.2dBm
CW	500Hz	-129.0dBm	-141.0dBm

The Minimum Discernible Signal (MDS) parameter is the signal level required to produce a signal-to-noise ratio (SNR) of 1 (0dB), which is equivalent to the receiver

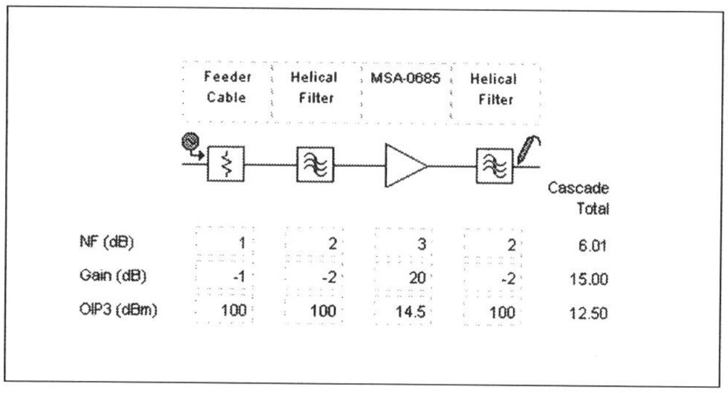

Fig 5.5. Example receiver front-end

noise floor. If the signal is at or below the noise floor we will not be able to detect it from the background noise. In this example the 'Sensitivity' parameter is the signal level required to produce an SNR of 12dB. For CW and SSB we can often tolerate a lower SNR, perhaps as low as 3dB for a skilled operator. For AM we can often get by with as little as 6dB SNR, for FM 10dB SNR is about the lower limit.

If the amplifier gain is reduced from 20dB to 10dB the ultimate sensitivity does not dramatically change, but the strong signal performance increases by a factor of 10dB. If the gain is reduced more the sensitivity drops a little more but the strong signal performance rises dramatically. The sensitivity achieved with different amplifier gains is listed below:

Modulation	20dB Gain	10dB Gain	3dB Gain
FM	-114.2dBm	-114.1dBm	-113.6dBm
AM	-118.2dBm	-118.1dBm	-117.6dBm
SSB	-121.2dBm	-121.1dBm	-120.6dBm
CW	-129.0dBm	-128.9dBm	-128.4dBm

Reducing the gain from 20dB to 10dB only reduces the sensitivity by 0.1dB. Reducing the gain by 17dB to 3dB only drops the sensitivity by approx. 0.5dB.

The difference in the signal handling for different amplifier gains is shown below for an input signal of -115dBm.

Gain dB	System Noise Figure dB	OIP^3 dBm	IIP^3 dBm	SFDR dB
20	6.01	+12.5	-2.5	82.5
10	6.13	+12.5	+7.5	89.1
3	6.60	+12.5	+14.5	93.4

The 'Spurious-Free Dynamic Range' (SFDR) parameter is a measure of how large the signal can become before significant 3rd order IMD products are generated. With a wanted input signal of -115dBm (0.3μV) the figure is 82.5dB, which is an interfering signal level of -32.5dBm (≈2mV) when the gain is 20dB, as the gain is reduced the SFDR level increases. At a gain of 10dB the allowable interfering signal level increases to -26dBm (≈5mV) and for a gain of 3dB the level becomes -21.6dBm (≈10mV).

As can be seen the reduction in the gain from 20dB to only 3dB makes very little difference to the sensitivity, but less gain brings with it far superior signal handling.

The strong signal handling can also be analysed by using the 'Output Third-Order Intercept Point' (OIP^3) parameter. In the example an Agilent Semiconductors device was chosen. The figure of OIP^3 is given as +14.5dBm; this is the power level at the output of the amplifier device where the 3rd order intermodulation generated spurious signal crosses the theoretical linear curve. The actual output power where the device deviates from a linear curve by 1dB is known as the '1dB compression point'. This is generally approx 11 to 15dB below the OIP^3 point. In the case of the chosen device this is a power level of +3.5dBm and is equivalent to an input signal of 19dB less, defined by the device gain. As the device has a small signal gain of 20dB the maximum input signal it can handle is 20dB below the 1dB compression point value, or approx -16dBm. This is the maximum allowable input signal level for wanted or unwanted signals.

Whereas it is normally advantageous to reduce the losses before the first active stage to as low as possible we also need to consider the strong signal handling aspects. As the prominent form of intermodulation is due to the odd order products the 3rd and 5th order are the ones that are the most destructive. If our receiver is for the 2m band and if there is a strong signal from a two-way radio or FM broadcast transmitter that is able to enter the receiver amplifier stage, the possibility of intermodulation products being generated are quite high.

Take the case of a 2m receiver and the FM broadcast transmitter as an example. The receiver is designed to cover the band of 144 to 146MHz and the FM broadcast transmitter can be anywhere between 88 and 108MHz. Normally there are many FM broadcast transmitters situated across the 88 to 108MHz portion of spectrum. If two of these have sufficient amplitude to drive the active device into a non-linear portion of its operating range, 3rd or 5th order products will be generated. If the frequency separation of the two FM broadcast transmissions is such to cause a spurious product to fall either into the 2m band or the subsequent down-converted IF, this signal will tend to block out any wanted weak signal. It must not be forgotten that the mixer in a receiver is a very non-linear stage - it needs to be in order to convert the input signal to the IF efficiently.

Intermodulation generation is very dependent on the unwanted signal amplitude. 3rd order products have a law that generates a spurious signal 3dB higher for each increase of signal level of 1dB. It can be shown that if we attenuate the unwanted signal by an extra 1dB then the spurious product will drop by 3dB. Similarly for the 5th order products the extra 1dB of attenuation causes the spurious level to drop by 5dB. If we attenuate the unwanted signal by an extra 10dB, the 3rd order spurious will drop 30dB and the 5th order product by 50dB. Sometimes placing a small value attenuator in front of the receiver will cause the IMD products to disappear for little loss of the wanted signal.

If the spurious generated is equivalent to a signal of -100dBm when we are attempting to receive a weak signal of -120dBm, the spurious signal will override the wanted one by 20dB, making it impossible to hear it. Hence, the design of the receiver filtering will need to provide adequate attenuation to any out-of-band signals that are likely to cause IMD products.

LOW NOISE AMPLIFIERS

When we require more sensitivity than the average receiver can provide we need to use an external preamplifier to raise the signal level to improve reception. However, there are several factors that need to be considered. The first involves the strong signal handling capability of the receiver. If the preamplifier gain is excessive the receiver may be 'pushed over the edge' of its linearity ability. Hence, the gain should only be sufficient to improve the noise performance and no more. Achieving greater sensitivity entails providing a lower noise figure with adequate gain to overcome system losses.

The noise figure of a receiving system is defined by the noise figures of the amplifier stages and any losses before the first amplifier. If we take an average 2m receiving system and examine the various components we can see the effect the noise figure of the amplifiers have. Suppose we have a length of coaxial cable between the antenna and the receiver with a loss of 3dB. The average noise figure for a commercially-manufactured 2m receiver is about 5dB for the better varieties and it can be as high as 8dB or more for mediocre versions. (Some Japanese 2m transceivers measured in the 1980s showed upwards of 15dB noise figure and can be considered very poor compared with what can be achieved with careful design.)

With the 3dB of feeder loss and a receiver with a noise figure of 5dB the system noise figure to a first approximation is 3 + 5 = 8dB, being the coax loss plus the receiver noise figure. If we place a low noise preamplifier of 3dB NF with adequate gain between the coax and the receiver at the shack end we obtain a new system noise figure of 3 + 3 = 6dB. This is hardly any better than the 8dB we had before and it will not make any significant improvement to weak signal reception. If we now use a lower noise amplifier, say 2dB, the system noise figure improves to 2 + 3 = 5dB, still not a huge improvement. If we use a very low noise preamplifier with a 1dB noise figure the system noise figure improves to 1 + 3 = 4dB.

However, if we now place the preamplifier at the antenna instead of in the shack we see that the system noise figure improves dramatically. With the 1dB noise figure preamplifier now situated at the antenna and feeding the receiver via the feeder cable of 3dB loss we achieve a noise figure close to 1dB. Exactly how close to 1dB we achieve depends on one thing: how much gain the preamplifier has in excess of the feeder loss following it. If the gain is above ≈20dB the system noise figure is so close to 1dB that it is academic. For an EME system operating on 2m 1dB NF is more than sufficient.

The exact system noise figure obtained for various preamplifier gains when using a preamplifier with a NF of 1dB and a 3dB feeder loss is shown below:

G = 10dB	NF = 2.53dB
G = 20dB	NF = 1.18dB
G = 30dB	NF = 1.02dB

From the above we can see that we require a certain minimum gain to overcome the feeder loss and the noise figure of the following receiver. If we now run the calculations for a really 'deaf' receiver we can see the effect this has on the system noise figure. Suppose our 2m receiver is a bad example and has a noise figure of 10dB. The system noise figure obtained with a 1dB noise figure preamplifier at the antenna will be:

G = 10dB	NF = 4.99dB
G = 20dB	NF = 1.61dB
G = 30dB	NF = 1.06dB

Hence, we need to use more gain for the poorer receiver to obtain a similar noise figure.

If we now consider the strong signal handling performance we will see that the higher the gain in the preamplifier the more likely the receiver will be driven into a non-linear mode if strong adjacent channel signals exist. Normally we would choose the lowest gain possible consistent with achieving an acceptable noise figure but also avoiding overloading problems. A simple way around this is to reduce the signal being fed to the receiver by increasing the loss in the feeder cable between the preamplifier and the receiver. This can be demonstrated by the next set of figures for the 10dB NF receiver case.

Feeder Loss dB	System Noise Figure dB
3	1.06
6	1.13
9	1.26
12	1.51
15	1.97
18	2.76
21	4.01

Before we continue we need to consider the ultimate sensitivity we can attain with the system noise figure. We saw earlier that the sensitivity largely depends on the bandwidth of the IF, which is dictated by the mode we are using. For SSB we need an IF bandwidth of about 2.5kHz and for CW we can usefully go as narrow as 300Hz with an appropriate IF filter. If we assume we can copy a signal 6dB above the receiver noise floor with CW and SSB and 10dB for FM this sets the sensitivity limit of our receiving system. For the various modes we can list the capability of our example receiver below:

Feeder Loss dB	System Noise Fig dB	Sensitivity CW dBm	Sensitivity SSB dBm	Sensitivity FM dBm
3	1.06	-142.1	-132.9	-123.9
6	1.13	-142.0	-132.8	-123.8
9	1.26	-141.9	-132.7	-123.7
12	1.51	-141.7	-132.5	-123.4
15	1.97	-141.2	-132.0	-122.9
18	2.76	-140.4	-131.2	-122.2
21	4.01	-139.2	-129.9	-120.9

Hence, we see that even an extra 18dB of feeder loss between the LNA and the receiver only changes the ultimate sensitivity by about 3dB. So we can usefully exploit this to improve the strong signal handling characteristics without throwing away much in ultimate sensitivity. Every 1dB of extra loss equates to a 3dB improvement in signal handling for the third order intermodulation products and 5dB for the fifth order products, so it is quite significant. 3dB is double the level of interfering signal.

Just to put this into perspective, here is what the 10dB NF receiver on its own with 3dB of feeder loss equates to in sensitivity:

CW = -130.2 dBm
SSB = -121.0 dBm
FM = -111.9 dBm

These figures aren't that bad compared with commercial two-way radio specifications. The normal benchmark for an FM receiver is about -115dBm. Note that we haven't considered the antenna gain in these equations. If a 10dB gain antenna precedes the LNA then the figures all improve by approx the same amount.

A simple method of attaining the required loss in the receiver feeder cable is to use lossy coaxial cable. If RG-58/U or URM-43 cable is used the loss is approx three times that of RG-213/U or URM-67. This cable is also far less expensive and easier to support due to its lower weight and smaller diameter. For the VHF bands up to 144 MHz this is a suitable method. For the higher bands such as 70cm and upward we often need to use cable such as RG-213/U or URM-67 because of the higher losses encountered.

Exactly how low a noise figure we can usefully use is dependent on the sky noise at the frequency of interest.

IMAGE REJECTION

Another aspect we need to consider is the image frequency. If the IF is a low one the image will appear close to the wanted band. For some simple short-wave receivers using a 455kHz IF when used on frequencies higher than about 15MHz the wanted signal can often be found at two tuning dial settings separated by 910kHz. If, however, the IF is chosen to be high the filtering aspects are less difficult.

Let us take a typical example to highlight this problem. We will choose a 2m receiver with an IF of 9MHz. We have two choices of the local oscillator injection; either we can use *low-side injection* or *high-side injection*. If we choose low side injection the local oscillator will be at a frequency 9MHz below the wanted channel frequency and hence the image will be a further 9MHz below the LO frequency. (Image = Wanted Freq - 2 x IF). For a channel frequency of 145MHz the LO will be 136MHz and the image at 127MHz. This is in the middle of the aeronautical communications band and any aircraft flying nearby which happen to be using 127MHz will be a very strong unwanted signal. Aeronautical transmitters are AM types and often in the 100W to 400W range.

If we choose high-side injection the LO will be 154MHz and the image at 163MHz (Image = F + 2 x IF). This is within the commercial VHF high-band allocation (136 - 174MHz) where there are many strong signals.

If we choose an IF of 10.7MHz the corresponding

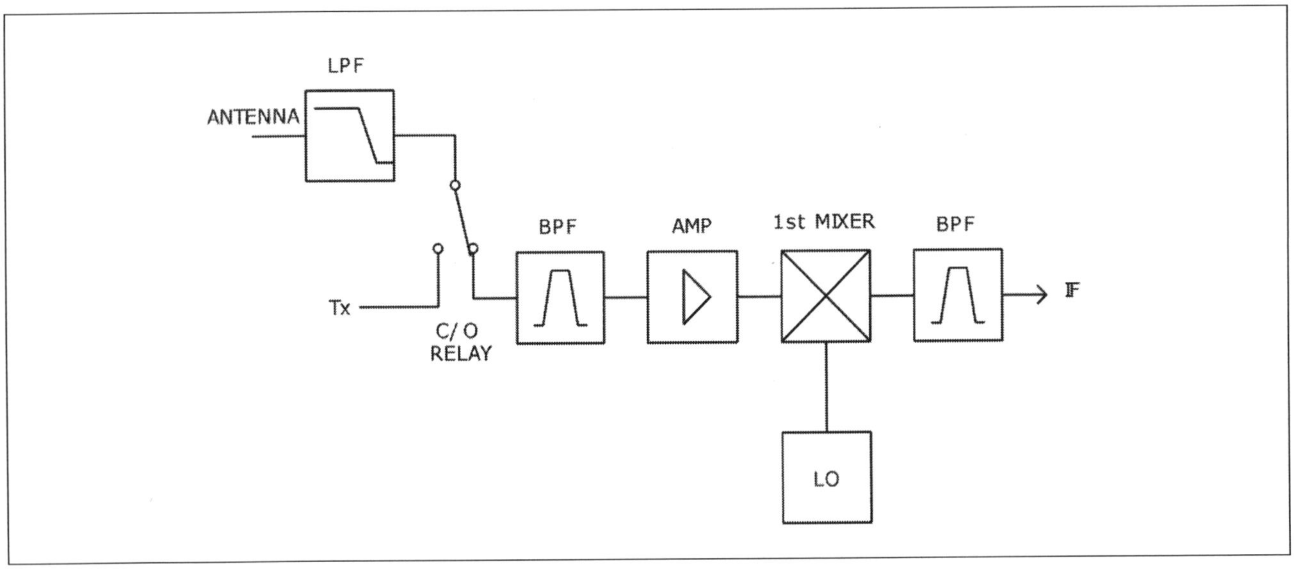

Fig 5.6. Harmonic low-pass filter used to improve receiver filtering

image frequencies would be 123.6MHz or 166.4MHz. 123.6MHz is within the aeronautical band and may be a problem, 166.4MHz is within the commercial VHF high-band allocation and is also likely to be a problem.

> A common problem with AM broadcast receivers is interference from amateurs operating in the 160m band. The RF amplifier stage in a typical AM broadcast receiver is not very selective and because of the local oscillator frequency and the common use of an IF of about 455kHz this causes a problem. The LO in the AM broadcast receiver is a high-side injection scheme in order to cover the whole AM broadcast band (540 to 1610kHz). The LO typically tunes over approx 995kHz to 1955kHz. The image falls within the 160m amateur band at a tuning dial setting of approx 900kHz.

If we choose an IF of 21.4MHz the image frequencies would be 102.2MHz or 187.8MHz. 102.2MHz is within the FM broadcast band and 187.8MHz is within the band-3 television band. Choosing an IF of 45MHz puts the image at 55MHz or 235MHz. With a large image offset this should be relatively easy to implement a filter with adequate image rejection and low insertion loss.

As we can see there is no simple solution, no matter which of the common IFs we select we are faced with the problem of an image frequency that may contain a strong unwanted signal, which will need to be attenuated by filtering.

In military radio communications where there may be several transceivers on a *platform* (typically a fighter aircraft or ship) these problems are sorted out by 'frequency planning', where the receivers and transmitters are programmed with frequencies that avoid potential self-jamming. For example, in an aircraft carrying one VHF and one UHF transceiver, the VHF transmitter will be programmed so that the second or third harmonic does not encroach closer than 10% away from any UHF receive channel frequency or image frequency.

Another factor is LO radiation. If the LO signal is allowed to leak out through the receiver front-end it will be radiated by the antenna, causing possible interference to other services. Often it is necessary to install a notch filter to reject any LO energy that manages to sneak out through the amplifiers and filters and limit the amplitude at the antenna connector. In some classes of radio equipment, for example FM broadcast receivers, television receivers and two-way radio receivers, there are laid down levels that may not be exceeded; fortunately most amateur radio equipment is not subject to this in most countries, except the USA.

> The FM broadcast receiver typically uses an IF of 10.7MHz and uses high-side injection to cover the band with a linear readout on the tuning scale. This places the LO between 98.7MHz and 118.7MHz, the upper portion of which is used by aeronautical systems for air-traffic control. Hence, any radiation can cause interference to this critical service and this is why stringent levels are imposed. Also a receiver radiating a signal at 98.7MHz will cause interference to other FM broadcast receivers in the vicinity because it falls within the FM broadcast band.
>
> A similar problem arises with UHF television receivers. The frequency allocation of the transmitters is normally such that a four channel spacing minimum between adjacent transmitters is adopted because of the potential of LO radiation from a receiver.

A further problem is that of multiple LO frequencies. It must not be forgotten that the mixer in a superhet receiver is a very non-linear device. Although the oscillator driving the mixer may be spectrally pure and contain no harmonics, the mixer will regenerate them up to the nth harmonic. For example, if the LO is at 100MHz, the mixer will generate equal amplitude carriers at 200, 300, 400, 500, 600, 700, 800, 900MHz etc. If any of these, with an unwanted signal that manages to get past the amplifier and filtering, has a frequency offset equal to the IF in use then it will generate a spurious IF signal. Therefore, we need to consider how much attenuation needs to be provided for these potential interfering signals.

Fortunately, once we have attained sufficient attenuation for the image frequency the filter often has enough attenuation to reject these other signals. However, the calculations should still be performed to highlight potential interfering signals. In two-way radio and military transceivers it is very common to find that the transmitter harmonic filter precedes the receiver bandpass filter. The insertion loss added by the harmonic filter is normally negligible, typically less than 0.25dB, but the extra attenuation afforded to high order signals is very worthwhile. This eases the design of the receiver filtering.

Very often when a top-coupled band-pass filter is used the slope attenuation on the high frequency side of the filter is less than on the lower-frequency side. Using the harmonic filter connected in series gives additional attenuation to the high side of the response.

ACCEPTABLE LEVELS FOR REJECTION

The image rejection is usually the most stringent parameter. A minimum value to aim for is 70dB below the channel frequency sensitivity. For those applications where more than 70dB of image rejection is required it is often not possible to achieve this with a normal band-pass filter. In such a case a notch filter placed ahead of the band-pass filter is a good approach. The insertion loss of the notch filter to the channel frequency is normally negligible, but an extra 40dB or more can readily be achieved at the image frequency. For UHF receivers where the inductor value or Q may become impractical a good alternative is a coaxial stub cut to the image frequency and connected in parallel with the receiver input. This can be a $\lambda/4$ open circuit stub cut to the image frequency.

The level for other unwanted parameters will depend largely on the physical location and what interfering signals exist in the nearby spectrum. In my case this is a serious problem. My QTH is situated on top of a mountain with a 360° radio horizon. Within 0.25km is a commercial two-way radio tower with high-band VHF and UHF transmitters. Within 0.5km is another commercial site with several UHF transmitters and a cell-phone tower. Approx 4km from the QTH is another mountain on top of which resides the local TV and FM broadcast transmitters. There are seven FM broadcast transmitters and five TV transmitters, three of which are Band III (VHF) and two are Band 4/5 (UHF). The FM broadcast transmitters are each rated at 100kW erp and the TV transmitters at 250kW erp. The signal levels from these are very strong as can be imagined and if adequate front-end filtering were not used it would be impossible to use the station on 2m or 70cm. I guess this is the penalty one pays for choosing a mountaintop QTH.

The strongest signal level present at the antenna port on 2m is from the Band III television transmitters and approx -10dBm, even with the non-resonant antenna for the unwanted signals. In my case 70dB rejection is totally inadequate and a figure closer to 120dB is required. This imposes a severe problem as the screening and shielding of commercial amateur receivers do not normally meet these criteria. Admittedly this is an extreme case, but it serves to highlight the design difficulties. For the average amateur living a fair distance from any strong interfering stations the rejection required will be proportionately less. However, if any contest operation is envisaged from a hilltop then this sets the target to aim for. In a contest there are many strong in-band and out of band signals that can cause receiver overload and spurious signals to be generated. Your receiver may perform perfectly well at home but completely fold up and die at a good contest site. Popular mountain-top contest sites are also used by commercial two-way radio repeaters and base stations, so the signal levels can be expected to be very high. Although these transmitters may be 50MHz or more from the amateur band, the levels of unwanted signals entering the antenna will be very high.

VARICAP TUNED BAND-PASS FILTERS

For operation where the signal levels are low, the varicap tuned band-pass filter offers a means of *tightening-up* the bandwidth. This means that the band-pass filter becomes what is known as a 'tracking filter', where a variable DC voltage is used to tune the filter to the wanted frequency. For bands such as 2m and 70cm they are not really necessary, it being very difficult to construct a filter with a bandwidth of less than the band. However, for 6m where the band extends from 50 to 52MHz (or 54MHz in some cases) the difficulty of constructing a filter that covers the entire band with very steep cut-off is more severe.

If commercial two-way radio receivers are considered, where the low-band extends from 66 to 87.5MHz and the high-band from 136 to 174MHz, the need for a tracking filter is paramount if the user has several frequency allocations scattered about the band. The same argument holds for HF receivers. In the case of the 80m band from 3.5 to 3.8 or even 4MHz it is very difficult to construct a band-pass filter to give full band coverage with adequate attenuation in the stop-band. On HF transceivers of yesteryears it was common to find a control on the front panel known as a *pre-selector* which the operator peaked on a signal for best S-meter reading. This is a form of tracking filter, but used an air variable capacitor. Today with solid-state transceivers it is almost never found.

A varicap tuned tracking filter to some extent is a good move, except where the level of unwanted signal is too high. The varicap diodes act like rectifier diodes above a certain RF voltage and can cause the DC bias voltage to be overcome. This then causes the filter to be slewed off-resonance with extra insertion losses. If varicap diodes are used on both the input and output band-pass filters and the input signal level increases, the output filter diodes will start to conduct before the input diodes because of the gain in the amplifier stage. This can be overcome to some extent by using dual diodes connected in series with the bias fed into the RF dead point via a high value resistor.

The varicap diode is normally connected across the high impedance portion of the parallel resonant tuned circuit. The RF voltage occurring here is determined by the *dynamic resistance* (R_D) of the tuned circuit at resonance. For example, if the inductor has a Q of 100 and the input impedance is 50Ω from the antenna then the value of R_D can exceed 20kΩ. The voltage 'step-up' is the ratio of the input impedance to the R_D. In this example the ratio is 20,000/50 = 400:1. If the DC bias on the varicap diode is approx 4V then it only requires about 10mV of

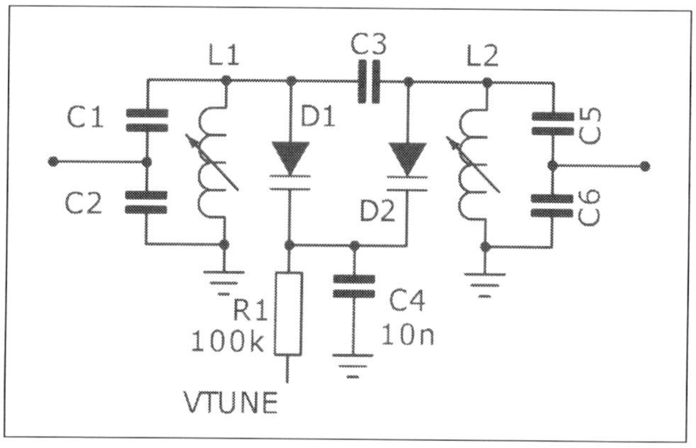

Fig 5.7. Varicap diode tracking filter

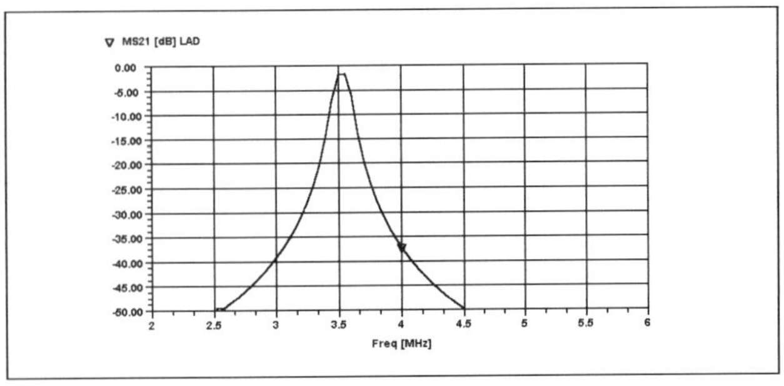

Fig 5.8. Pass-band response of 80m tracking filter

RF signal to cause the varicap to start to detune the circuit.

If the signal causing the detuning is due to an in-band wanted signal (within the IF pass-band) this is of little consequence, the AGC will already have reduced the front-end gain to a low value. However, if the signal is a strong in-band signal not within the IF pass-band the receiver may never reach full sensitivity under these conditions. Under crowded band contest conditions this can be disastrous as any signal greater than ≈10mV will cause the receiver to lose sensitivity. During VHF contest operation from a hilltop site with large antennas this is a common problem.

It is a good plan to have a switchable step attenuator so it can be switched in or out of circuit between the antenna and the receiver. If a 6dB attenuator is used the greatest loss in sensitivity will be 6dB. However, if a strong in-band signal close to the wanted channel frequency causes the filter to detune by 20dB you will be 14dB better off with the attenuator switched into circuit. An extra amount of attenuation also reduces the IMD products, for a 6dB attenuator the 3rd order IMD performance will improve by 18dB, which is about 70 times as a power ratio.

For the HF bands where very strong signals can be present the tracking filter front end is a good move. By making the filter very narrow the levels of adjacent channel signals can be reduced considerably. Tuning is via a variable DC voltage and so the filter can be remotely situated from the controls eliminating the mechanical problem of shafts that need to traverse the equipment from the front panel to reach conventional tuning capacitors.

IMPROVING STRONG SIGNAL HANDLING

The weak link in any chain is the part that will cause the greatest problem. In a superhet receiver the weak link is normally the mixer. To recap, the mixer needs to be a non-linear stage to convert from the RF signal to the IF signal efficiently. A mixer is a *multiplier device*, it acts by multiplying the LO signal with the RF signal to generate several new frequencies, one of which is the wanted IF signal. As well as the IF there are products of 2 x LO, 3 x LO etc and LO + RF, LO - RF, 2 x IF, 3 x IF, 2 x RF, 3 x RF etc. So a whole host of other signals exist at the output of the mixer, only one of which is the desired one.

A way to improve the strong signal-handling characteristic is to use a mixer and LO signal running considerable power. The higher the power of the LO signal the better the mixer is able to handle the strong unwanted signals. In days gone by the receiver was constructed using valves, today these are considered old fashioned but in actual fact were excellent in handling strong signals because they were running considerable power. With a heptode mixer (six-electrode valve) and an LO drive of 10V rms there is very little using solid-state devices that come close to it in performance. Many amateurs still regard the Drake R4C receiver as one of the best ever made for its excellent signal handling performance. Even equipment made during WWII, such as the AR88 series, are still better than modern amateur equipment using solid-state devices. One of the advantages of valve receivers is the longevity of the valves; some of these wartime receivers are still giving good service with valves made over 60 years ago.

> Valves in receivers rarely fail because they are being very much under-run and the problems that do arise with age are normally due to capacitors failing short circuit, resistors going high in value, or lack of maintenance and corrosion on the valve pins causing intermittent contact. In many cases if the other faulty components are replaced we find the valve is still in good working order. The most common failures are the rectifier and the audio output valves because these are dissipating the greatest power. These can be replaced with solid-state devices with little effort. The advantage of a valve as a rectifier is it inherently protects the equipment should a short circuit appear across the HT rail. This is because they have a limited anode current. Also they give a gentle start as the heater warms up, so the HT rises slowly while the other valves filaments are coming up to full temperature. Replacing a valve as a rectifier with silicon diodes gives an instant HT supply at switch on and can cause the smoothing capacitor to have a hard life. It is normally necessary to insert a current limiting resistor between the rectifier diodes and the smoothing capacitor to protect the capacitor from high surge currents.

A modern approach to high signal handling would be to use solid-state where it performs best, normally this is in the RF amplifier stages where the lower noise figure they attain out performs valves. (Although as we have demonstrated the average HF receiver does not need such low noise figures to perform satisfactorily and valves can achieve the required noise figure with good signal handling.) But for mixers and LO generation valves still take a lot of beating. Valves are also a better choice for IF amplification because they inherently are high impedance devices and we can obtain higher working Qs from the IF band-pass filters. Not only that, but the stage-gain a valve can attain is normally greater than solid-state; two valves in cascade can deliver about 90dB of gain with good stability. AGC is simpler to achieve with valves because the grids can be biased more towards cut-off with a negative voltage. Many valves contain two or even three valves inside one glass envelope so reducing the number required. A high performance receiver could be built using just five valves. So the ultimate would be a hybrid design using valves and solid-state devices.

> The Yaesu FT-200 SSB transceiver was almost the ultimate receiver. It used transistors for the VFO and crystal oscillators and valves elsewhere. In comparison to the then new FT-101 MkI solid-state transceiver the FT-200 out-performed the newer design by a large margin. On 40m in Europe the FT-101 was unusable with an average dipole antenna due to intermodulation products generated in the solid-state receiver mixer. The FT-200 was unaffected by the many strong signals and large portions of the band had unused channels. On the FT-101 the whole band *seemed* to be full, but this was caused by intermodulation signals. The later versions replaced the MosFet mixer with a double-balanced ring mixer to improve the signal handling characteristics.

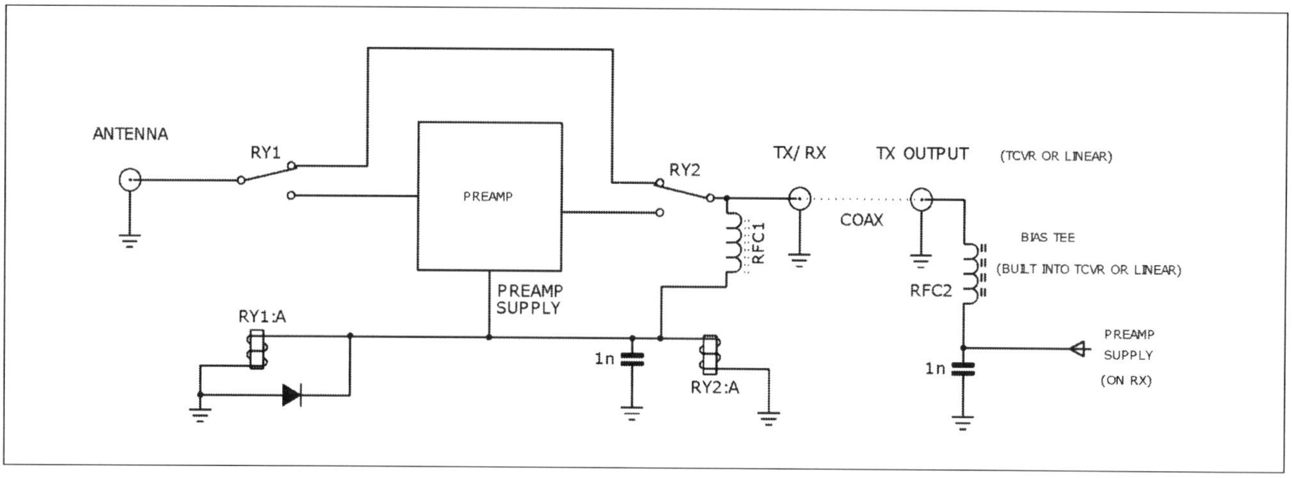

Fig 5.9. Method using two coaxial relays. Note: relays energised during receive. Both relays need to be rated at full RF power

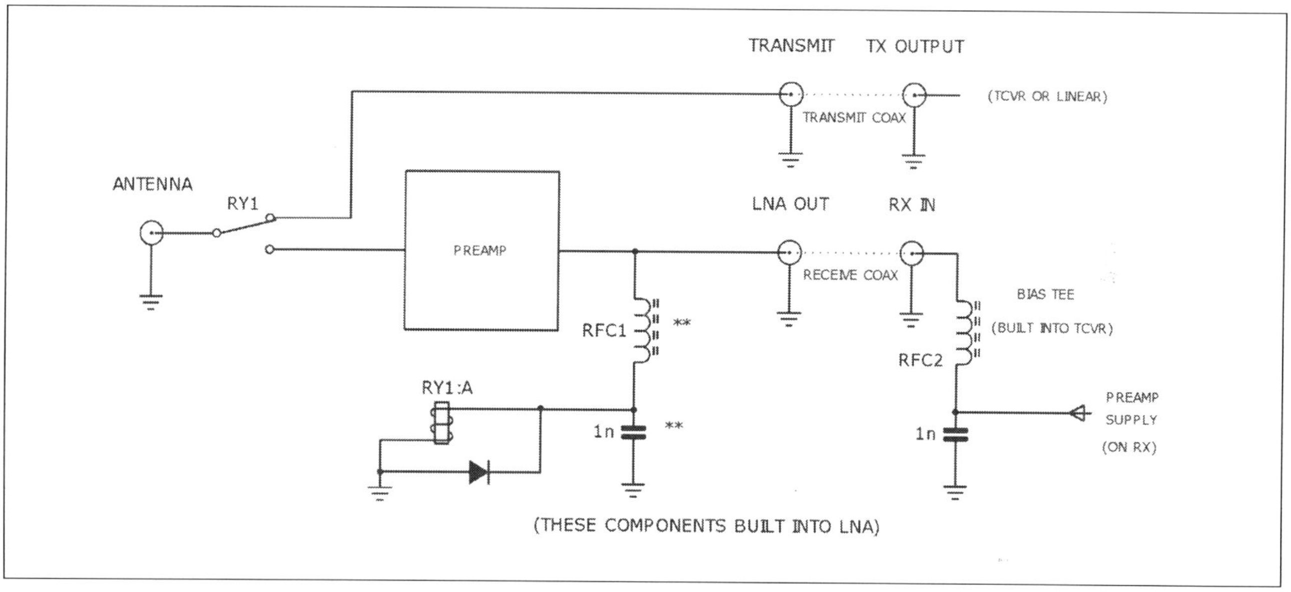

Fig 5.10. Alternative switching method. Note: relay is energised during receive. Relay must be able to handle full RF power

SWITCHING MASTHEAD LOW NOISE AMPLIFIERS

When a low noise amplifier is mounted at the antenna we can save some expensive coaxial relays and cable by simplifying the system. There are two methods to switch the antenna between the transceiver and the antenna. One method uses two coaxial relays and the other employs just one. The traditional method is to use a single low-loss coaxial cable between the transceiver and the antenna but this requires two expensive coaxial relays and suffers from greater insertion loss than the alternative method.

Fig 5.9 shows one method of powering a dual relay system and LNA using a bias tee connected to the coax feeder. An alternative method used by some amateurs is a separate DC control cable with three conductors. One conductor serves as the common ground and the other two switch the relays and the LNA. In this system the relays are normally energised on transmit and the LNA during receive. But this system has little in the way of 'fail-safe' should something go wrong. The method shown in Fig 5.9 is a safer method and it eliminates the control cable. The only important components are the two RFCs connected to the feeder cable. These need to be able to withstand the transmitter potential peak RF voltage when high power is used and have sufficient reactance to prevent loss of power. An alternative to the RFCs is $\lambda/4$ 'shorted stubs' cut for the operating frequency.

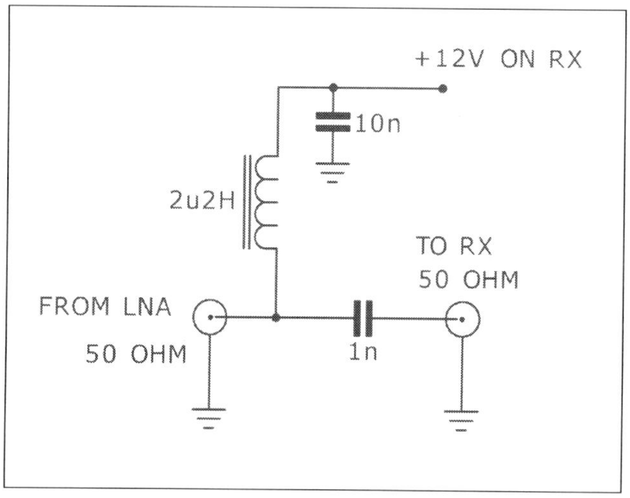

Fig 5.11. Bias-tee circuit

In the alternative method we require a second coaxial cable to carry the receive signal, but this can be low cost cable and it is even beneficial if this cable has some loss so as to improve the strong signal handling performance of the system. The cable for receive also carries the DC control signal for the relay and the LNA power. The advantage of this system is that when the station is powered down the antenna is disconnected from the LNA and this provides a fail-safe mode. During receive the DC supply is supplied to the *bias-tee* and the relay and LNA are energised. During transmit the DC supply is removed and the LNA is switched off and the relay drops out.

The cable for receive is fed to a bias-tee circuit to allow the DC and RF to use the same coaxial cable. This can be built into the transceiver or linear amplifier or made as a separate unit. The basis of a bias-tee is shown in **Fig 5.11**. The RFC can be a small moulded choke as the power level is virtually zero.

SEQUENCING

When using a masthead-mounted LNA and coaxial relays we need to use a reliable 'sequencer circuit' to control the operation. With the alternative method we can eliminate several other coaxial relays from the system saving cost and power loss. In the transceiver there is normally a changeover relay for the transmitter and receiver, this can be removed and a connector fitted to take the receiver coaxial cable. If an external linear amplifier is used this often has two coaxial relays. One is a low power type connected to the input and serves to divert the power to the amplifier input or to the output when the amplifier is bypassed. The second relay is connected between the amplifier output and the antenna connector. This relay needs to be rated at the power output of the amplifier output and hence is a large and expensive item.

MEASUREMENTS ON SOME TYPICAL AMATEUR 2m TRANSCEIVERS

As an indication of the figure of merit for some of the popular amateur 2m transceivers, the results determined by DF9IC using very good test equipment are listed in **Fig 5.12**. As will be apparent from the table, the older generation 2m transceivers are noticeably superior to the more modern types. Equipment made approx 20 to 30 years ago is still generally better in terms of strong signal handling and blocking. The outstanding model is the FT-225RD fitted with the expensive Mutek front-end module. Although this raises the IP3 considerably to a very creditable +7dBm it doesn't improve the wide-band blocking characteristics which are still poor compared with the older TS-700S. Note also how generally the better the noise figure the lower the value for IP3 due to excessive gain before the mixer. The IC-746, IC-821H, IC-901H, FT-817 and FT-857D are particularly poor performers. The IC-821H and IC-901H are 2m handheld radios so we would expect these to have inferior performance, whereas the IC-202 although regarded by many as an inferior performer is actually not too bad in comparison with some of the other models when the wide-band blocking performance is examined. The later generation TS-790 that was manufactured after the TS-700 was phased out is quite a bit inferior to its predecessor.

TRX	Owner	NF	IP3	RX Blocking in USB mode (3 dB S/N reduction in typ. 2,5 kHz BW) dB		
		dB	dBm	20 kHz offset	50 kHz offset	200 kHz offset
IC275E	DF9IC	5.6	-7.5	98	110	117
IC7000	DD9WG	5.6	-7.5	91	94	102
IC706 - measured by DL2KCK	DL2KCK	-	-	-	-	-
IC746	DJ0QZ / DK1VC	3.6	-7.5	87	95	109
IC821H	DK9VZ	3,4	-9	80	90	101
IC910H	DK9IP	3.7	-8.5	81	89	100
IC202	DL3IAS	7.7	-14	100	104	107
Hohentwiel	DL3IAS	11.4 (?)	-5.5	96	97	100
FT225RD + MuTeK + mods	DK9VZ	6.1	+7	90	90	110
FT817	DK2DB	5.4	-12	87	96	106
FT857D	DK9VZ	6.1	-2	86	96	106
TS700G mod. with GaAsFET	DK8SG	4.9	-13	100	108	111
TS700S (preamp off)	DB6IR	6.6	-7	100	107	111
TS790E	DJ5IR	4.5	-14.5	103	104	109
DK2DB homemade 1976	DK2DB	-	-11	109	110	112

Fig 5.12. Comparative performance of typical 2m transceivers *(courtesy of DF9IC)*

Note that the average noise figure is of the order of 6dB and this would be adequate for all but the most exacting weak signal modes such as EME.

The TS-700 series can be improved even further than a standard device. The main limitation is the LO filtering and the receiver mixer, both of which are less than optimum and can be improved considerably with careful engineering. As an example of how the TS-700 local oscillator can be improved we have included the following section which details the changes required. These entail a new oscillator board and a different mixing technique to filter the mixing products better and to prevent spurious signals.

2m TRANSCEIVER LOCAL OSCILLATOR DESIGN PROBLEM

The Kenwood TS-700G 2m multi-mode transceiver uses a single conversion on transmit and receive with a 10.7MHz IF. The tuning is performed by a Mixer-VFO system with a variable frequency oscillator tuning between 8.2MHz and 9.2MHz. The band is split into two segments, 144 to 145MHz and 145 to 146MHz. The mixer for the LO generation uses a dual junction Fet balanced mixer with the VFO fed into one port and the conversion oscillator into the other. Selectable crystal oscillators provide for the two band segments. A further requirement is for repeater offset and reverse operation. Hence, there are three crystals that can be selected to cover all possibilities.

The band pass filtering of the local oscillator mixer output is very critical. Because of the frequencies chosen, dictated by the VFO module, the separation between the crystal oscillator and the final output frequency is very small. The local oscillator needs to tune over the range of 133.3MHz to 135.3MHz to cover the whole 2m band. When the band segment is 144 to 145MHz the VFO output is mixed with a 125.1MHz crystal oscillator and this provides coverage of 133.3MHz to 134.3MHz. For the 145MHz to 146MHz band segment the crystal oscillator is 126.1MHz and the output covers 134.3MHz to 135.3MHz. The elimination of the 125.1 or 126.1MHz carrier in the band-pass filter stages after the mixer is poor and this causes spurious products to be generated in the transmitter final mixer. This also degrades the receiver spurious response performance and causes 'bird-

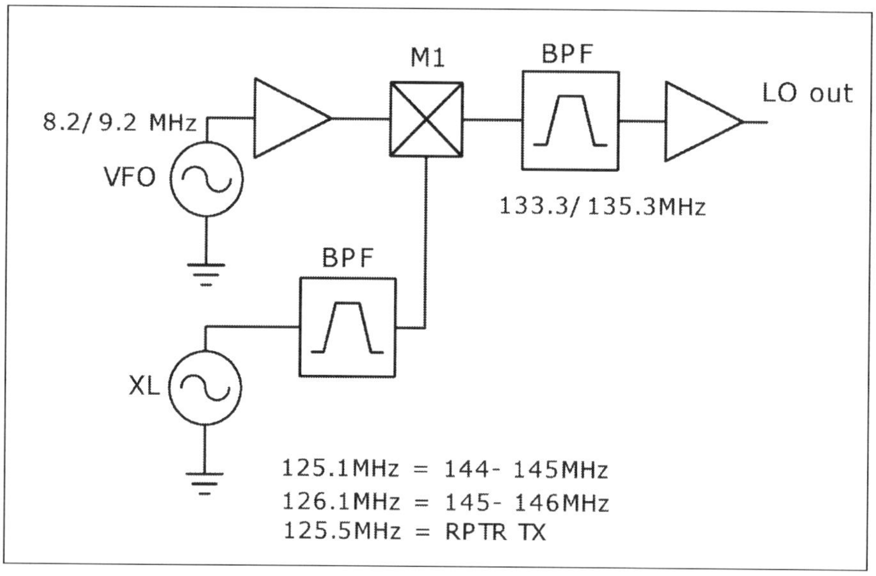

Fig 5.13. Existing TS-700G LO block diagram

ies' to occur at certain tuning points on the main dial.

The block diagram of the local oscillator module is shown in **Fig 5.13**.

The major difficulty is achieving a narrow enough band-pass filter that can reject the 125 or 126MHz oscillator signal. The percentage bandwidth required is approx 1.5% and the limited Q of the slug-tuned inductors is the major reason the required narrow bandwidth cannot be achieved.

One of the contributing factors to the poor selectivity is the use of single-sided printed circuit boards with insufficient grounding for sensitive portions of the circuitry. Another problem is the purity of the VFO signal. The VFO used is a sealed unit with good low-pass filtering incorporated to reduce the levels of the harmonics. However, a portion of the circuitry on the local oscillator board amplifies the VFO signal and this is used to drive a carrier level detector. This uses a rectifier diode to switch on a transistor. The rectifier diode regenerates the harmonics, which have been so carefully eliminated in the VFO module. Because of the high level of this signal, and the poor layout and grounding on the board, the spurious radiation allows harmonics of the VFO to enter the main mixer circuitry and this causes 'birdies' at certain tuning points. These show as very fast reverse tuning carriers.

The band-pass filter response required is shown in **Fig 5.14** with the main signals.

As well as the LO carrier at approx 125MHz and the wanted local oscillator output at 133MHz the image signals also appear at the output of the mixer. This occurs between 114.9 and 116.9MHz. The level of the 125MHz carrier is some 10dB greater than the VFO carrier presented to the push-pull mixer, as required for best mixing efficiency. The gates of the Fets are driven in push-push by the 125MHz signal and in push-pull by the VFO signal. Although a balance pot is used in the sources of the Fets the 125MHz carrier cannot be completely nulled out. Therefore it is a sizeable signal appearing at the output of the mixer.

The first filter stage consists of a critically coupled inductor pair with a large physical separation. These inductors are unshielded and prone to pick up and radiate spurious signals. The loss through this coupled pair is approx 10dB. A BJT NPN transistor with tuned base and tuned collector using two shielded inductors performs the final filtering and LO amplification.

From measurement made on the system the best case is for the spurious products to be suppressed some 50dB below the wanted local oscillator carrier. Attempting to adjust the tuned circuits for more LO output power degrades the spurious rejection to as little as 30dB. Hence, the tuning is very critical to attain good suppression of spurious products consistent with adequate LO power output. For this class of 2m transceiver the spurious rejection to aim for should be greater than 60dB and even 70dB is not enough in some cases where high level signals may be present on receive. It must not be forgotten that because the same LO is used for transmit the potential of generating spurious products that may affect other spectrum users exists.

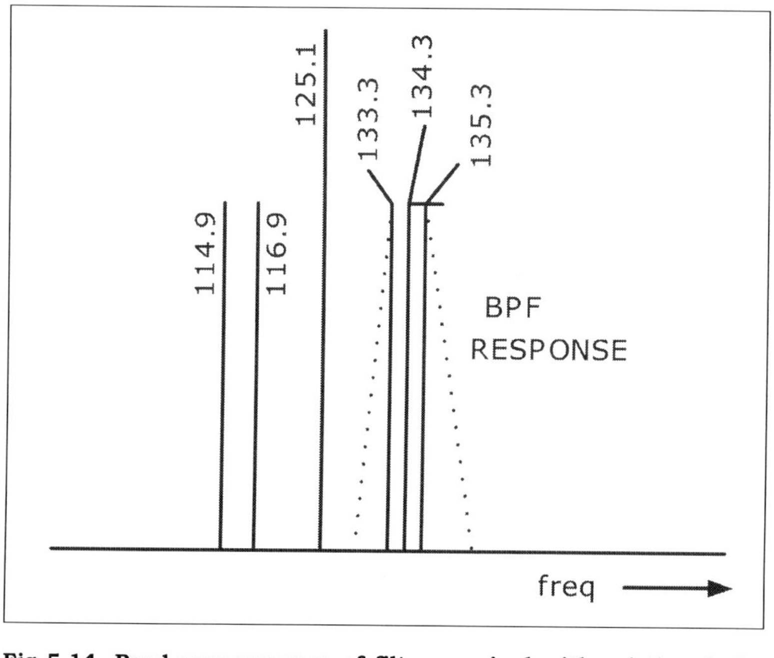

Fig 5.14. Band-pass response of filter required with existing design

ALTERNATIVE DESIGN

In order to obtain better performance with available off the shelf tuneable inductors an alternative design is now offered. The VFO cannot be changed but the frequency plan can be altered to achieve better performance.

The block diagram of the alternative scheme is shown in **Fig 5.15**. This uses two mixers and band-pass filters. In principle the use of a wider frequency separation between the wanted LO output and the conversion oscillator dictates the scheme. As in a re-

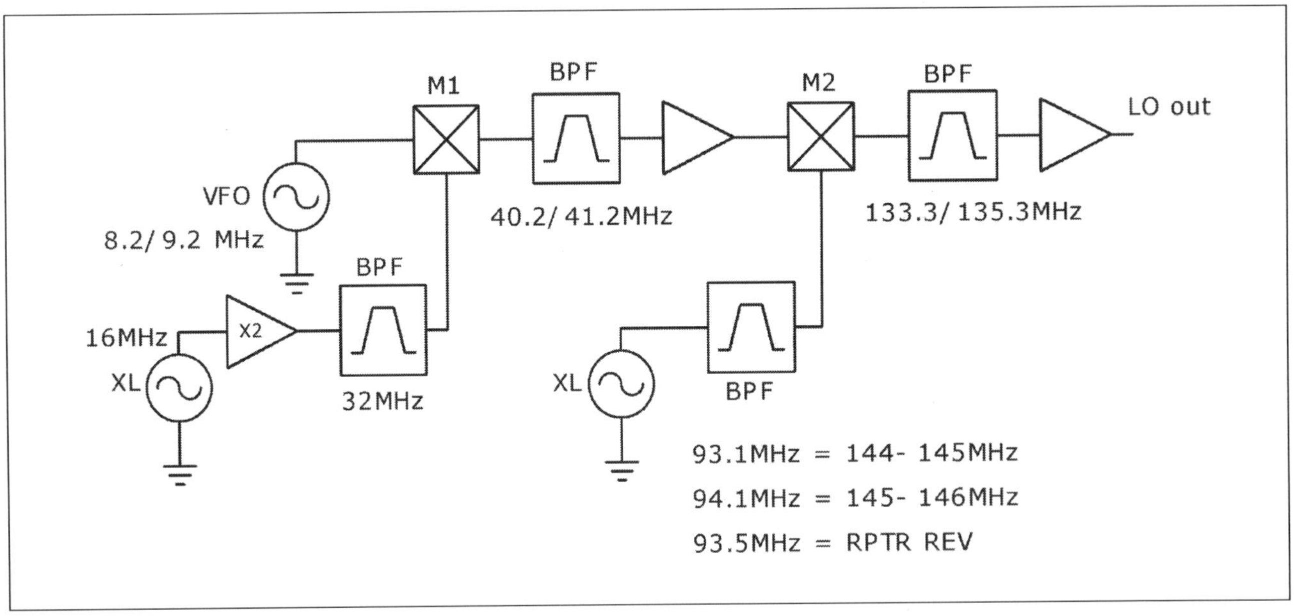

Fig 5.15. Alternative LO system block diagram

ceiver using a low IF the image suppression is a problem; the same argument holds for this scenario.

The VFO is first mixed with a 32MHz carrier to produce a signal that varies between 40.2 to 41.2MHz. The 32MHz signal is obtained from a 16MHz crystal oscillator that is doubled by a multiplier stage and then selected by a narrow band filter. The reason 16MHz was chosen will be explained later.

The ≈40MHz signal is filtered and amplified before being applied to the second mixer stage. The carrier for the second mixer is derived from one of three different crystal oscillators. These allow full band coverage plus the repeater offset function. The required oscillator is powered up when required.

The mixer products are now the sum and difference of a 40MHz carrier and one of approx 93MHz. The sum product is the wanted signal at about 133MHz. The difference product at about 53MHz is rejected by the band-pass filter, as well as the ≈93MHz carrier. Because of the wide frequency separation the Q of the tuned inductors required to construct the band-pass filter is not as high as for the original design. Using standard Toko 10mm inductor with a nominal Q of 100 provides adequate rejection in a two inductor top-coupled pair. An amplifier stage follows with extra selectivity to raise the local oscillator level to that required. With careful construction and alignment spurious rejection of 80dB should be achievable.

The choice of 16MHz as the starting point was for several reasons. Firstly a suitable crystal is available at low cost from a local supplier intended for microprocessor clocks. This is designed for fundamental oscillation (parallel resonant mode) and it is easy to pull on to frequency with a trimmer capacitor. It would be possible to start with a 32MHz crystal (3rd overtone), but this would need to be made to special order at far greater expense. A benefit of starting with 16MHz is due to the frequency multiplication. 144MHz can be derived from a 48MHz crystal by multiplying three times. Similarly as 16MHz is one-third of 48MHz it will generate a weak marker carrier at 144.000MHz when correctly set. Hence, it provides a useful additional feature to identify the lower band edge. By utilising the 100kHz marker which is built into the transceiver the 16MHz crystal can be accurately set on to frequency by tuning for zero beat.

The *receiver incremental tuning* (RIT) feature requires the local oscillator to be varied slightly in frequency when receiving an SSB or CW signal. The original design allowed approx 5kHz shift about the centre frequency. In the original design the RIT was achieved by varying the DC voltage on a varicap diode connected across the 125MHz crystals. This required three varicap diodes and gave varying frequency shifts as the three crystals had slightly different pulling characteristics. By using only one varicap connected to the 16MHz crystal the shift should be the same for any band segment. The 125MHz crystals are 5th overtone types and therefore have a limited pulling range and a non-linear characteristic versus capacity applied in shunt with the crystals. This gives approx 4kHz lower but only 2.5kHz higher when the varicap bias is varied. Another method would have been to use a varicap diode within the VFO but this was not allowed for in the original design.

BAND-PASS FILTERS REQUIRED

To implement the alternative design several band-pass filters are required that ideally should use off the shelf Toko variable inductors. As the Q of these is limited to approx 120 careful consideration of the topology needs to be made. The 32MHz band-pass filter needs to be narrow but also reject high order products. Here a bottom coupled Cohn filter is the best option. A suitable Toko 7mm inductor is available with approx 1μH inductance (see the chapter on RF filter design for details).

The 40MHz band-pass filter needs to pass 40.2 to 41.2MHz signals and hence only needs a bandwidth of approx 2.5%. The consideration of the mixing products and the carrier dictates the use of a capacitor top coupled filter to provide greater rejection to the image and carrier products. A suitable Toko 7mm inductor has a Q of approx 120 at 40MHz and an inductance value of ≈350nH.

The final band-pass filter for 133MHz can now be

quite wide as the frequency separation between the wanted carrier and the image is now ≈40MHz and not 10.7MHz as in the original design. This allows lower insertion loss and lower inductor Q. Again we need more suppression on the low frequency side and hence a capacitor top coupled filter is required. The Toko 10mm inductor available has a nominal value of 100nH. All of these filters can be realised with just two coupled inductors to provide adequate suppression of the spurious products.

The final band-pass filter is another 133MHz type after the amplifier stage. The top coupling capacitors have a very low value (0.5pF) and may present a problem in implementation. This can be attained by series connection of 1pF capacitors.

MIXERS REQUIRED

Although Mini-Circuits ASK-1 mixers would be acceptable, they are quite expensive and require 50Ω terminations. Because the frequency separation is now quite high, single-ended Dual Gate MosFet mixers can be employed with the tuned circuits in the drain network. These offer good spurious free dynamic range and offer a conversion gain of approx 6 to 10dB when correctly biased. This topology also suits the Toko inductors, as none of the mixers now require push-pull tapped inductors.

The plots that follow are for the prototype band-pass filters, simulated using ARRL software. The MS21 value is the filter insertion loss when correctly terminated. The insertion loss at 32MHz is approx 6dB and the attenuation at 48MHz is greater than 50dB.

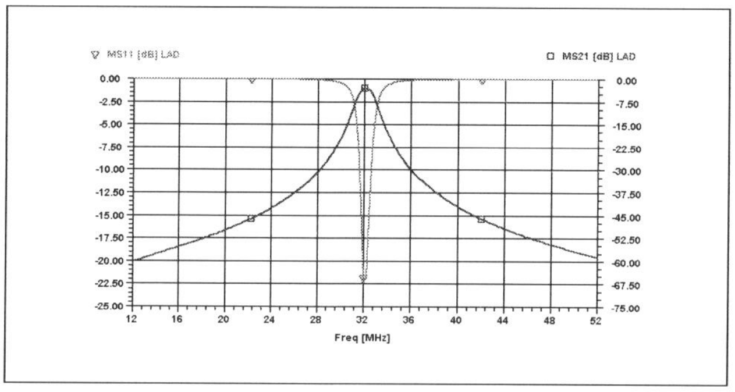

Fig 5.16. 32MHz multiplier band-pass filter (bottom coupled design)

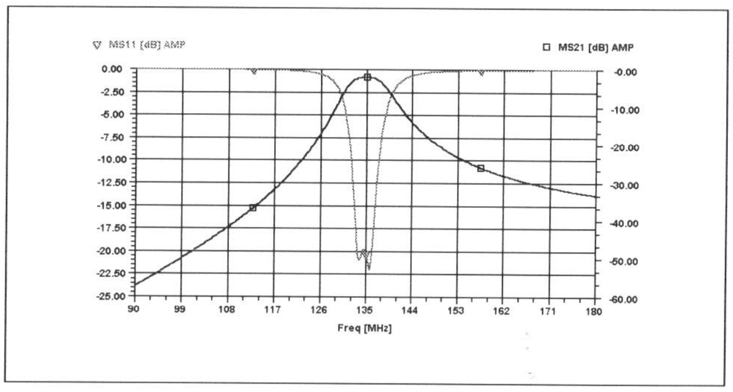

Fig 5.18. 2nd mixer 133MHz band-pass filter (top coupled design)

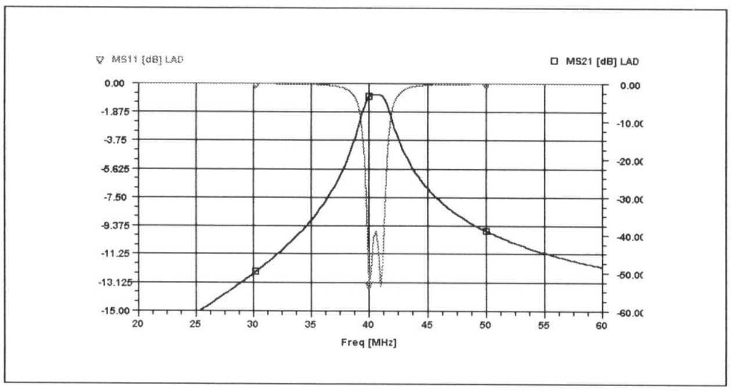

Fig 5.17. 1st mixer 40MHz band-pass filter (top coupled design)

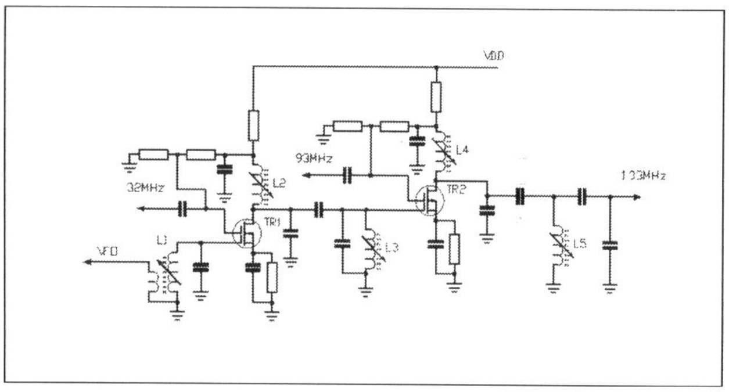

Fig 5.19. Basis of new LO mixer system

Chapter 6

Design of Radio Frequency filters

THE average amateur transceiver contains many different filters to shape the RF spectrum. Some of these are low pass filters, used for harmonic suppression, and some are high pass or band pass filters. Of the three main types the band pass filter causes the most problems. Low pass and high filters are simple to design and make, by consulting tables, such as those found in Zverev's *Handbook of Filter Synthesis,* [published by John Wiley & Sons Inc. ISBN: 0-471-74942-7]. Anatol I Zverev, an engineer and a formidable mathematician, moved to the USA after the WWII and took up a position at Westinghouse Electric Corporation. His book, published in 1967, is a classic, now selling for about $300. Secondhand copies, in any condition, command a very high price.

The intention of this chapter is to show the amateur constructor the best topology to use for a particular application. By following these simple guidelines an amateur constructor will be able to design and build a filter with minimal effort and a good certainty of success. Having covered the basic LC filter types we will later look at some other filter types in common use.

LOW-PASS FILTERS

There are myriad different types in common use. They include the *Butterworth*, *Chebyshev* and the *Cauer-Chebyshev*, normally called the *Elliptic Filter.* Of the three main types the Chebyshev types allow the designer to select the ripple in the pass band and hence the ultimate slope of the filter beyond the cut-off frequency. The Butterworth filter is almost never used. It features the lowest ripple in the pass band but the poorest roll-off in the stop band. The Chebyshev types offer much steeper cut-off slopes and hence greater attenuation to the harmonic signals at the expense of some ripple in the pass band.

For low-pass filters intended for transmitter harmonic suppression the greater the ripple allowed in the pass band the steeper the slope of the filter after the cut-off frequency is exceeded. In practice the ripple should not be allowed to become excessive or else the attenuation loss in the filter can reach unacceptable levels. Generally, a maximum ripple of 0.5dB is the figure most designers aim for, although many would prefer a lower value such as 0.1dB.

It must not be forgotten that most LC filters are what are known as 'reflective types', that is to say the filter prevents the unwanted (harmonic) radio frequency power from passing through the filter by generating a high VSWR at the input and output ports. For a transmitter this can give rise to instability at frequencies outside the pass band if the transmitter design does not tolerate this high VSWR.

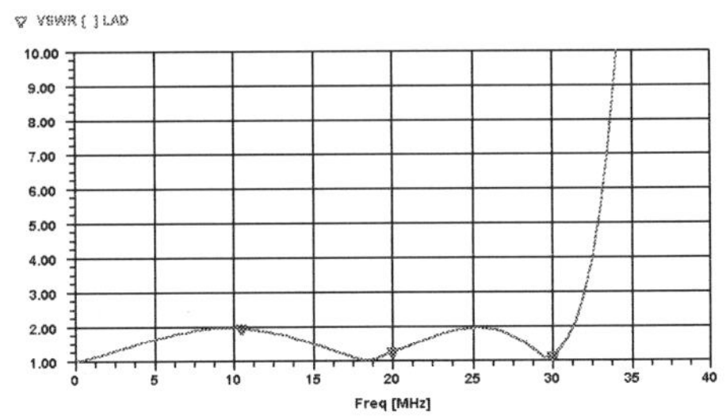

Fig 6.1. Typical VSWR plot for a low-pass filter

At frequencies far from the pass band the filter reflects all the power back towards the transmitter output stage and presents an infinite VSWR to the transmitter. There are certain types of *non-reflective* low-pass filters we will examine later. These maintain a low VSWR irrespective of the frequency presented to the filter. However, the design of such a filter is more complicated.

HOW A LOW-PASS FILTER IS CHARACTERISED

There are several parameters that define a low-pass filter. The first is the impedance the filter is designed to work into. For harmonic antenna filters this is normally 50Ω, although some transmitters are designed for a different impedance.

The second parameter is called the *stop band.* This is the frequency above the normal band that will pass radio frequency power with little attenuation. The point at which the *cut-off* frequency is placed determines the transition between the *pass band* and the *stop band.* This cut-off frequency needs to be carefully selected. For most harmonic filters the choice of the cut-off frequency is critical. If the cut-off point is placed too close to the upper frequency range of the transmitter, the insertion loss at this point can become excessive and a great deal of radio frequency power will be absorbed by the filter, causing excessive heating and a large drop in transmitted power. In the Butterworth filter the cut-off point is defined as 3dB down from the pass-band loss (in the Chebyshev types it is not so simple).

The third parameter is of course the *pass band*, within which the transmitter is able to deliver maximum power with little insertion loss due to the attenuation in the filter. Because of the topology of the standard low-pass filter this normally extends from DC to the cut-off frequency.

Finally, the *pass-band ripple* defines how much additional attenuation is added to the inherent insertion loss of the filter.

If the pass-band ripple is excessive, say 1dB, we can expect the output power to drop by about 20% when the filter ripple increases the insertion loss to maximum. Most designers prefer to minimise the ripple to allow maximum power to be transmitted. But as already mentioned this has an effect on the ultimate cut-off region and the attenuation it is possible to obtain in the stop band. To achieve high values for the stop band attenuation requires more components to be added, which increase the pass band insertion loss.

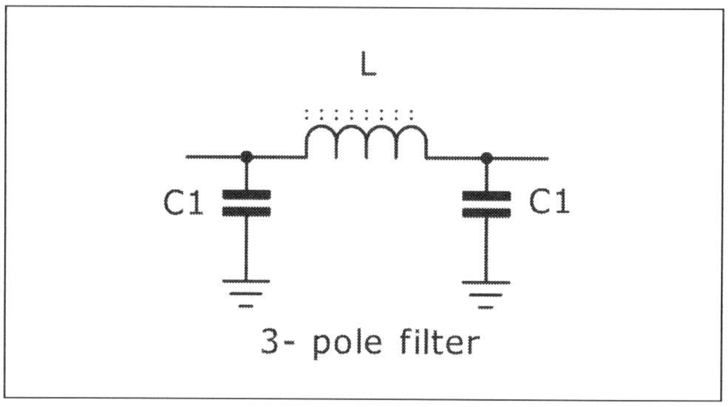

Fig 6.2. 3-pole low-pass filter (π section)

COMPONENT Q VALUES

In an ideal filter the inductors and capacitors used to construct it would have Q values approaching infinity. In the real world this is often not realisable. The component with the lowest Q will dictate the insertion loss and the ultimate roll-off after the cut-off frequency is exceeded. In practice the component that normally limits the response is the inductor. The Q value of the capacitor is normally at least a factor of 10 greater than the inductors used. So if the inductor Q is poor the insertion loss will be higher and the ultimate attenuation to the harmonics will be less than is possible.

In order to obtain high Q values for the inductors we need to use a wire that is substantial and has low DC resistance and *skin resistance*. In general, for high power low-pass filters, above about 50W, we will need to use 16SWG (1.6mm diameter), or thicker, copper wire to manufacture the inductors. In some cases thin wall copper tube is used.

As an example: an inductor for an HF transmitter harmonic filter will exhibit an unloaded Q of not much more than 500 if wound from a substantial gauge of wire. To achieve this Q value requires a large diameter, single layer, air dielectric solenoid winding, with the turns spaced at least one-wire diameter from the next turn. The optimum unloaded Q occurs when the length to diameter ratio is approx 1:1. So if the coil is 1 inch in diameter (≈25mm) the length should not exceed approx 1.2in (30mm) or be less than 0.8in (20mm). Should the inductance value be too low, the coil diameter should be increased to obtain the correct value and optimum Q.

If the inductor is placed within two coil-diameters of a metallic shield it will severely *de-Q* the inductor. The high voltage RF capacitors used will normally have unloaded Q values of more than 2000. If the capacitor Q is poor they will heat up and in severe cases will explode, so it is important to select them carefully. It must not be forgotten that very high values of circulating RF current will be flowing in both the inductors and the capacitors, in some cases as much as 50A for a 400W transmitter. The VSWR the antenna presents to the filter also increases the voltage stress if the VSWR is high.

SELECTING THE CORRECT FILTER ORDER

The number of components used in a filter is known as the *filter order* (N). So if a low-pass filter consists of three inductors and four capacitors there are a total of seven components, and this is known as a *7-pole filter*.

The greater the number of poles, the steeper the cut-off slope will be, but at the expense of a slightly greater insertion loss. For standard Chebyshev filter types the best performance is obtained when the filter has an odd number of poles, so 3-pole, 5-pole, 7-pole etc are the order of the day. It is possible to design an even number of poles, but the component values become impractical beyond a certain point.

Fig 6.2 shows the most basic low-pass filter using 3-poles (N = 3). This is often referred to as a *π section*, and a typical filter will be built up from several sections connected in series to achieve the required stop-band attenuation.

For a low-pass filter where the input and output impedance are the same, the value of C1 will be the same at both ends of the filter. The value of C1 and L will depend on the operating frequency, system impedance, pass-band ripple and the desired cut-off frequency. The filter is symmetrical and because the input and output impedance are the same it does not matter which way around the filter is placed in the circuit. For the rarer case where the input and output impedance are not the same, the values of C1 will be different as will the value of L for a particular frequency, and the filter must be connected the correct way around. It is not proposed to cover this type here, but the corresponding values can be obtained from Zverev's tables if needed.

The frequency response of a 3-pole filter is shown in **Fig 6.3**. You will notice the rather poor roll-off above the upper pass-band frequency. This due to the limited number of poles being used. The filter is designed to pass a band of frequencies from 0 to 10MHz. The attenuation at 50MHz is about 38dB. In this example the attenuation in the stop band is 38dB at a point that is five times the

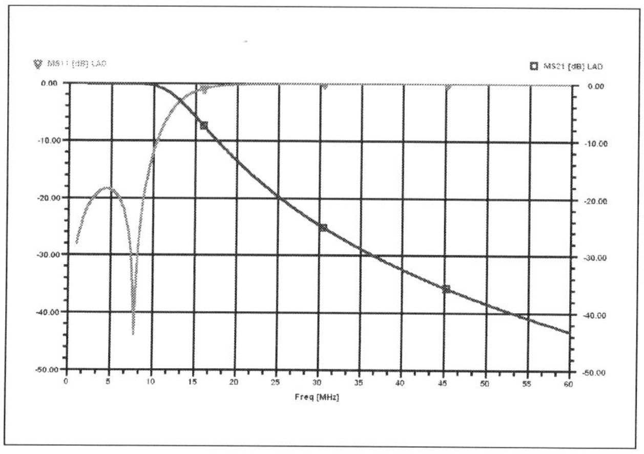

Fig 6.3. Response of a 10MHz 3-pole filter

cut-off frequency. Hence, we can calculate the attenuation as being 38/5 = 7.6dB/octave. So at 20MHz we would expect to see an attenuation of approx 15dB, at 40MHz we would expect to see ≈30dB etc. So if the frequency in the stop band extended to, say 100MHz, which is about nine octaves, the attenuation to be expected is approx 60dB. In practice it is often not possible to make these assumptions as the inductors and capacitors at some very high frequency may exhibit self-resonant characteristics and become either parallel or series resonant.

In the Butterworth filter (known as the *maximally flat* filter because it inherently has no ripple in the pass band) the attenuation is approx 6dB per pole per octave. So if the filter consisted of 6-poles it would have an attenuation of 36dB at twice the cut-off frequency and 72dB at four times the cut-off frequency. The Chebyshev types offer much greater attenuation in the earlier portion of the stop band than the Butterworth types, for the same number of poles.

HOW MANY POLES ARE NEEDED?

This very much depends on how much attenuation is required and how far from the cut-off frequency. Many constructors choose the number of poles rather unscientifically; even RF design engineers are sometimes guilty of this! The number of poles required depends on the spectral purity of the transmitter and with what legislation it is necessary to comply.

Let us take a transmitter and run it without a low-pass filter and measure the harmonic products. Suppose we use a 30MHz transmitter and drive it to full output power and with a spectrum analyser measure the harmonic products. The second harmonic at 60MHz is likely to be the worst offender and the products at 90MHz, 120MHz and higher are usually less of a problem. The harmonic energy tends to fall off in a fairly linear manner. If the output stage of the transmitter consists of a push-pull topology the suppression of the even harmonics is normally better than the odd-order harmonics. Hence, the second harmonic at 60MHz is usually better suppressed than the third harmonic at 90MHz. In a single-ended stage the suppression of the even- and odd-order harmonics is about the same and expected to drop off in a fairly linear manner as the harmonic number rises.

Suppose we measure the second harmonic at 60MHz to be 30dB below the fundamental at 30MHz and the third harmonic at 90MHz to be 45dB below the fundamental. If the legislated specification is, say, below 60dB for all unwanted products then we can see that we require an additional 30dB at 60MHz but only an additional 15dB at 90MHz. If we simply took the 60dB specification and designed a filter with a minimum of 60dB attenuation at the second harmonic the filter would have more poles and hence a greater insertion loss than is required.

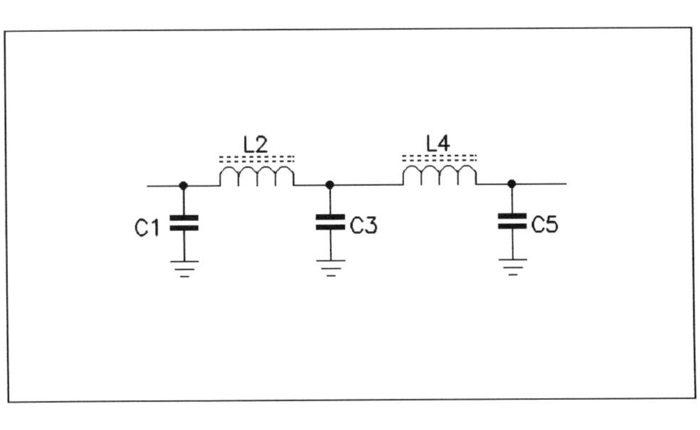

Fig 6.4. 5-pole Chebyshev filter

The order of the filter required then is not that high, and a 5- or 7-pole filter will probably fit the bill. It is preferable initially to select the lower order filter if possible to reduce the insertion loss at the 30MHz portion of the spectrum and to reduce the complexity. Let us therefore design a 5-pole filter and see what response we get. We will choose a 0.5dB ripple factor as this will ensure a faster roll-off above the cut-off point, and hence more attenuation at the 60MHz point on the response.

You will notice that the components are numbered from left to right and start with C1. This is a convention in a shunt first element filter. The capacitors are given odd numbers and the inductors assigned even numbers. An alternative topology is to have a series inductor as the first element and the first capacitor in shunt to ground after the first inductor. In this case the inductors would be assigned as L1, L3 and the capacitors C2, C4 etc. Hence, there would be three inductors and only two capacitors. As the inductors generally need to be wound by hand, often it is better to choose the *coil-saving design* with three capacitors and only two inductors.

> The component numbering convention is due to Zverev's tables, which can be read from top to bottom or bottom to top. For the standard shunt first element being a capacitor the capacitors have odd numbers and the table is read from top to bottom. When an inductor is used as the first series element the numbering needs to be changed so that the correct values in the table are used.

The cut-off frequency for a 5-pole filter needs to be placed at about 11% above the highest operating frequency, so in this case this puts it at 33.3MHz. The cut-off frequency is the point where the insertion loss curve and the input VSWR curve cross. The cut off point varies depending on the filter order; higher order filters require a smaller percentage. Typical percentage values are: for a 3-pole filter the cut off percentage is approx 15%, for a 5-pole filter it is approx 11%, for a 7-pole filter it is approx 9% and a 9-pole filter approx 7%.

> If **Fig 6.3** is examined you will see the cut-off occurs at about 13MHz. If the cut-off where chosen to be 10MHz the filter would already be rolling off at 10MHz and therefore the highest frequency we could use with low attenuation would be about 8.5MHz.

By selecting the correct page in the Zverev's tables we get the following values:

N = 5. 0.5dB ripple. Z = 1Ω . ω = 1rad/sec

C1 = 1.8068 Farad
C3 = 2.6914 Farad
C5 = 1.8068 Farad
L2 = 1.3025 Henry
L4 = 1.3025 Henry

Note that these are the *normalised* values and are in Farads and Henrys at a frequency of 1radian/s and impedance of 1Ω. To convert to the design frequency we need to do some mathematical juggling. To convert the values to 50Ω we *divide* the capacitor values by 50 and *multiply* the inductor values by 50. The table values are derived from the reactance values in ohms to obtain the correct response. This gives the following:

C1 = 36.136 x 10^{-3} Farads
C3 = 53.828 x 10^{-3} Farads
C5 = 36.136 x 10^{-3} Farads
L2 = 65.125 Henrys
L4 = 65.125 Henrys

To convert to the required cut-off frequency of 33.3MHz we need to find the value of 2πf. This is in this case 209.23 x 10^6. The rule is that the capacitors' and inductors' values are then *divided* by the frequency constant.

So for our filter we end up with the following circuit values:

C1 = 172.8pF
C3 = 257.3pF
C5 = 172.8pF
L2 = 0.311μH
L4 = 0.311μH

As you can see the filter is symmetrical and C1 and C5 are the same value, as are L2 and L4, because the input and output impedance are the same. It is important to use the values calculated and not to make large value substitutions. The 172.8pF may be made up from a 150pF and a 22pF in parallel. The 0.8pF error is not going be an issue. If, however, the 172.8pF value was substituted by a 180pF this would cause the response to be incorrect with more insertion loss, ripple in the pass band and a less steep roll-off.

Generally, for high-power filters when low-cost ceramic disc or ceramic plate type capacitors are used paralleling several lower value capacitors to obtain the correct value is a better method than using a single capacitor, even if it happens to be the correct value. The reason for this is that paralleling capacitors reduces the effect of the lead inductance and increases the total current handling rating, both of which are important. If the capacitor type is a clamped silver mica (Unelco) or ATC 100B the capacitor normally has sufficient current handling and low enough lead inductance not to upset the performance. The 257.3pF may be made up from a 220pF and a 39pF in parallel; or with two 120pF and 18pF, the small error resulting is not going to be too serious. The inductors in any case will need to be wound to the exact values and then adjusted in circuit, preferably with the prototype filter measured on a sweep generator or vector network analyser to get the correct response.

When constructing a filter it is also important to minimise the mutual inductance between adjacent inductors. The best method is to mount the inductors so that they are at 90° to each other. Shielding plates fixed between the inductors will also reduce the mutual coupling to very small values. Alternatively for a low-power filter we could use inductors wound on low permeability ferrite toroidal cores. As toroidal cores are inherently self-shielding this make the filter compact.

If we now analyse the filter with the above values we see that the response looks like that shown in **Fig 6.5**. The insertion loss remains substantially flat from DC to 30MHz with a very small amount of ripple. Here we see that we have achieved our objective of 30dB attenuation at 60MHz, in fact the response passes through 30dB at 50MHz, so we have some attenuation in hand. The attenuation at 90MHz is greater than 50dB, although we only required 15dB to meet the harmonic specification. The rippled trace shows the input match as the magnitude of S_{11} (input reflection coefficient or input VSWR), between 0MHz and 30MHz. It is very low and as to be expected becomes very high beyond the cut-off frequency. The value at 30MHz is -23dB, which is a VSWR of 1.15:1. The ripple in the pass band can be seen and the response kicks up to minimum attenuation towards the top of the pass band. This is typical of a Chebyshev response filter.

If we had estimated the required filter order as being higher, say a 7-pole filter, we would have more insertion loss in the pass band, hence less available output power into the antenna and much more attenuation in the stop band, which we did not really require.

The inductor Q assumed in this simulation is only 50. The capacitors are assumed to have infinite Q. In practice we would choose a much higher Q for the inductors. With a Q value of 50 the maximum insertion loss is approx 0.6dB, which may be too high for some applications. If we now model the filter with a more practical Q value of 250 we can see the maximum insertion loss will go down. It drops to 0.5dB (10%) because this is the maximum ripple value chosen; the minimum insertion loss is 0.018dB. If we were concerned about power loss in the filter we could select a lower ripple figure, say, 0.1dB.

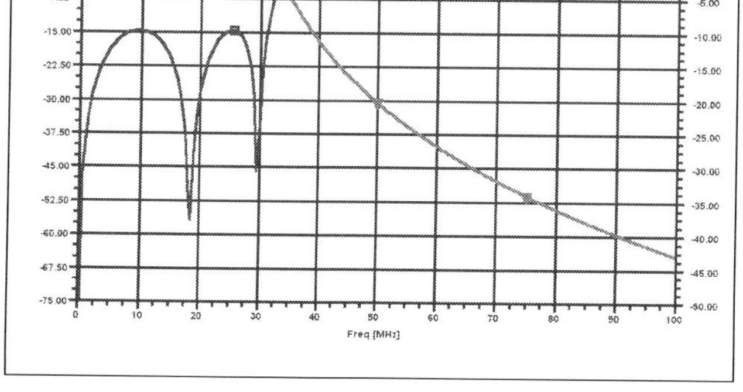

Fig 6.5. 30MHz 5-pole low-pass filter response

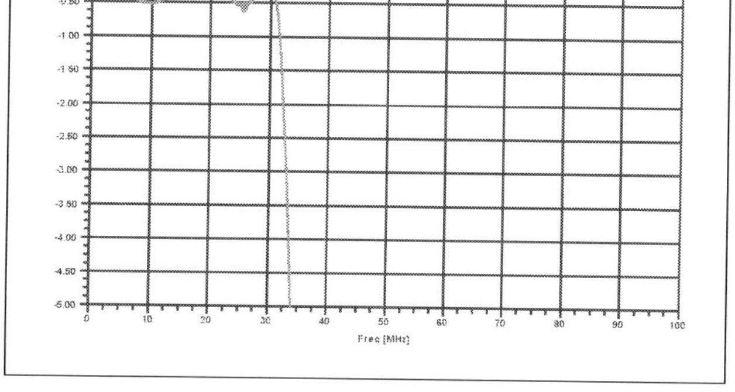

Fig 6.6. Expanded plot of pass-band ripple

Then the maximum insertion loss would be approx 0.1dB in theory, but the filter would require more poles because the lower ripple filter has a less steep roll-off above the cut-off frequency. (It is impractical to increase the *pole order* in units of one, as this would become a 6-pole filter and as has already been explained this tends to lead to impractical values. Hence, the next logical filter would be a 7-pole.)

In practice as long as the 11% cut-off rule is observed the lowest insertion loss will normally occur just before cut-off, which in this case is between approx 28 and 30MHz so suiting the 10m-band transmitter perfectly.

In order to examine the pass-band ripple more closely, **Fig 6.6** shows it on a 0 to 5dB vertical scale. As is to be expected the ripple in the pass band is 0.5dB.

SELF-RESONANT FREQUENCY OF CAPACITORS AND INDUCTORS

In constructing any LC filter it is well to remember that the capacitors inherently have some self-inductance, apart from the lead lengths. If the leads are excessively long the lead inductance cancels out part of the capacitance and makes the capacitor appear to have a lower value. A good estimate is 1nH of inductance for each millimetre of lead length. So if we used a lead length of 2mm on each lead of the capacitor it amounts to an additional 4nH apart from the internal inductance, which might be as much as 5nH for a ceramic disc capacitor.

For high-power filters the best choice of capacitor is either compressed mica types such as Unelco, Arco or Semco [Silvered Electronic Mica Company, Willimantic, CT, USA] or ceramic high-voltage types made by ATC [American Technical Ceramics Corp, Huntington Station, NY USA]. These are available in working voltages up to 600V DC and the RF voltage rating can be safely assumed to be of the same order. Both of these capacitor types are usable to the upper UHF spectrum, indeed they need to behave as real capacitors in a harmonic filter if they are going to act like a short circuit to the higher order harmonic energy. The inherent self-inductance of the Unelco, Semco, Arco and the ATC types is for all practical purposes zero, provided the capacitor is soldered directly to a low inductance ground plane. Silver mica disc capacitors are hard to find today and are a poor substitute for Arco or ATC types. They have much higher self-inductance than the other types and relatively poor Q values.

Similarly the inductor will at some very high frequency exhibit a parallel resonant characteristic as the stray inter-turn capacity causes the inductor to become resonant.

STANDARD LOW-PASS FILTERS

Whereas it is relatively simple to design a low-pass filter from scratch using the Zverev's tables, my usual method is to calculate for a standard frequency of 1MHz and then to scale the filters accordingly. Hence, once the tedious calculations have been done for each filter order and ripple factor, it is only necessary to make some simple calculations to ascertain the values for any other cut-off frequency. If, for example, a filter is required for a cut-off of 100MHz, the 1MHz values for each component are simply divided by 100.

TABLE OF STANDARD LOW-PASS FILTERS

3-Pole Chebyshev, Fc = 1MHz, Z = 50Ω

Ripple dB	C1 & C3 pF	L2 µH
0.5	5932.023	10.189
0.1	4560.744	12.682
0.01	3759.558	14.494

5-Pole Chebyshev, Fc = 1MHz, Z = 50Ω

Ripple dB	C1 & C5 pF	C3 pF	L2 & L4 µH
0.5	5751.223	8566.992	10.365
0.1	4142.167	7133.643	12.381
0.01	3108.614	6482.700	13.408

7-Pole Chebyshev, Fc = 1MHz, Z = 50Ω

Ripple dB	C1 & C7 pF	C3 & C5 pF	L2 & L6 µH	L4 µH
0.5	5696.474	8650.708	10.314	14.241
0.1	4015.479	7127.595	12.093	13.372
0.01	2905.214	6372.882	12.690	14.884

9-Pole Chebyshev, Fc = 1MHz, Z = 50Ω

Ripple dB	C1 & C9 pF	C3 & C7 pF	C5 pF	L2 & L8 µH	L4 & L6 µH
0.5	5672.919	8645.932	8828.006	10.282	11.079
0.1	3961.685	7072.845	7307.440	11.950	13.392
0.01	2818.316	6243.330	6594.426	12.345	14.814

Note: If the need is for different system impedance than 50Ω, then you need to recalculate the *de-normalised values*. If the system impedance was 75Ω, the capacitor values would be *divided* by (75/50) = 1.5 and the inductor values *multiplied* by the same figure.

NON-REFLECTIVE HARMONIC FILTER

For applications where the input VSWR presented to the source must remain low, another circuit topology is used. In this case the filter actually consists of two filter types. One is the conventional Chebyshev low-pass filter and the other is a Chebyshev high-pass filter. The two filters are connected together at the input of the filter and each filter feeds a separate load. The low-pass filter connects to the antenna port, and the high-pass filter connects to a terminating load resistor.

The RF spectrum within the low-pass filter pass band is all passed to the antenna port. The RF spectrum containing the harmonic energy is passed only via the high-pass filter and is dissipated in the terminating load resistor. The cut-off point of the two filters is chosen to be the same frequency. As the low-pass filter rolls off the power passing through it, the high-pass filter begins to roll down in insertion loss towards zero. Hence, the worst

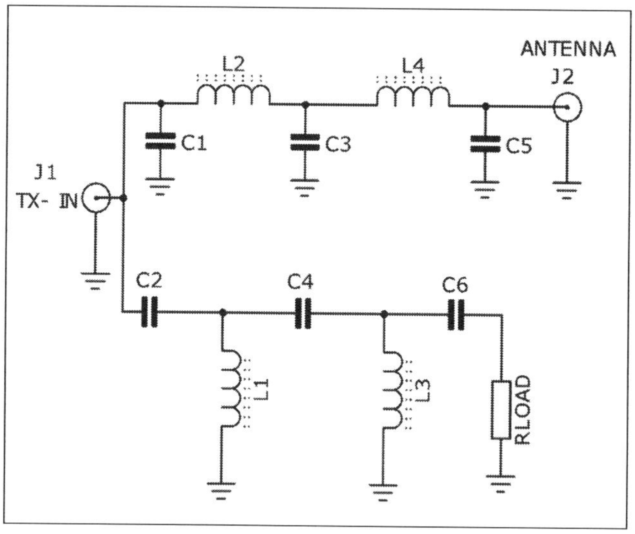

Fig 6.7. Non-reflective harmonic filter

input match occurs at the cut-off point and returns to a low VSWR at frequencies removed from the cut-off point. With correct design the input match can be kept below 1.5:1 VSWR across the band from DC to some very high frequency. This circuit configuration is often called a *diplexer*. The basic configuration is shown in **Fig 6.7**.

The low-pass filter consists of a 5-pole type and the high-pass filter also consists of a 5-pole type. However, they do not have to be the same order. The high-pass filter consists of C2, C4, C6 and L1 and L3. The load resistor is the same as the system impedance, typically 50Ω. It need only have a power dissipation equivalent to the total harmonic power contained in the unwanted products. If the transmitter was rated at 100W and the second harmonic is 20dB below the fundamental, the power contained in the harmonic is approx 1W. Hence, a 2W resistor would suffice in most cases.

HIGH ATTENUATION LOW-PASS FILTERS

In many applications the need is for more attenuation at the harmonics than a standard Chebyshev filter can provide. In this case a modification of the standard Chebyshev filter is used. This filter is known as the *Cauer-Chebyshev* or 'Elliptic Filter'. In essence it is a normal Chebyshev response but additional *zeros* are placed in the response with parallel resonant traps. The basic configuration is shown in **Fig 6.8**.

In the diagram C2 and L2 resonate to provide a parallel resonant trap at one of the harmonics and C4 and L4 provide another trap at another harmonic. In practice the second and third harmonics are often chosen as they

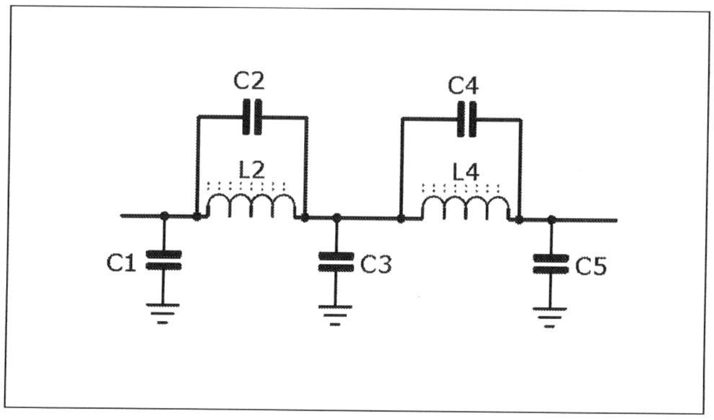

Fig 6.8. Cauer-Chebyshev elliptic filter

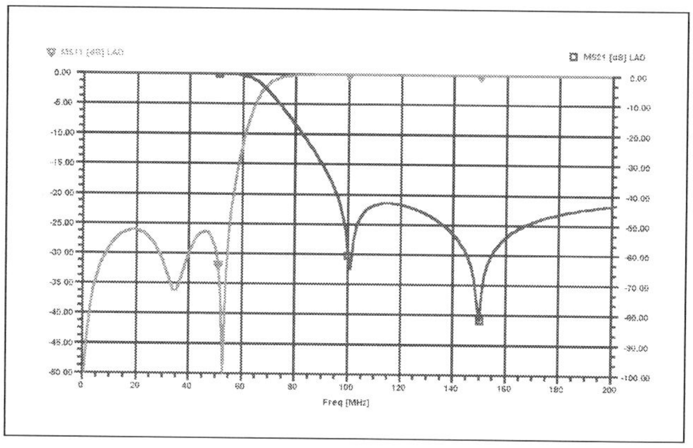

Fig 6.9. Elliptic filter response

contain the greatest harmonic energy. The response of the elliptic filter is a very steep roll-off until the first notch frequency is reached then a more gradual re-entry of the attenuation curve. The response of a typical elliptic filter is shown in **Fig 6.9**. This filter is designed for a 50MHz transmitter where the possibility of interference to FM broadcast receivers operating at about 100MHz is a common problem. The attenuation at 100MHz is approx 65dB. The filter is a conventional 5-pole Chebyshev with two traps, so there are seven components used. The second notch is placed at 150MHz. With the elliptic filter attenuation values of approx 80dB per octave are readily achieved up to the first notch frequency. A normal 5-pole filter would only achieve about 40dB.

DESIGN OF HIGH-PASS FILTERS

The high-pass filter is simply an inverted mirror image of the low-pass filter. In the high-pass filter the inductors are connected in shunt to ground between the series signal path. The series signal path consists of capacitors connected in series between the input and output.

A basic high-pass filter with 3-poles is shown in **Fig 6.10**. We can work out this is a high-pass filter from first principles if we remember how the different reactances work. For capacitive reactance at DC the reactance is infinite, falling towards zero as the frequency increases. For inductive reactance it is zero at DC and rises towards infinity with increasing frequency.

As with the low-pass filter, if the input and output impedances are the same the value of C will also be the same at both ends of the filter. The frequency response of the 3-pole HPF is shown in **Fig 6.11**.

In order to demonstrate the various types of filter order we will also use a 25MHz cut-off for all the following plots. We will choose 0.1dB ripple in the pass band and design for 50Ω impedance. You will notice that the curves of S_{11} and S_{21} cross at the cut-off frequency. In this filter the pass-band attenuation at the cut-off point is exactly 3dB below the zero insertion loss portion, the same as

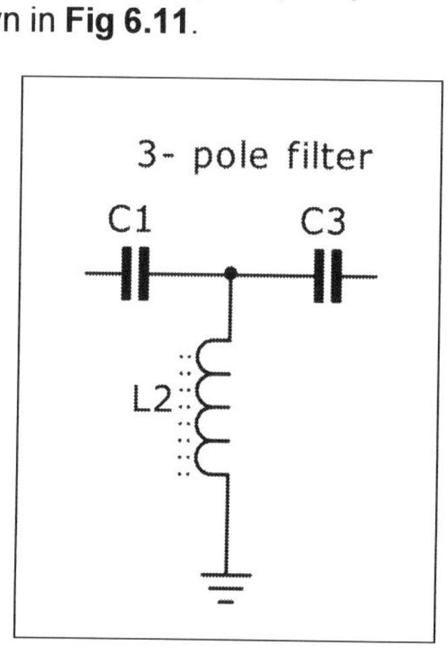

Fig 6.10. 3-pole high-pass filter

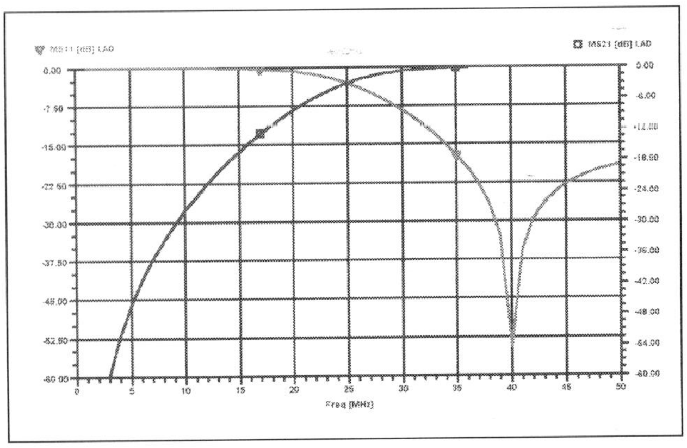

Fig 6.11. 25MHz cut-off 3-pole HPF

a Butterworth type. The point at which the insertion loss drops to zero in this example is approx 33MHz. This is about 13% higher in frequency than the cut-off point. This agrees quite well with the 11% rule discussed in the section on low-pass filters. As the filter order rises the cut-off point in percentage drops towards zero. In practical filters (those below 11th order) it will always be above approx 2%.

Now look at **Fig 6.12** and **Fig 6.13** showing the plots for 5-pole and 7-pole versions of the HPF.

You will notice that the curves of S_{11} and S_{21} all cross at the cut-off point of 25MHz. The main difference between the three plots is the slope from the stop band to the pass band. All the vertical scales are a maximum of 60dB attenuation. The 5-pole filter reaches zero attenuation at approx 29MHz and the 7-pole at approx 26MHz. The stop band 50dB points are 4MHz for the 3-pole version, 10.5MHz for the 5-pole version and 15MHz for the 7-pole version. If we were to use a 9-pole version, the 50dB point would be reached at approx 20MHz.

The pass-band insertion loss is essentially zero,

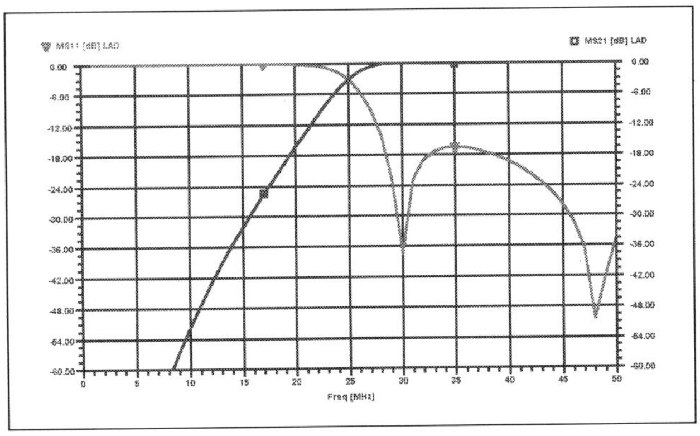

Fig 6.12. 25MHz 5-pole HPF

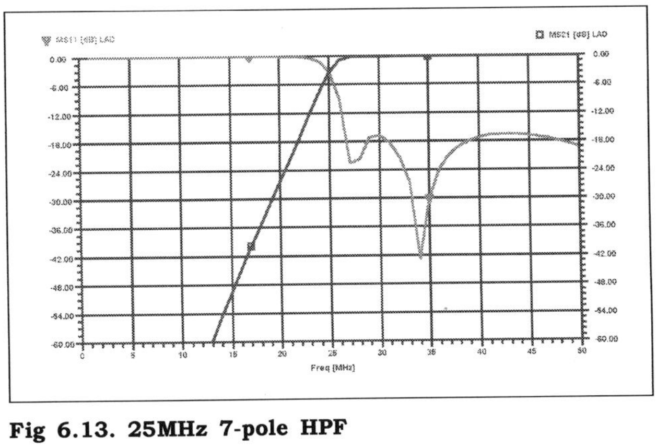

Fig 6.13. 25MHz 7-pole HPF

with a maximum value only slightly above the ripple factor chosen of 0.1dB. The pass-band insertion loss is dependent on the Q of the inductors and capacitors. If the Q is low the insertion loss will be higher. The pass band theoretically extends to infinity, in practice at some very high frequency the inherent self-inductance of the capacitors and the stray capacitance of the inductors will modify the response.

In the stop band the attenuation at DC is infinite. For the 3-pole version it reaches 90dB below the pass band at approx 1MHz. For the 5-pole version 90dB down is reached at approx 4MHz and the 7-pole version reaches the same point at approx 8MHz.

The high-pass filter can be easily designed with the use of Zverev's tables. Again, I prefer to design standard filters with a 1MHz cut-off and then scale them to the required frequency. These standard filters are listed below in tabular form. All have a 1MHz cut-off and impedance of 50Ω.

To scale to a higher frequency simply *divide* all the values by the new frequency. For example, for 100MHz, divide all the values by 100. To scale to another impedance, modify the values by the *ratio* of the new impedance to 50Ω. For example, for an impedance of 300Ω we would *multiply* all the inductor values by six and *divide* all the capacitor values by six.

3-pole

Ripple dB	C1 & C3 pF	L2 µH
0.5	1708.038	6.215048
0.1	2221.593	4.993253
0.01	2695.029	4.369028

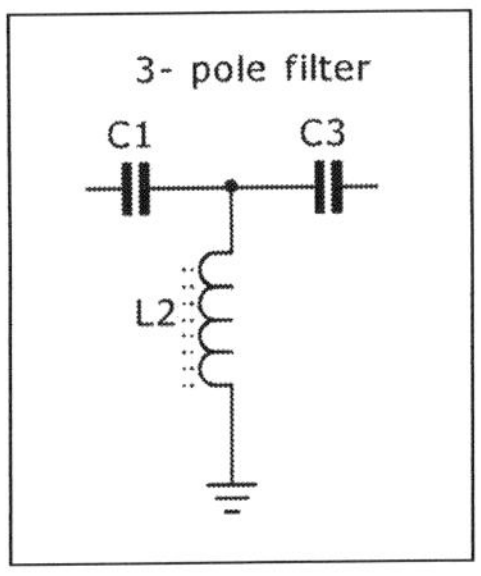

(Right) Fig 6.14. 3-pole filter

5-pole

Ripple dB	C1 & C5 pF	C3 pF	L2 & L4 µH
0.5	1761.733	1182.693	6.109595
0.1	2446.091	1420.329	5.114562
0.01	3259.368	1562.947	4.722979

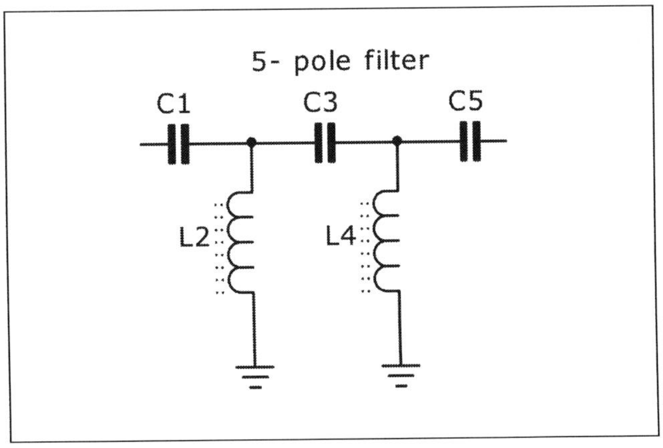

Fig 6.15. 5-pole filter

7-pole

Ripple dB	C1 & C7 pF	C3 & C5 pF	L2 & L6 µH	L4 µH
0.5	1778.665	1171.247	6.139764	5.746495
0.1	2523.265	1421.534	5.236738	4.735627
0.01	3487.564	1589.880	4.990122	4.25457

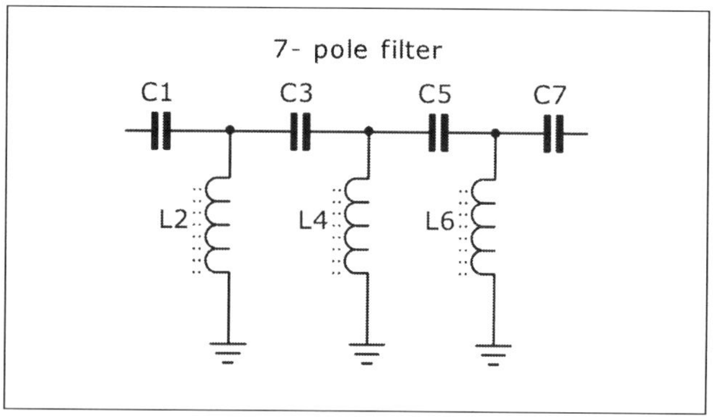

Fig 6.16. 7-pole filter

9-pole

Ripple dB	C1 & C9 pF	C3 & C7 pF	C5 pF	L2 & L8 µH	L4 & L6 µH
0.5	1786.050	1171.894	1147.724	6.158770	5.715951
0.1	2557.528	1432.538	1386.548	5.299159	4.728592
0.01	3595.097	1622.871	1536.467	5.129728	4.274682

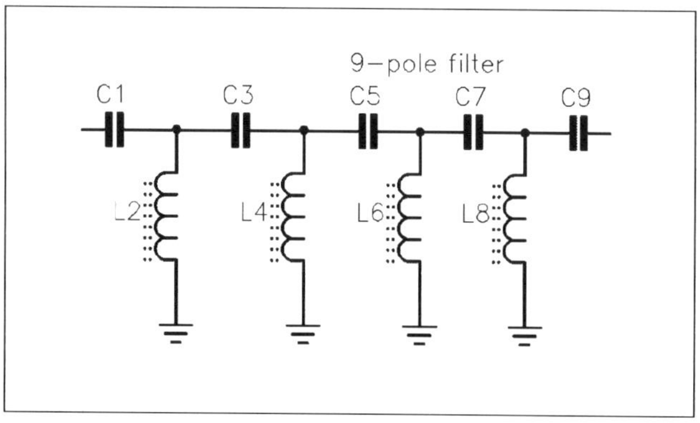

Fig 6.17. 9-pole filter

CONVERTING A LOW-PASS FILTER FILE TO A HIGH-PASS FILTER

Often only values for low-pass filters are published. It is very simple to convert from one type to the other. The example below illustrates this.

The values for a 5-pole low-pass filter with a 0.1dB ripple are:

C1 = 1.3013
C3 = 2.2411
C5 = 1.3013
L2 = 1.5559
L4 = 1.5559

These are the *normalised* values in Farads and Henries at 1Ω impedance and a frequency of 1radian/sec. To convert to a high-pass filter we take the reciprocal of the low-pass values. For example the capacitors in the high-pass filter are C2, C4 and C6 and the inductors L1 and L3.

C2 = 1/C1 = 0.7684623 Farad
C4 = 1/C3 = 0.4462095 Farad
C6 = 1/C5 = 0.7684623 Farad
L1 = 1/L2 = 0.6427148 Henry
L3 = 1/L4 = 0.6427148 Henry

Now we convert to 50Ω by dividing the Cs by 50 and multiplying the Ls by 50:

C2 = 0.0153692 Farad
C4 = 0.0089242 Farad
C6 = 0.0153692 Farad
L1 = 32.13574 Henry
L2 = 32.13574 Henry

Now all we need to do is to convert to the required cut-off frequency. We do this by dividing the values by 2π f. For a 100MHz cut-off this is 6.2832×10^{8}. The values become:

C2 = 2.4461×10^{-11} = 24.461pF
C4 = 1.4203×10^{-11} = 14.203pF
C6 = 2.4461×10^{-11} = 24.461pF
L1 = 5.1146×10^{-8} = 51.146nH
L3 = 5.1146×10^{-8} = 51.146nH

BAND-PASS FILTERS

Intuitively the reader might presume that a simple way to make a band-pass filter would be to cascade a high-pass and low-pass filter. Whereas this approach is possible it will not normally work if the filter needs to have a narrow bandwidth. For cases where we require a very wide bandwidth within the pass band it can be made to work.

However, when using this approach one often finds that the low-pass section requires considerably more poles than the high-pass section to obtain a symmetrical response. Consider the plot in **Fig 6.18** which is a filter constructed by cascading a low pass and high pass. The high-pass filter consists of 7-poles and the low pass 9-poles. The attenuation of the high frequency signals is not as sharp as the low frequency end of the spectrum and would require more poles to get a symmetrical response.

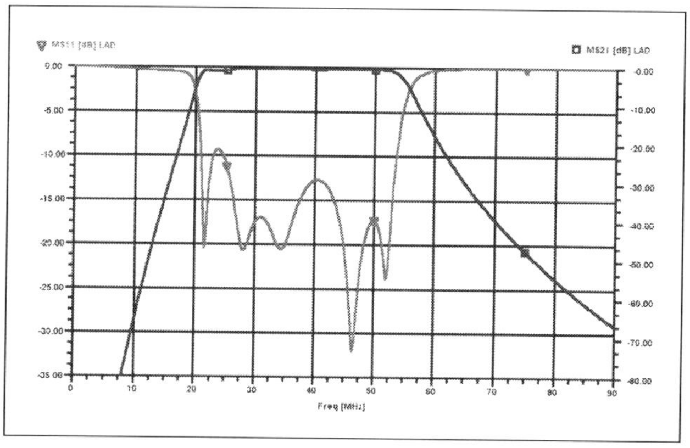

Fig 6.18. Combined low-pass and high-pass filter

ADVANCED BAND-PASS FILTER DESIGN

This is a rather specialised application and for many applications we require a relatively narrow band-pass filter. We will take the example of an oscillator multiplier chain for an UHF transmitter to illustrate this approach. We might start with a crystal oscillator at 12MHz and multiply by 36 to get to 432MHz. The possibility of doing this in just one stage is very remote. To be able to make a band-pass filter that can pass 432 to 434MHz but also reject signals only 12MHz either side is next to impossible.

So we need to attack the problem in small bites. Starting at 12MHz means we have several choices where the first multiplier stage can provide an output. It is prudent not to exceed a multiplication factor of about 4. The choice of multiplication is limited to integer numbers and the various multiplier outputs need to be readily multiplied to 432MHz.

We might choose the first multiplier to be a tripler stage and the output will then be at 36MHz, this can work as we need a further multiplier of times 12 to attain the final frequency. This will require either a times 3 and a times 4 stage or two times 2 and one times 3. It is generally better to use the highest multiplication for the first stage because the design and construction of a filter is easier at the lower frequencies. Therefore we choose a times 4 multiplier as the first stage. So our first multiplier will pass the wanted 48MHz product but it also needs to reject the other products generated at 12MHz, 24MHz, 36MHz, 60MHz, 72MHz, 90MHz etc. The required bandwidth of this filter is not very wide. If the transmitter covers the 432 to 434MHz portion of the band the filter only needs to pass signals between 48.0MHz and 48.22MHz, a bandwidth of 220kHz. In percentage bandwidth terms the unwanted signals are 25% either side of the wanted signal.

This filter is the most critical because if any 12MHz products get past it to the next multiplier we will end up with spurious signals 12MHz either side of the final signal. Hence, it will require several sections to achieve at least 80dB of spurious suppression. In practice at least 3 or 4-sections will be needed, meaning three or more inductors and resonating capacitors. As the next multiplier stage needs a certain drive power to be efficient the best approach would be two 2-section filters cascaded with an amplifier between them to provide the gain.

The second multiplier stage will raise the signal frequency to 144MHz, a multiplication factor of 3. The filtering here is a little easier as the unwanted products are now 48MHz either side of the wanted signal and the most destructive signals occur at 96MHz and 192MHz. In percentage bandwidth terms they are 33.3% either side of the wanted signal. This filter needs a bandwidth of 660kHz. A 3-section filter followed by an amplifier stage will normally suffice.

The final multiplier will raise the frequency to 432MHz; again a factor of 3 is required. Now the unwanted multiplier products are 144MHz away from the wanted, but the filter bandwidth required is now 2MHz. To prevent interference to other users of the spectrum requires this filter to have a very steep cut-off and be able to reject signals down to at least 70dB. The insertion loss in the filter is not too important as we will have a lot of amplification following the final multiplier stage. So if we can tolerate quite a lot of loss in this filter, perhaps as much as 10dB, then the easier it is designing a steep-sided filter slope.

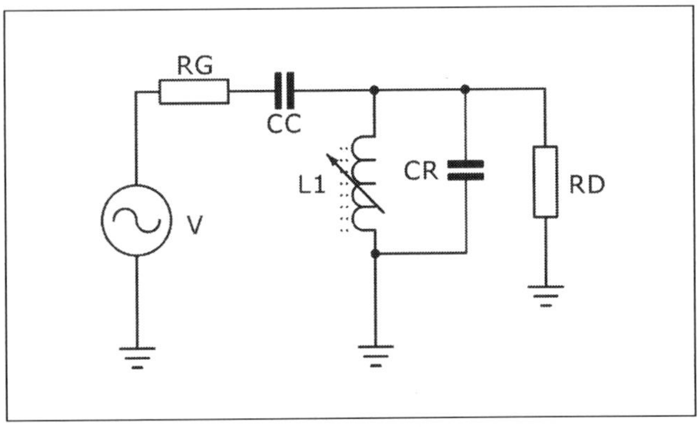

Fig 6.19. Basic band-pass filter

Before we continue let us look at the various filter types and the pros and cons of each type.

The basic circuit of a band-pass filter is shown in **Fig 6.19**. Here only the input network is shown. Again the components that go to make up a filter are known as *poles* and the individual building blocks are called *sections*.

The filter is connected to a signal source, which has characteristic impedance, RG. The filter section consists of a parallel resonant network consisting of the capacitor CR and inductor L1. At resonance the network develops an impedance across it called RD. At resonance this impedance is non-reactive, hence this is known as the *dynamic resistance* and it behaves as a pure resistance. The coupling into the first network is performed by a reactance with a very high value. In this case it is achieved by CC, which will be a low value capacitor. (We could also use an inductor in which case it would have a high value.)

As we saw in the Tuned Circuits chapter, to attain a high RD the value of the inductor needs to be high and CR needs to be low (a high L-C ratio). If the inductor L1 has a high Q the value of RD will also be high (assuming the capacitor Q is much higher than the inductor, which is almost always the case). In some cases RD can be as much as 100kΩ. Hence, any loading on the parallel resonant network by CC or the following sections will dampen the response and increase the bandwidth and the insertion loss.

Where the input source impedance is low, such as 50Ω, a better method of coupling into the network is by an *impedance converter*. This can be a tapped inductor acting as an *auto-transformer* or a pair of capacitors connected in series as a potential divider. Of the two methods the tapped capacitor method is much easier to analyse and design with. Tapped inductors do not easily lend themselves to accurate computer modeling or network analysis. They are also difficult to make changes to as the design evolves. So at present we will only look at

> A typical Toko tunable inductor at 21.4MHz has a measured Q of 110 and when resonated by the parallel capacitor the value of RD is about 20kΩ. If the Q or the inductor value is doubled the value of RD also doubles. For example if L is raised to four times the current value and CR reduced to resonate at the same frequency, with the same Q of 110 the RD goes up to 80kΩ.

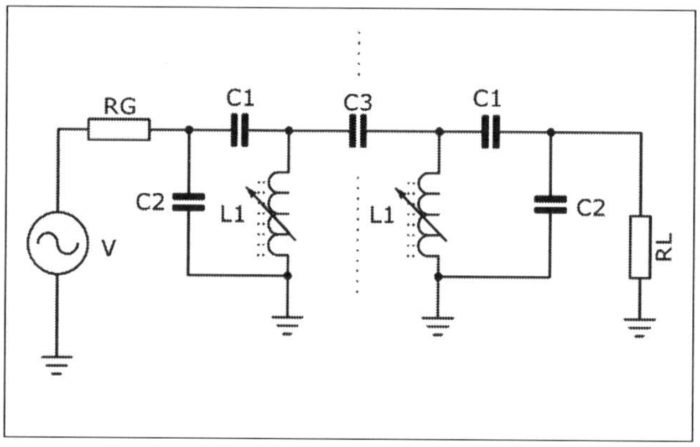

Fig 6.20. Two section band-pass filter

capacitor tapped matching to the network.

Fig 6.20 shows a simple 2-section band-pass filter using tapped capacitors to feed the signal into and out of the filter. The inductors L1 are the same value and the filter is symmetrical about the centre line of C3. The two sections are *mirror images* and hence, because the input and output impedance is the same, the component values for either side will be exactly the same, i.e. C1 and C2 on the input will be same as C1 and C2 on the output etc.

Note: in all the filters to be described it is assumed the individual inductors are completely shielded from each other, either by placing them into screening cans or orientating the inductors to achieve minimal mutual inductance. If some mutual inductance exists then the response will be severely skewed. Pre-wound variable inductors in shielding cans, such as those made by Toko or other manufacturers, are the best option, although hand-wound inductors using ferrite toroidal cores are also a good method. The most important thing is to achieve the correct inductance value and acceptable Q values.

The value of capacitance that resonates the inductor is the product of C1 and C2 in series. If the source and load impedance is low the value of C2 will normally be much greater than C1. If the inductor Q is low the ratio of C1 to C2 will be different to when the inductor Q is high; this is because the value of RD changes directly with the inductor Q. If the value of the source and load impedance equal half the value of the circuit RD value then C1 and C2 will be equal values. It is the *reactive potential divider* properties that make the tapped capacitor matching such a flexible method.

The coupling between the two sections is by capacitor C3, which has a very small value. Half the value of C3 is in fact added to both sides of the filter, so the resonant frequency of the two sections appears slightly lower in frequency. This can be corrected by a slight reduction in the value of L1.

> Tables of band-pass filter values in some publications often give incorrect results because the tables assume the inductor and capacitor Q values are infinite, which is never the case in a practical filter. Consequently it is not so easy to design a standard filter at, say, 1MHz and then scale to another frequency as we can for low-pass and high-pass filters. The inductor Q and hence RD values directly affect the capacitor and inductor values. If the inductor Q values are the same then a scaling technique will work for a given source and load impedance, but equivalent Q values are difficult to achieve over a wide range of frequencies.

All the radio frequency power has to pass through C3 and so it contributes additional loss over and above the inherent insertion loss of the two sections. For a filter handling substantial RF power this causes the capacitor to heat up due to its inherent series resistance and this causes the capacitor to change its value. This limits the power that can be handled.

The other limitation is the *voltage rating* of the parallel resonating capacitors and if this is exceeded the capacitor will fail. As the voltage developed across the inductor / capacitor combination is the ratio of RD to source or load impedance the voltage multiplication becomes very high when high Q inductors are used. For example, a filter handling a signal of 1V rms at 50Ω at the input might have a RD value of 10kΩ. This is a voltage step-up of 200 times. Hence, we would expect to measure a signal across the inductor of 200V rms, a pk-pk voltage of 565V. If the capacitors are not adequately rated they will fail. These factors normally limit the power handling capabilities of conventional band-pass filters to a few tens of milliwatts.

UNDER-COUPLING AND OVER-COUPLING

The value of C3 also affects the bandwidth and pass-band ripple. If C3 is very small the slopes of the filter will be very steep and the bandwidth the minimum possible, but the insertion loss will also be the highest. This condition is known as being *under-coupled*.

If C3 is increased in value the insertion loss reaches a minimum but at the expense of the slopes becoming less steep. This condition is known as *critical coupling*. If C3 is further increased in value the slopes start to rise further and the bandwidth broadens and at the same time the pass-band response becomes double humped with a distinct dip at the centre frequency. This is known as *over-coupling*. For some applications where the bandwidth is the more important factor an over-coupled filter is often used, for example in *wide-band couplers* in old-fashioned HF transmitters. To make a band-pass filter covering the entire 80m band from 3.5 to 3.8MHz often requires an over-coupled type filter.

To illustrate the effect C3 has we can look at three simulation plots of a typical band-pass filter. I have chosen a filter for a 2m receiver front-end amplifier stage, which needs to cover 144 to 146MHz. The first response plot, **Fig 6.21**, shows the under-coupled response.

As you can see, the slopes are very steep, the pass band very narrow - too narrow to cover the band - and the insertion loss is quite high, almost 8dB, which

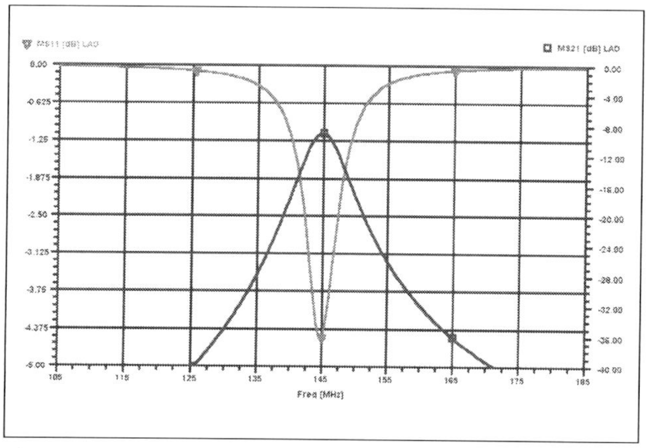

Fig 6.21. 2m band-pass filter: under-coupled

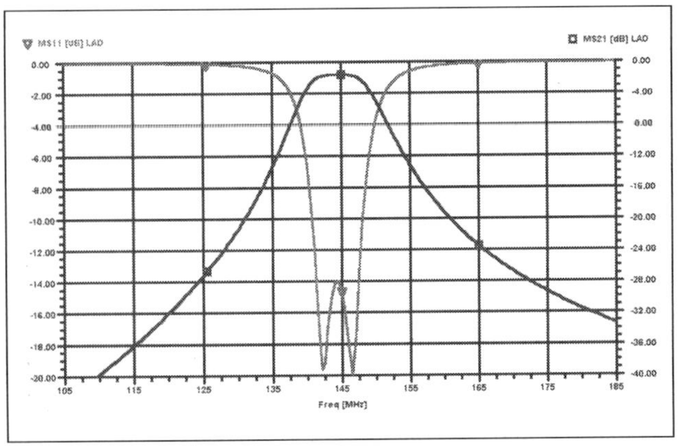

Fig 6.22. 2m band-pass filter: critical coupling

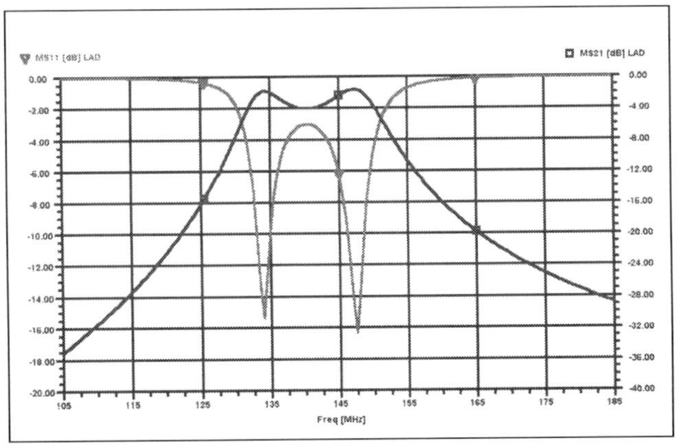

Fig 6.23. 2m band-pass filter: over-coupled

will degrade the receiver noise figure. The coupling capacitor value required is 0.12pF, which is an impractical value.

The next plot, **Fig 6.22**, is the filter optimised to cover 144 to 146MHz by increasing the coupling capacitor value and slight adjustment of the inductors to bring the response back to the centre frequency.

Here, the slopes are much less steep but the filter now covers the whole 2m band. The insertion loss has dropped to less than 1.5dB at mid-band. The coupling capacitor value is now 0.4pF, which is still a bit impractical, but could be achieved by connecting two 0.82pF capacitors in series.

Finally let us over-couple the filter to see the effect it has: **Fig 6.23**. Here we can see the distinctive dip in the pass-band response and the poor rejection to adjacent frequencies. Also, notice that the centre frequency has moved slightly lower in frequency because the inductors

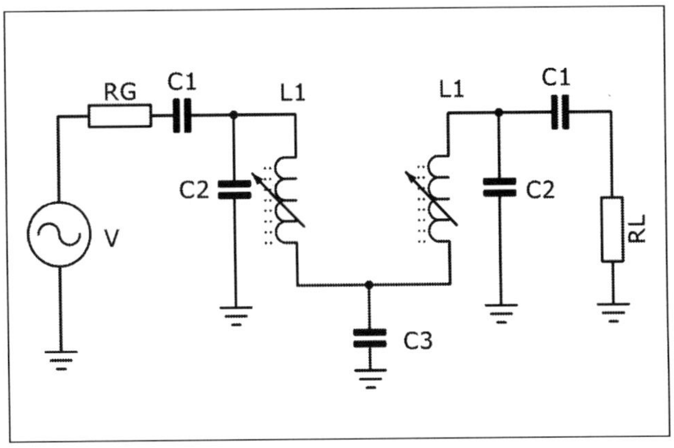

Fig 6.24. Bottom-coupled band-pass filter

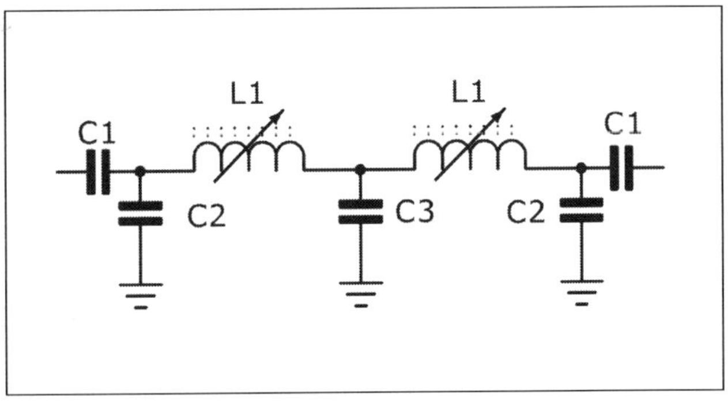

Fig 6.25. Redrawn Cohn filter circuit

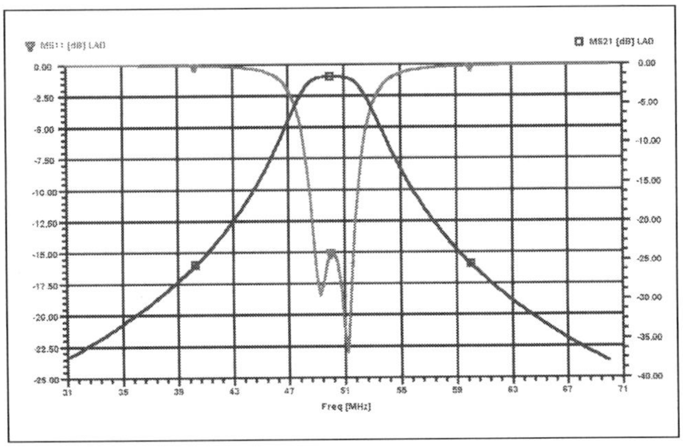

Fig 6.26. 6m band-pass filter

are still the original value. The coupling capacitor is now 1.2pF, which is still a very small value, although obtainable as a standard E12 capacitor value.

COHN FILTERS

The above highlights another problem with the top-coupled band-pass filter. If you look at the slopes at either side of the centre frequency you will see that the one on the low frequency side of the centre is steeper than the one on the higher frequency side. This is because the filter is tending to behave like a high-pass filter. If you look at the circuit topology you can see it bears a resemblance with the top-coupled inductors used in a high-pass filter. For very high frequency band-pass filters, where the top coupling capacitor can become too small to be practical, a better topology is the *bottom-coupled filter*. This is also known as a *Cohn filter*, after its inventor [S B Cohn *Direct-coupled-resonator-filters*, Proc IRE, Vol 45, pp187-196, Feb 1957]. This filter is shown in **Fig 6.24**.

The inductors are now connected at the ground end to a common coupling capacitor C3. Also the capacitor tapping has been dispensed with and top coupling used to feed the signal in and out. The combination of these does two things. Firstly, it makes the value of C3 a practical value and secondly it makes the response either side of the centre frequency more symmetrical. The symmetry is achieved by combining the high-pass effect of the input and output-coupling capacitors with the low-pass filter characteristic of the bottom coupled filter. If the filter is redrawn as in **Fig 6.25** this becomes obvious.

In **Fig 6.26** the response of a Cohn filter designed to cover 50 to 52MHz is shown. This filter still only uses two sections but has superior performance to a top-coupled version using an identical number of components.

The bottom-coupling capacitor has a practical value of 680pF.

The pass-band response shows the filter to be slightly over-coupled; this is because the nearest E12 capacitor value of 680pF was chosen for C3 rather than the exact calculated value, which was about 80pF higher. By connecting an additional 82pF in parallel with C3 the response can be optimised. Notice how symmetrical the filter response is and the rejection either side of the pass band, more than 35dB at only 20MHz offset. The band centre insertion loss is only 2dB. This filter uses Micrometals Inc ferrite toroidal cores (T37-12) with an unloaded Q of 160, which is easily achieved. The required inductance is 350nH and the input and output coupling capacitors are 6.8pF, the resonating capacitors are 22pF.

An alternative to a capacitor bottom-coupled filter is the inductor bottom-coupled filter. This, as the name suggests, uses an inductor to couple the two resonating inductors together. However, the inductor value required can also become quite small and is often not practical. This filter is shown in **Fig 6.27**.

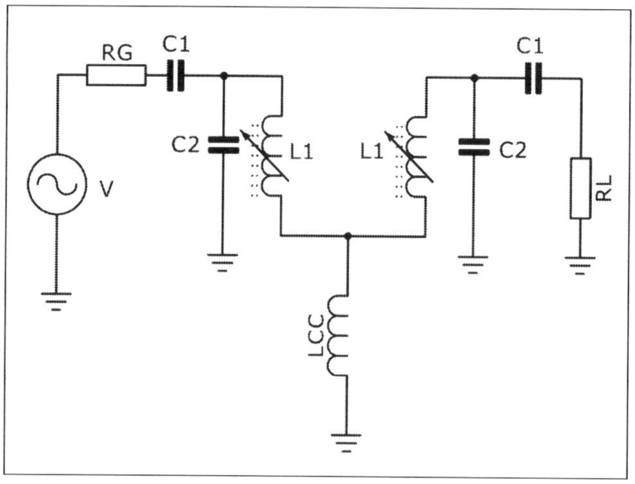

Fig 6.27. Inductor bottom-coupled filter

HIGHER ORDER BAND-PASS FILTERS

Where the need is for more attenuation in the stop band or a faster roll-off we can increase the number of sections. In doing this we use the same type of section for the input and output but the centre section or sections now have a different topology. If we choose a three-section filter the centre section consists of a pure parallel resonant combination of a capacitor and inductor resonated to the filter centre frequency and the two end sections are resonated one higher and one lower to spread the pass band out to the required value. Of the three-section types the best approach is to choose all the sections to be top-coupled with low value capacitors, despite the fact that these may reach impractical values. This is shown in **Fig 6.28**.

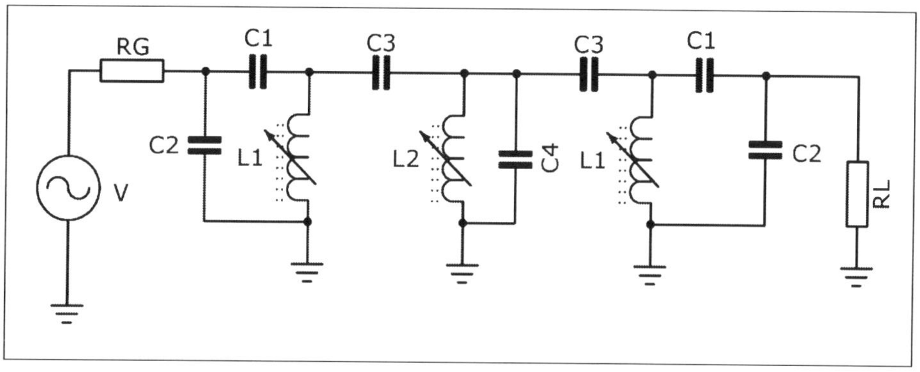

Fig 6.28. 3-section band-pass filter

Only when we choose an even number of sections, such as a 4-section filter, can we utilise the bottom-coupled technique. The centre two sections, however, still need to utilise the top-coupling capacitor. A typical 4-section band-pass filter is shown in **Fig 6.29**. This uses bottom coupling for the end sections and top coupling for the two centre sections.

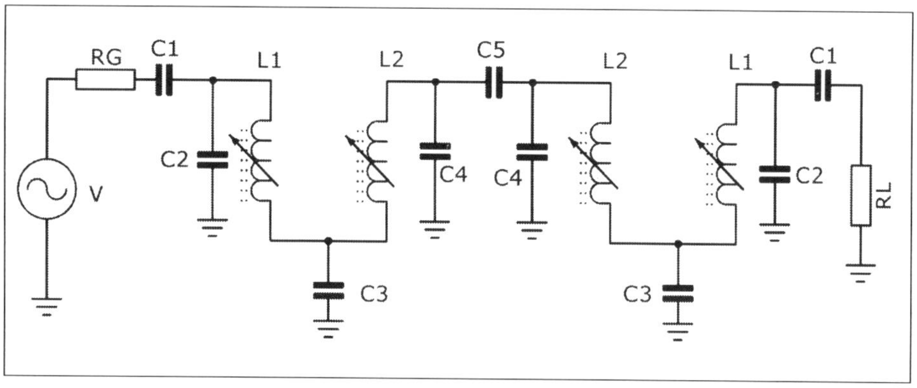

Fig 6.29. 4-section band-pass filter

In the filter shown in **Fig 6.29** the values assigned to the various components are the same for each end of the filter, provided Zin = Zout. Hence, if C1 is a 10pF at the input end it will also be a 10pF at the output end. The value of C3 will quite large and C5 will be quite small. C4 will be sum of C1 and C2 plus half the value of C5. As C5 is likely to be very small for a high frequency filter this simplifies to C4 ≈ C1+C2. The inductors L1 and L2 will be approx the same value, within a few percent, for the narrowest pass-band response.

With adequate Q values for the inductors and the

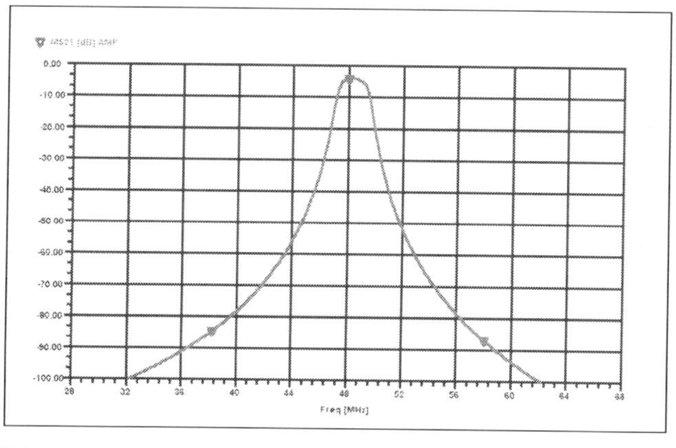

Fig 6.30. 48MHz band-pass filter response

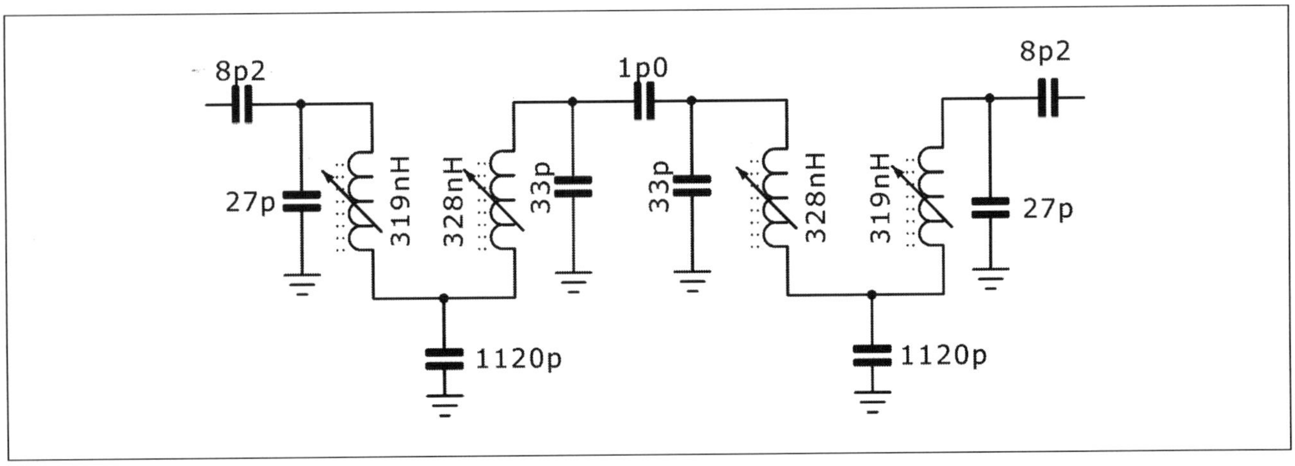

Fig 6.31. 48MHz band-pass filter schematic

capacitors the response at 10% off the centre frequency will be of the order of 50dB or more down from the pass-band response when the band-pass portion is critically coupled. If the coupling is slightly under-coupled this value can attain values of more than 70dB.

Again, if the filter is looked at carefully, it can be split into two mirror images about C5.

Going back to our original requirements for a multiplier chain for the 432MHz transmitter, we can now look at a simulation of a 4-section band-pass filter for the 48MHz filter stage. **Fig 6.30** shows the response achieved. As you will recall, we needed about 80dB attenuation at 36MHz and 60MHz. The response meets this with some attenuation to spare. The component values are very practical and have been optimised for E12 capacitor values.

As you can see, we achieve 92dB attenuation at 36MHz and approx 94dB at 60MHz. The pass-band insertion loss is 4dB, so our attenuation is actually (92-4) = 88dB, which is still better than we require. The filter component values are shown in **Fig 6.31**. The filter is designed for 50Ω impedance.

All the capacitors are E12 values. The bottom coupling capacitors are made up from two 560pF in parallel. The inductors are wound using Micrometals T37-12 cores and have a nominal Q of 160 at 50MHz.

STANDARD BAND-PASS FILTERS

Standard 2-section band-pass filter prototype
Z = 50Ω.
Attenuation = ≈40dB at 20% from centre frequency.

F MHz	C1 pF	C2 pF	C3 pF	L1 μH Q=100min
10	39	68	2200	2.472
20	22	39	1120*	1.213
30	15	27	820	0.706
40	10	22	820	0.518
50	8.2	22	820	0.348
60	5.6	15	680	0.350
70	4.7	12	560	0.32
80	3.9	12	560	0.255
90	3.9	10	470	0.230
100	3.3	8.2	390	0.227

* = 2 x 560pF. Inductance tuning range ±20%

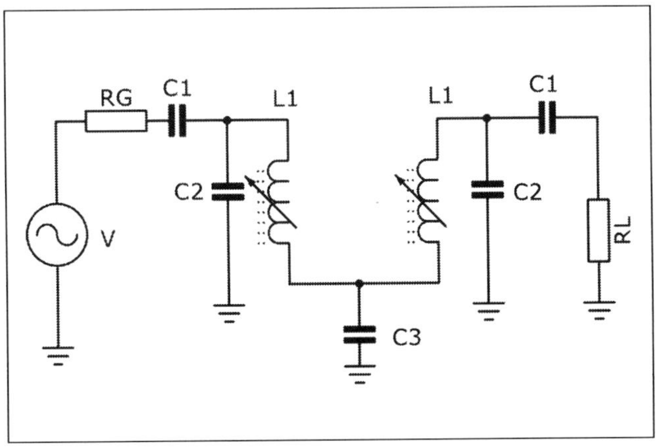

Fig 6.32. Standard 2-section band-pass filter prototype

Note: All filters optimised for best pass-band insertion loss, input return loss and stop-band attenuation, using standard E12 capacitor values.

To scale to a new frequency, divide or multiply all values by the ratio of the new frequency to the old. For example, to make a 1MHz filter multiply all the values by 10 for the 10MHz design. To make a 200MHz filter, divide all the values by 10 for the 20MHz design, or divide the 100MHz values by 2. For example, we need a filter with a pass-band centre of 15MHz. We would take the 10MHz filter values and divide all the values by 1.5. As we prefer to use E12 values if possible, this gives us the following new values.

C1 = 26pF use 27pF
C2 = 45.33pF use 47pF
C3 = 1466pF use 1500pF
L1 = 1.648μH use value calculated.

Alternatively, take the 30MHz design values and multiply by 2. This gives the following values:

C1 = 30pF use 27pf or 2 x 15pF
C2 = 54pF use 2 x 27pF or 56pF
C3 = 1640pF use 2x 820pF
L = 1.412μH use value calculated.

For frequencies in between the design standards we can interpolate the nearest E12 values. For example, we need a filter for 35MHz. By assessing the values at 30MHz and 40MHz we can choose suitable values:

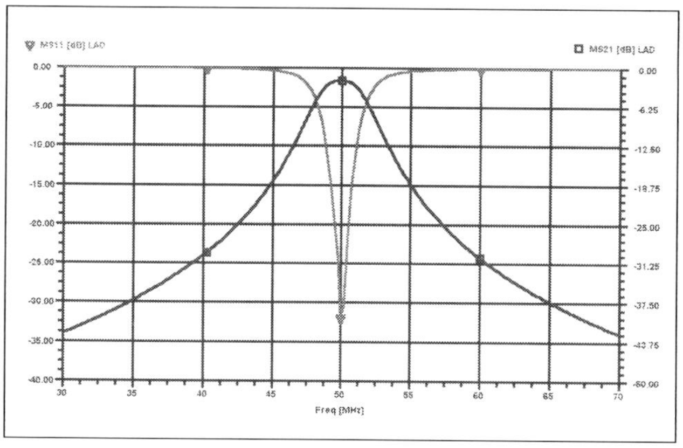

Fig 6.33. Typical attenuation response for the standard filters

C1	= 15pF at 30MHz	10pF at 40MHz	use 12pF
C2	= 27pF at 30MHz	22pF at 40MHz	use 2 x 12pF
C3	= 820pF at 30MHz	820pF at 40MHz	use 820pF
L1	= 0.706μH at 30MHz	0.518μH at 40MHz	use ≈0.6μH

Alternatively use the 70MHz values and divide by 2. The typical attenuation response for these standard filters is shown in **Fig 6.33**.

STANDARD E12 CAPACITOR VALUES

Base value (pF)	x10	x100	x1000	x10,000	x100,000
0.1	1	10	100	1000	10000
0.12	1.2	12	120	1200	12000
0.15	1.5	15	150	1500	15000
0.18	1.8	18	180	1800	18000
0.22	2.2	22	220	2200	22000
0.27	2.7	27	270	2700	27000
0.33	3.3	33	330	3300	33000
0.39	3.9	39	390	3900	39000
0.47	4.7	47	470	4700	47000
0.56	5.6	56	560	5600	56000
0.68	6.8	68	680	6800	68000
0.82	8.2	82	820	8200	82000

Note: Values below 0.47pF are not normally available in standard capacitor types. Where very low value capacitors are required either a 'gimmick' capacitor made from two insulated wires twisted together and pruned to length may be used, or a high value carbon-composition resistor. A 1MΩ 1/8W resistor has a parallel capacity of approx 0.2pF.

SPECIAL FILTERS

In some applications we require a far better filter than the conventional types. A typical case would be a repeater transmitter to reduce the energy transmitted in the receiver pass-band from 'desensing' the receiver. Normally we would use a type of cavity filter or a helical filter. A helical filter is a coiled-up quarter-wave and occupies much less volume than a full quarter-wave line. The length of transmission line is less than a true quarter-wave if it were uncoiled and this is because the top end has some fringing capacitance and we apply even more with a tuning disk to bring the line to resonance.

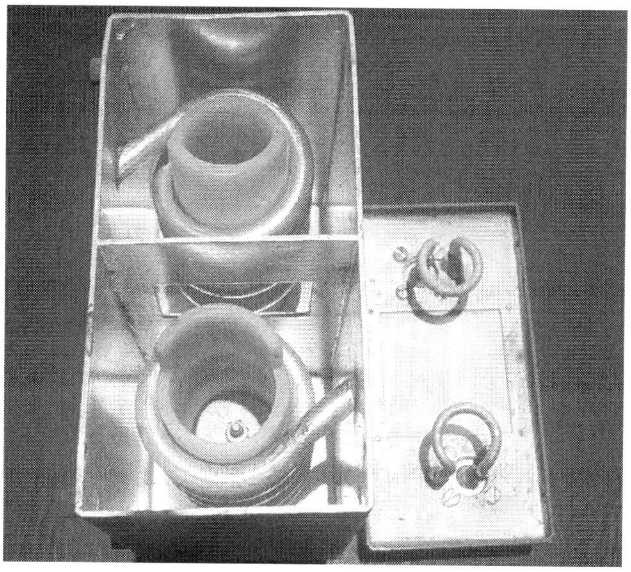

Fig 6.34. VHF helical filter

A typical helical filter for a 100W VHF transmitter and made by Philips Telecommunications (South Africa) is shown in **Fig 6.34**. This is a 160MHz narrow-band filter used to reduce the level of transmitter broad-band noise. The insertion loss is quite high, being nearly 0.8dB, so some of the transmitter power is lost in the filter. The filter is very compact and the coupling to the helical resonator is by two loops at the ground end of the inductors. The coupling between the two sections is by an iris plate with a slot cut to obtain the optimum coupling. The helix is wound using 8mm copper tube and the ground point is made to the wall of the resonator cavity. The filter is heavily silver plated to lower the insertion loss. The tuning capacitors are two copper slugs that are attached to threaded rods and these run inside the PTFE tubes and support the helix. The 3dB bandwidth is approx 2MHz.

Pye Telecommunications used a variation on this method in their early VHF equipment: **Fig 6.35**. It was a bottom inductively-coupled band-pass filter similar to the Cohn design described earlier. It used an additional compensating inductor inductively coupled to the main inductor

Fig 6.35. Pye Telecommunications antenna filter. 66 - 88MHz version shown for 25W transmitters

Fig 6.36. Comb line filter for 500MHz (cover removed)

that reduced the coupling. The two inductors forming the main resonators were shielded from each other in screened cans and the coupling wire entered via a ceramic feed-through insulator.

For higher frequency operation we can use either a quarter-wave resonator or we can use a different method. For frequencies above about 430MHz a good filter is a type using coupled shortened quarter-wave lines. There are two distinct types in common use, although many amateurs incorrectly call them both 'inter-digital filters'. The second type is known as a 'comb filter', because the quarter-wave resonators are mounted like the teeth of a comb with all the grounded ends on the same side. An inter-digital filter has the resonators mounted on opposite sides and every other line is inverted. (If you hold both hands out and spread the fingers on both hands and then slip the fingers between the opposite hand you get an 'inter-digited' filter.) A typical comb line filter is shown in **Fig 6.36** and an inter-digital filter in **Fig 6.37**.

The filter shown in Fig 6.36 is all copper construction with air-spaced tuning capacitor 'slugs' fitting over

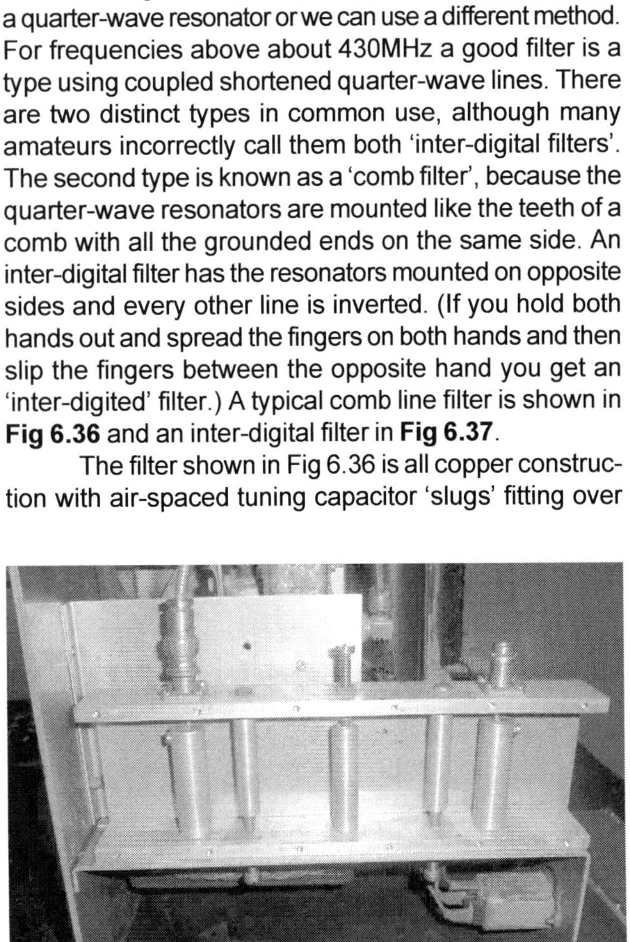

Fig 6.37. Inter-digital filter for 1.3GHz (cover removed)

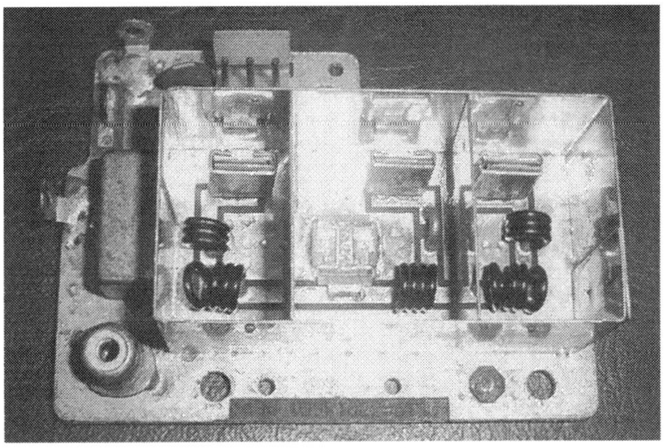

Fig 6.38. UHF harmonic filter

the round ends of the two square section resonator lines. The 50Ω coupling lines are rectangular section and terminate in TNC connectors. The insertion loss at 500MHz is approx 0.5dB. Rejection at 100MHz off the centre frequency is approx 50dB. The filter measures approx 75mm in length.

The inter-digital filter shown in **Fig 6.37** is a 5-pole design based on an article that was originally published by DL3NQ in *UKW-Berichte* (*VHF Communications*). It has an insertion loss of 0.6dB, a 3dB bandwidth of 30MHz at 1.3GHz and the rejection to signals spaced 100MHz either side is approx 60dB. It was used in my original 23cm power amplifier to clean up the 2C39 power-mixer up-converter from 144MHz to reduce the 1152MHz LO injection to insignificant levels. The power amplifier is a WB6IOM design producing 150W output with a gain of 13dB using dual 2C39BAs with water-cooling. The filter can handle as much as 30W with no difficulty. The construction is all aluminium with critical coupling between the resonator lines and tuning screws to bring the resonators on frequency. The two thick lines couple from 50Ω to the resonator lines. The aluminium section acts as a waveguide and the ends are left open as no radiation occurs from the open ends.

Fig 6.38 shows a Motorola harmonic filter used in a 450MHz 10W mobile transceiver. It uses conventional lumped LC networks and strip-line inductors for trap sections. The capacitors are clamped silver mica (Unelco). Note the connector used: an RCA (phono) socket, which is a very good 50Ω RF connector even at UHF.

SPECIAL MICROWAVE FILTERS

If a low-pass filter is required at a frequency much above 1GHz the component values become very small and difficult to manufacture. A technique that overcomes this is the 'Rod-Disk' filter (sometimes called the 'Rod-Bead' filter). These can be manufactured up to at least 10GHz with high attenuation in the stop band and very low insertion loss in the pass band. Because the series element is a low DC resistance they are often used for feeding supply voltages into a network as an RF choke.

The inductors are made from round rod of appropriate diameter and length and the capacitors from round disks of metal. The value of the capacitor is determined by the air gap between the outer edge and the outer coaxial sleeve. The thickness and length of the disk can be made to produce the required capacity value with adequate

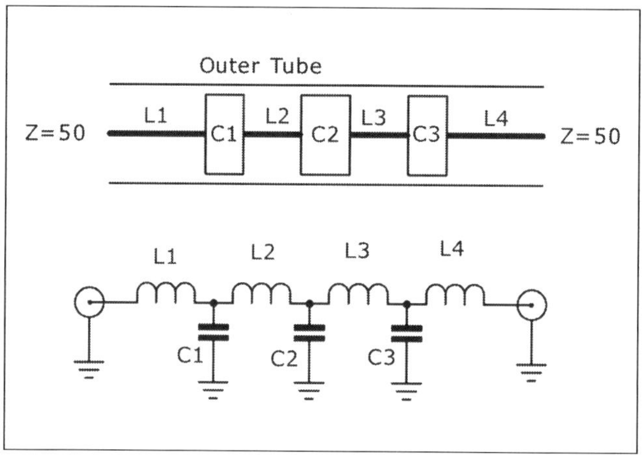

Fig 6.39. Rod-disk 7-pole low-pass filter and equivalent circuit

breakdown voltage. Sometimes the dielectric space is filled with a sleeve of PTFE made from a rolled up sheet of thin material. This raises the dielectric constant by 2.12 over that of air and so larger gaps may be used to obtain higher breakdown voltages.

Fig 6.39 shows the cross section of a typical rod-disk filter. The limitation is that the first element is a series inductor, but this is of no concern in most cases. Manufacturers such as Microlab offer styles with a shunt capacitor as the first element, as shown in **Fig 6.40**. The two ends of the outer tube have suitable RF connectors attached to connect to the rest of the equipment.

The central rod is formed from a single piece with the various capacitor disks attached. The spacing between adjacent disks sets the inductor value. Large inductance values require more distance between the capacitor disks. Using this technique low-pass filters of up to 21-poles may be made in a very compact manner. A typical use is for the low-pass filter preceding the 1st mixer of a spectrum analyser where the 1st IF is above the highest swept range. This is often about 2.5GHz and the low-pass filter is placed before the mixer to eliminate spurious signal from the input passing to the 1st IF.

The Q of the inductors and capacitors is very much higher than when conventional components are used and a capacitor Q of over 5000 is quite practical. The inductor Q is of the order of 1000 to 2500 when silver or gold plating is used. Some years ago I made a 15-pole low-pass filter with a cut-off at 2.2GHz for a piece of test equipment. The series inductor was made from 1.6mm brass rod and the disks parted off from a piece of 12mm diameter brass rod. The disks were drilled 1.6mm through the centre and slid on to the inductor rod and soft soldered in place. The outer tube was a piece of 15mm OD copper water pipe fitted with end caps and SMA connectors. The insertion loss below the cut-off frequency was measured as being less than 0.2dB with approx 0.25dB ripple in the pass band. The attenuation in the stop band at above 2.5GHz was greater than 80dB. The test equipment available could only cover up to 20GHz so the response above this frequency is unknown, but probably still acceptable for the purpose.

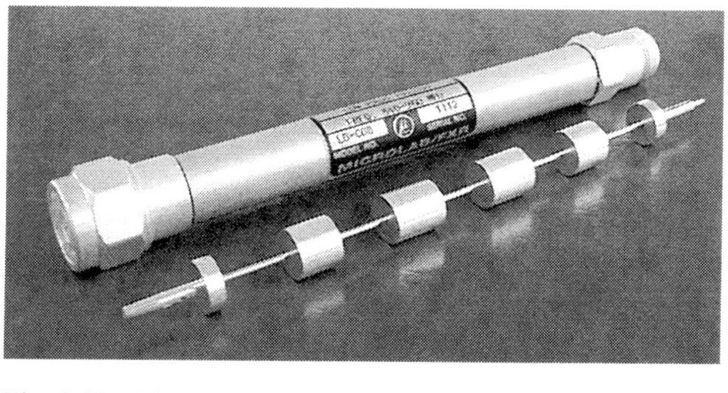

Fig 6.40. Microlab rod-bead 11-pole filter construction (*used with permission of Microlab / FXR Inc*)

QUARTER-WAVE CAVITY RESONATORS

One of the misconceptions about this type of band-pass filter is that they are a 'single-frequency' type (monotonic). This is not so and this can cause some embarrassing results. A quarter-wave resonator has 'overtone' modes that occur at odd multiples of the resonant frequency.

In days gone by many writers advocated the use of 'High-Q Breaks' to insert between a transmitter and the antenna to eliminate harmonic power from passing to the antenna. These are one or more coupled quarter-wave lines with input and output matching networks, often just a tap placed low down on the first and last quarter-wave line.

Included in the quarter-wave resonator category are types such as the helical filter that mimic a true quarter-wave line but with more compact dimensions.

Because the quarter-wave line is not monotonic it works just as well at three, five and all other odd harmonics of the input frequency. At the third harmonic the line is effectively 3λ/4. So if a 144MHz transmitter is fitted with a High-Q Break the 3rd harmonic at 432MHz is hardly attenuated at all. In fact given the wrong length of coaxial line connecting the transmitter to the filter the harmonic power at 432MHz can be as much as 20dB higher than when the transmitter is run into a true 50Ω dummy load. This occurs when the phase angle of the reflected wave is such to cause added 'gain' when the transmitter is operating into a severe mismatch at the harmonic. So a High-Q Break is a poor method and a far better option is a monotonic filter such as a proper low-pass filter.

Another severe limitation is the breakdown rating of the tuning capacitor or the end of the line and the surrounding shielding enclosure. Because the Q is very high the RF voltage at the high impedance end of the line is also high. In some cases the RF voltage may reach values as high as 10kV even with a transmitter of only 100W. If the atmospheric humidity is high this can cause severe arcing when high power is applied. Normal air-spaced tuning capacitors often have a limited breakdown voltage rating and a better option is the use of disks with a large distance between the fixed and moving disk. All the edges should be rounded to reduce the corona discharge effect as much as possible.

The High Q break shown in **Fig 6.41** uses top capacity coupling for the input and output connectors and if analysed degrades to a high-pass filter at some very high frequency and offers little attenuation to the higher harmonics.

COAXIAL STUBS FOR FILTERING

The λ/4 stub can be used for filtering. If we have an interfering signal at, say, the image frequency we can eliminate it simply with a short piece of coax cable. Suppose the interfering signal occurs at 100MHz. If we make a

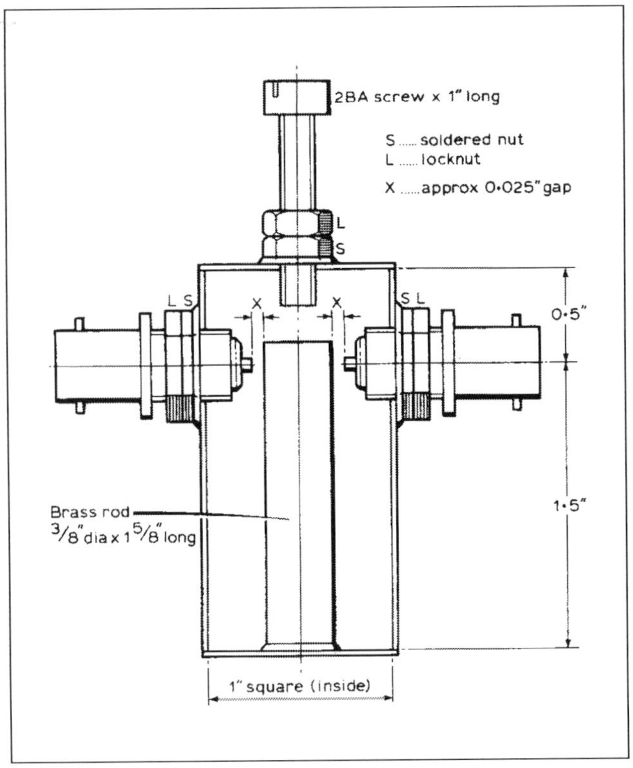

Fig 6.41. High-Q Break for 1.3GHz (from *RSGB VHF/UHF Handbook*)

λ/4 open circuit stub cut for 100MHz and attach it in parallel with our receiver input it appears as a short circuit to signals at 100MHz. The insertion loss is practically zero. But there is another phenomenon with the λ/4 stub. At a frequency where the stub is a length of 3λ/4 or 5λ/4 and all other odd λ/4 lengths it behaves in the same manner. So a stub cut for 100MHz will work at 300MHz and 500MHz.

If we now short-circuit the inner and outer of the stub the properties reverse. At 100MHz the stub appears as an open-circuit and at other frequencies it appears as a short-circuit to the signal. This will also occur at 300MHz and 500MHz. The shorted λ/4 stub is a good way of protecting the front-end of a receiver from lightning damage as the DC condition is a short. When cutting the stub, allowance must be made for the velocity factor of the coaxial cable used.

The same effect happens if we use a λ/2 stub. In this case the impedance appearing at either end is the same. So if we place a λ/2 stub across the receiver input and leave the far end open it behaves as an open circuit across the receiver at that frequency and a short circuit at other frequencies corresponding to when the coax is resembling a λ/4 or odd multiples. If we short-circuit the far end it appears as a short across the receiver input.

The graph in **Fig 6.42** shows insertion loss versus frequency. At the operating frequency the stub appears as an open circuit. At 150MHz and 250MHz it behaves in a similar manner. At 100MHz, 200MHz and 300MHz it appears as a short circuit across the input.

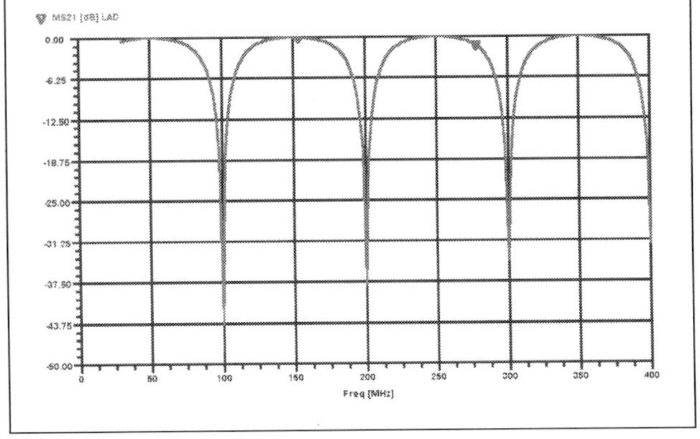

Fig 6.42. Simulation of a λ/4 shorted stub cut for 50MHz

STUBS AS HARMONIC FILTERS

Another useful property of the coaxial stub is that that it can perform as a harmonic filter with virtually zero loss. A common problem with 50MHz transmitters is the second harmonic that falls in the FM broadcast band. By using a λ/4 stub cut for 50MHz with the far end shorted the impedance the transmitter sees is an open circuit, so no power is lost. But at 100MHz the stub behaves as λ/2-length and appears as a short circuit. In many 50MHz amplifiers using a Pi tank network the need exists for an RF choke across the output to blow the high voltage fuse should the anode blocking capacitor fail short circuit. The shorted λ/4 stub can replace the normal RF choke and this will also provide additional harmonic suppression with less loss. Although the length of cable required is quite long it can be coiled up into a small volume. If RG-58/U or URM-43 cable is used with a velocity factor of approx 0.66 the total length required is about 1m.

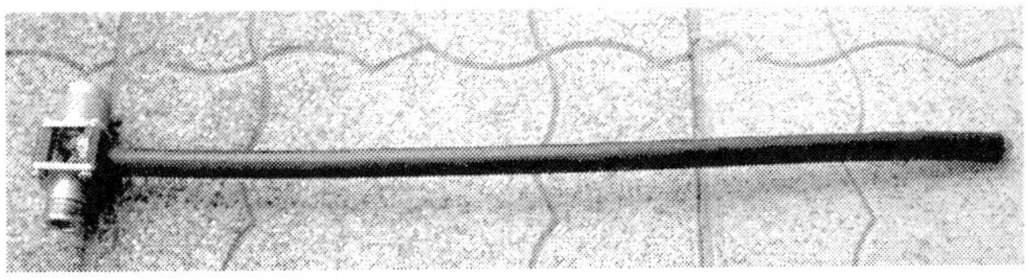

Fig 6.43. Quarter-wave shorted harmonic filter for 144MHz using RG-213 coax

MORE COMPLEX RESONATOR FILTERS

A variation of the normal coupled quarter wave resonator is the so-called 'half-wave' resonator. Theoretically the unloaded Q it is possible to achieve with this type is twice that of a quarter-wave resonator, but in practice the Q realised is often much less due to the construction. In the simplest form the half-wave resonator is two quarter-wave sections connected end to end with the outer ends being grounded (**Fig 6.44**). In the centre a tuning capacitor or disk is connected to ground. Input and output coupling is by taps, coupling loops or capacitive tapping between a point near the ground end and the connectors.

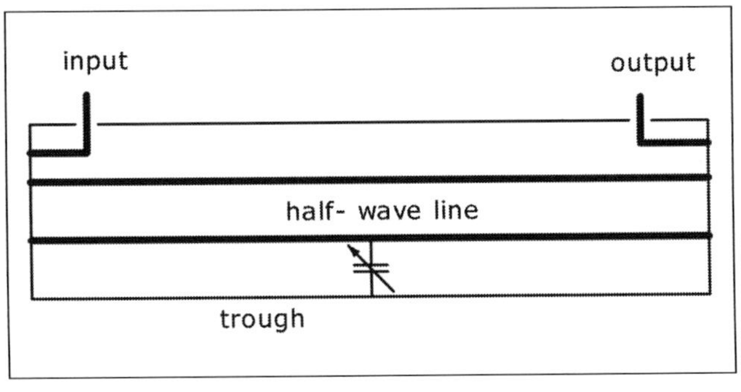

Fig 6.44. Half-wave resonator

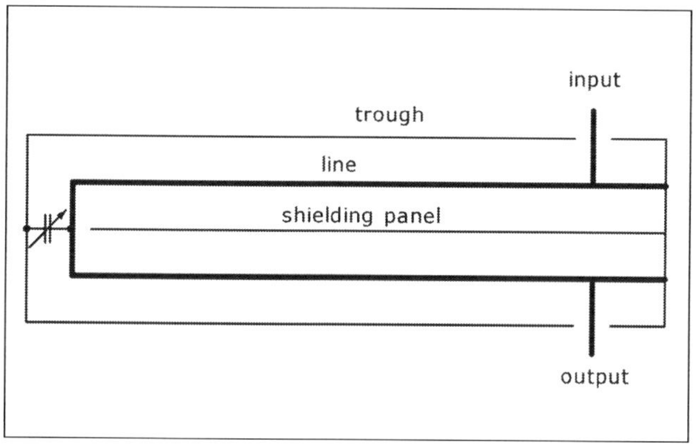

Fig 6.45. Folded up half-wave resonator

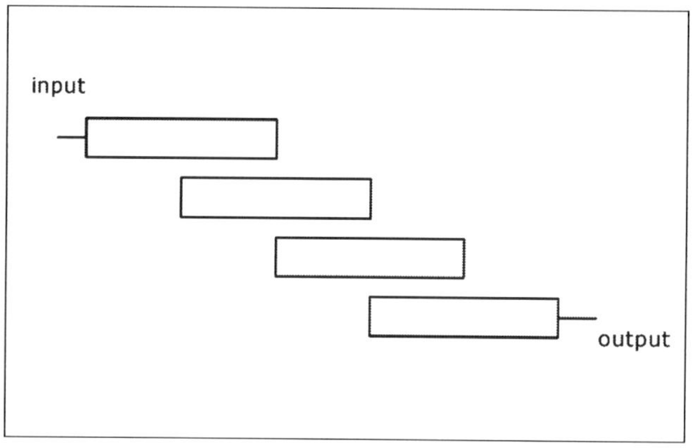

Fig 6.46. Basis of edge-coupled line band-pass filter

A further variation, to reduce the length, is to bend the two lines from one piece of tubing or rod into a U-shape and fit a screening panel between the two halves (**Fig 6.45**). In concept the half-wave resonator is very similar to the Cohn bottom-coupled filter where the tuning capacitor becomes the coupling capacitor as well.

The half-wave lines may also be printed on fibreglass or other printed circuit board material. This concept is extensively used in the range of microwave transverters sold by Down East Microwaves for 1.3GHz and above. The limitation of the board material, particularly the loss tangent, makes these filters less than optimum in an effort to keep the cost down. However, if a better board material were used the performance would improve considerably.

A further variation is the 'Edge-Coupled Line' band-pass filter printed on microwave substrate. These are tricky to get right without some software, but offer a compact low-loss band-pass filter up to many GHz. They consist of several quarter-wave lines placed parallel and slightly overlapping with a small gap between adjacent lines. Coupling between adjacent lines is a mixture of inductive and capacitive coupling. With the correct line overlap and gaps the response can be very good but a lot depends on the accuracy of the etching process and they can be frustrating filters to design for this reason. The tolerances of the gaps and line lengths need to be held to a very high accuracy for the filter to work correctly and be on the correct centre frequency. Very small dimensional errors can skew the filter far off the centre frequency and affect the coupling dramatically. Normal G-10/FR-4 substrate varies too much for accurate results and a more exotic substrate is normally required such as Teflon or RT Duroid. Even so it is not uncommon to have to make several filters with minor tweaks to the artwork before a correct response is obtained. Even then minor trimming of the line lengths with a craft knife is normally needed to optimise the response due to variations in the substrate from batch to batch, so this needs to be factored into the artwork when drawing it up.

Chapter 7

Frequency multipliers

SEMICONDUCTORS of various types can be used for multiplier applications. Many previously published articles have omitted details of how to extract the optimum performance from simple multipliers. There are two types of frequency multipliers in common use, one is the passive type using diodes and the other is the active type using transistors. We will firstly look at various diode multipliers.

> Note: This chapter is based on a two-part article that I wrote and which was originally published in *VHF Communications* magazine edited by Andy Barter, G8ATD. It is used with permission of the publisher and has been updated and combined into one chapter for easier reading.

DIODE FREQUENCY MULTIPLIERS

Some of the designs previously published give details of expensive diodes but often a cheaper diode can be pressed into service. Exotic diodes such as Step Recovery Diodes or Varactor diodes can often be changed for more mundane types with little loss in performance. For amateurs involved with VHF / UHF and microwave construction the need to multiply up from a base oscillator frequency to reach a higher frequency is a common task. Whereas diode multipliers may not be the best approach they are still the basis of many designs. With an understanding of the limitations a better insight into the best method is possible.

SIMPLE DIODE MULTIPLIER

A typical diode multiplier circuit is shown in **Fig 7.1**. This circuit is often used where a medium step in multiplication is required; say three times the input frequency.

The diode D1 is either a hot carrier or Schottky diode. The input drive comes from a low impedance source such as the emitter follower TR1 to drive a large current into the diode. The diode is connected in shunt across the input and output and the high circulating current in the diode generates harmonics of the input signal. The output of the multiplier is then connected to a selective band-pass filter via C3. If the spectrum of **Fig 7.1** circuit is examined the response is typically that shown in

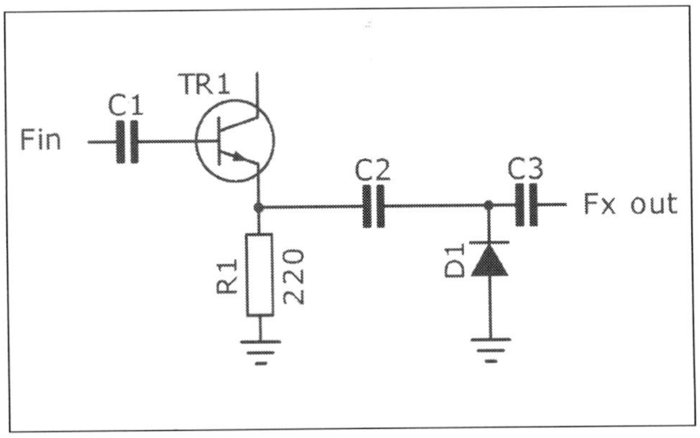

Fig 7.1. Basic diode frequency multiplier

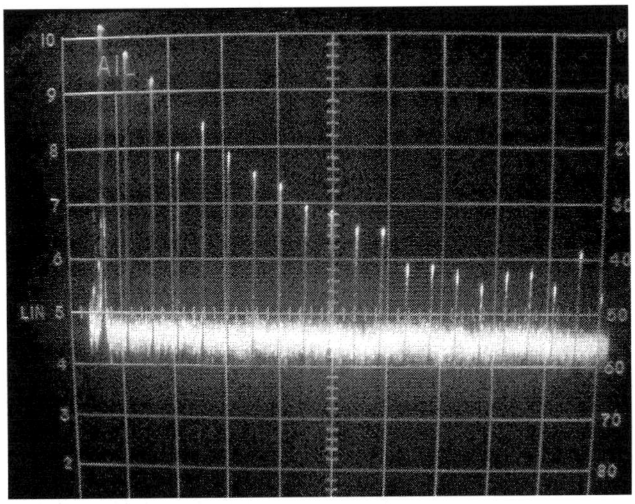

Fig 7.2. Spectrum generated by simple shunt diode multiplier

Fig 7.2. This was taken from a 50MHz crystal oscillator driving a very low impedance buffer amplifier feeding a hot carrier diode multiplier.

The extreme left-hand signal is the zero Hz marker on the spectrum analyser, the first spike after this is the 50MHz oscillator. The next signal is the 100MHz (2nd harmonic) product and the next is the 150MHz (3rd harmonic) product. The drive level into the diode is +17dBm (50mW) at 50MHz; the spectrum analyser is set to +20dBm (100mW) full scale. The third harmonic signal is lower in power level than would be extrapolated by the other harmonics and is approx 17dB below the 50MHz fundamental. The vertical scale is 10dB per division and the horizontal scale covers from 0Hz to approx 1GHz. So we can see that for 50mW input power the 3rd harmonic is only 0dBm (1mW), or 1/50th of the input power. Hence, the efficiency as a tripler is very poor at approx 2%.

This multiplier requires a severe amount of filtering to reject the fundamental, 2nd, 4th and higher harmonics. The order of filtering required is approx three narrow-band tuned circuits to reject the unwanted products to less than -60dBc. Even with a very good filter the output power of the wanted 3rd harmonic is going to be quite low and will require considerable gain at 150MHz to bring it back to a usable power level for the next multiplier stage. If the specification demands all unwanted products, apart from the 3rd harmonic, to be at least 60dB down, then the attenuation required for the fundamental signal is at least 77dB and the attenuation for the 2nd harmonic is at least 73dB. This order of filter will have considerable insertion loss to the wanted 3rd harmonic, perhaps as much as 6dB in a practical filter, so the output power will be of the order of -6dBm after the filter.

The basic problem is that a small signal hot carrier or Schottky diode is not inherently a multiplier, rather it is a divider. If a simple half-wave rectifier is used the diode produces one output pulse for every two pulses applied; the diode only operates on one half cycle of the input sine wave, hence it makes an efficient divide by two but a poor multiplier. To enhance the multiplication we need to do a few modifications to the simple circuit.

From Fourier analysis we know that we can construct a square wave signal by adding together the various sine wave harmonics in the correct amplitude. The value of the 3rd harmonic content is hence quite small. Even though we are driving the diode very hard to generate a quasi-square wave current in the diode the amplitude of the higher harmonics is expected to tail off in a more or less linear manner. This can be clearly seen in the spectrum plot in **Fig 7.2**.

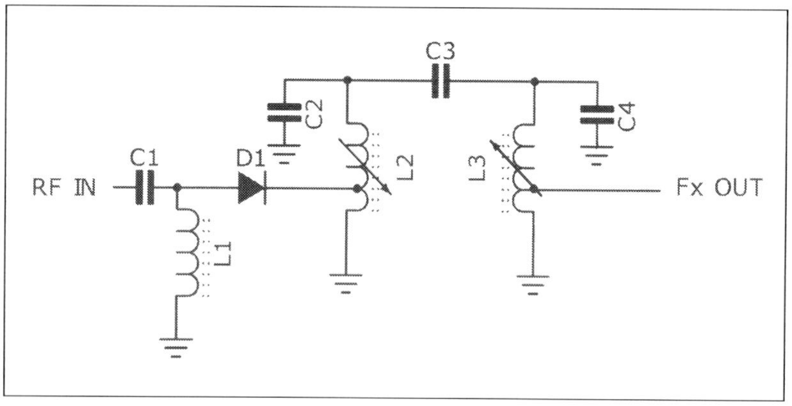

Fig 7.4. Improved diode multiplier

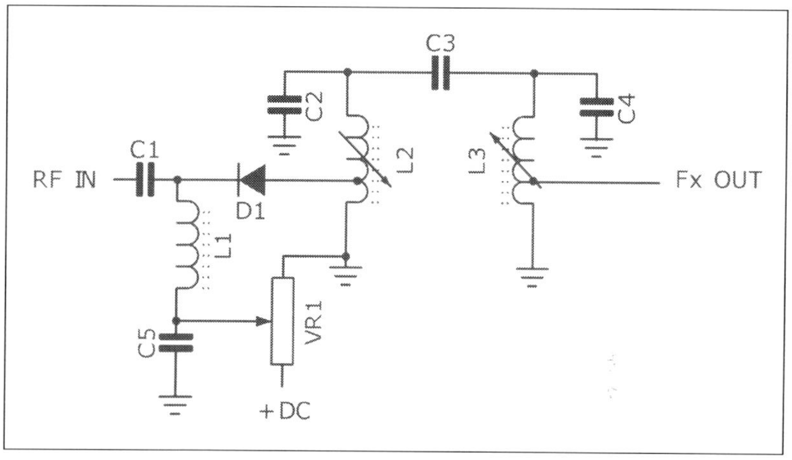

Fig 7.5. Alternative diode multiplier circuit

The apparent lower amplitude at the 3rd harmonic is probably due to some other resonance occurring in the circuit parasitic inductance and stray capacitance. As the input drive level is varied these show as peaks and troughs at different parts of the spectrum. The spectrum plot shown is that of a 50MHz calibrator for the spectrum analyser to allow alignment. It is known as a 'comb generator' and ideally all the output products should be of similar amplitude. This can be achieved with an exotic diode such as a Step Recovery diode, but these are expensive.

The harmonic generation is dependent on the conduction angle of the diode. To improve the harmonic generation we need to reduce the conduction angle. A simple way of achieving this is to reverse bias the diode with a standing DC bias voltage. This is shown in **Fig 7.3**. The variable resistor VR1 and a high value resistor R3 acting as a RF choke applies the reverse bias to D1. No static current flows in D1 apart from the reverse bias leakage current, which is negligible in most cases.

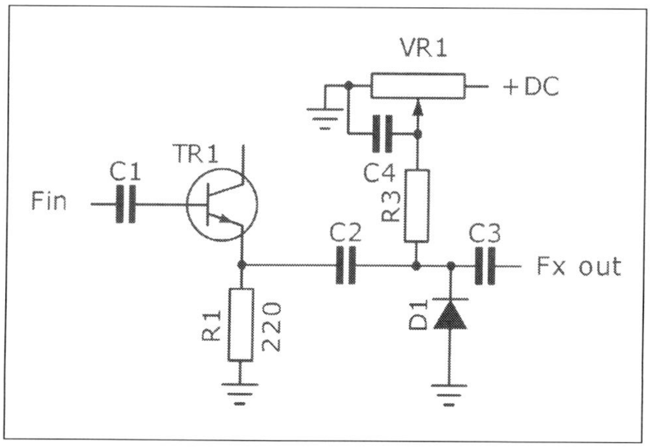

Fig 7.3. Reverse biased diode multiplier

Another version of the single ended multiplier is shown in **Fig 7.4**. This is somewhat better but still not optimum when high multiplication values are required. This circuit is often used for the final multiplication of a microwave local oscillator chain, for example a 384MHz to 1152MHz chain for a 23cm converter using a 144MHz IF. The filter is often two $\lambda/4$ lines with the multiplier diode tapped well down on the input tuned line.

Diode D1 is pumped by a high level signal at the fundamental frequency. The diode is tapped well down on the input inductor L2. Inductor L1 is the DC return for the circulating DC diode current and needs to represent a high impedance to the input frequency but a low DC resistance for the diode current. The fundamental energy is dumped into the low inductance tap point on L2 so suppressing the power contained. The top coupled inductors L2 and L3 which form the band-pass filter act as a high pass filter to the higher harmonics and so the spectrum roll off on the high frequency side is not as good as it could be.

An improved circuit is shown in **Fig 7.5** where a pre-bias is applied to the diode to enhance the multiplier efficiency. By varying the bias voltage the circuit can be made to peak at a particular input drive level. This in many cases can deliver up to 10dB more output power than an unbiased diode. The exact reverse bias needs to be found empirically and a good starting point is about 1 to 2V reverse bias. This of course assumes adequate RF input power is available from the preceding multiplier stages. A good rule of thumb is that the peak RF voltage swing is at least the same as the DC reverse bias, which in many cases means we need more than +10dBm of drive power.

LIMITATION OF SIMPLE DIODE MULTIPLIER

All practical simple diode multipliers have conversion loss. In many cases this can be overcome by simply increasing the input drive level, up to the power dissipation of the diode being used. If the input drive level exceeds the safe diode dissipation the diode will heat up internally and may fail.

The basic problem with all these circuits is the type of diode used. A Schottky or hot carrier diode is not the best choice for a multiplier because of the small junction capacity, although they make good receiving mixers with a reasonably low noise figure. We know from a typical varactor diode power tripler that an input of 10W at 432MHz can yield approx 6W at 1296MHz, an efficiency of 60%. What distinguishes the varactor diode from other types is the large junction capacity.

Energy pumped into the diode on one half cycle of the RF waveform is stored as charge and is released on the next half cycle with virtually no losses. Step Recovery diodes are similar but able to multiply to greater frequencies because the stored charge is released very much quicker than a normal varactor diode, so producing harmonics to a far higher order. This '*parametric conversion*' is the clue to making high efficiency multipliers. If the junction capacity of a Schottky diode is examined against applied reverse bias voltage it appears like the graph in **Fig 7.6**.

The amount of energy that can be transferred is directly proportional to the junction capacity. The curve in Fig 7.6 is that of the BAT-85 Schottky diode. If a typical small signal varicap diode characteristics is examined we see a very different picture.

Fig 7.7 shows the junction capacity versus reverse bias voltage of the BB-809 varicap diode. With zero applied bias voltage the junction capacity is more than 50pF and with 10V bias it is approx 13pF. By swinging the input RF voltage between approx 1V and 10V we have a total capacitance variation of 30pF. Compare this to the Schottky diode curve in Fig 7.6 and we can see that it only swings approx 5pF for the same DC and RF conditions. Hence, we can intuitively see that the BB-809 diode has about 6 times the capacitance variation and hence will perform better as a multiplier than the BAT-85. This can be borne out in experiments where a greater harmonic power is generated when the varicap diode is used. The exact junction capacitance required is a function of the frequency being used, for microwave multiplier stages the junction capacitance will be proportionally less than for a low frequency multiplier. There are many low cost varicap diodes available from several manufacturers that can be used. For low frequency applications the BB-809 or the BB-132 in SMD are a good choice and for higher frequencies a good choice is the BBY-31 which is a SMD device.

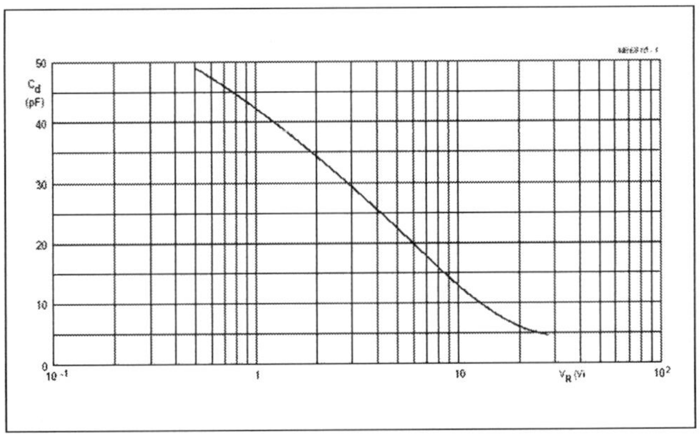

Fig 7.7. Typical small signal varicap diode characteristics

SPECIAL MULTIPLIERS

Often we require to multiply by a fixed factor, for example times 2. For this application we can utilise the dividing property of the typical diode. If two diodes are arranged as in **Fig 7.8** the two half cycles of the input sine wave are used and the predominant current pulses at the output are even harmonics. This configuration is known as a *push-pull* doubler. Because only the even harmonics are used the circuit offers a greater rejection to odd harmonic multiplication factors. It works well at 2x and 4x but very poorly at 3x and 5x because of the balanced configuration. If the input transformer has sufficient bandwidth the circuit can work over an octave or more of bandwidth. Normally the transformer is wound on a small ferrite bead or core with good high frequency characteristics. The typical transformer is wound with tri-filar wire on a small two-hole ferrite core.

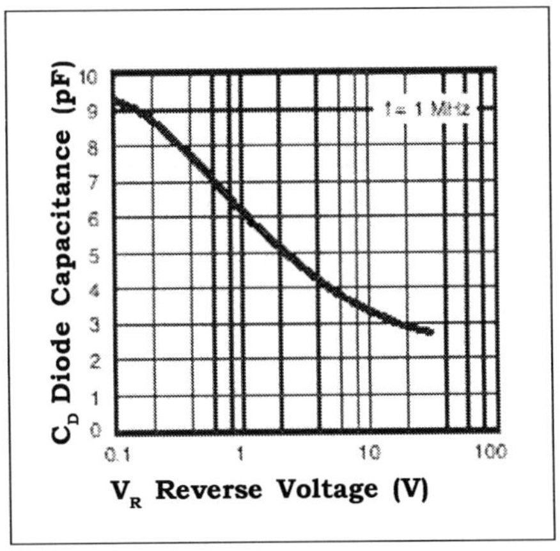

Fig 7.6. Typical small signal Schottky diode characteristics

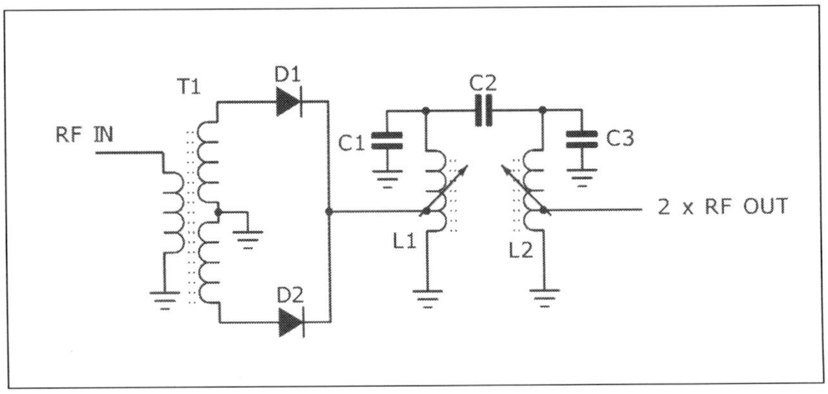

Fig 7.8. Basic diode doubler circuit

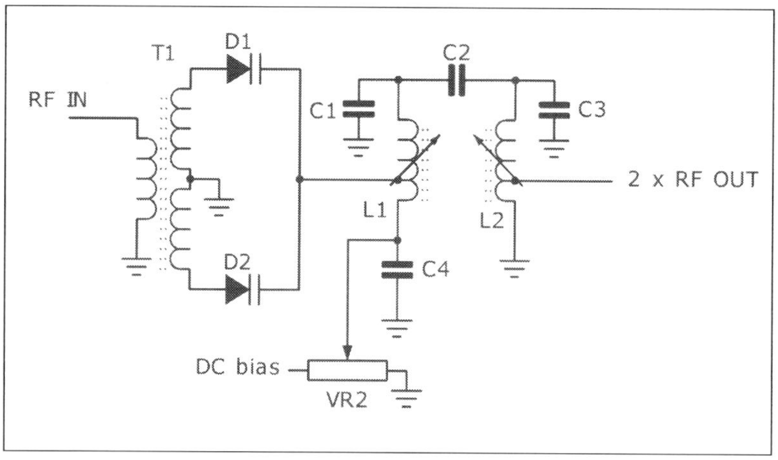

Fig 7.9. Improved diode doubler

The input power is applied to the transformer T1, which develops a push-pull drive to two diodes. The summation of the two diodes at the tap point on L1 is rich in even harmonic power and the band-pass filter selects the required harmonic, in this case the second harmonic. By selecting suitable values for L1, L2 and the tuning capacitors the circuit can be used as a 4x, 8x multiplier etc. However, the output power as a 4x multiplier is not as much as a 2x multiplier, typically it will be 6 to 12dB less than the 2x multiplier.

By substituting varicap diodes for D1 and D2 with some pre-bias the output power can be greatly increased over the simple Schottky diodes types shown. This circuit is shown in **Fig 7.9**.

Diodes D1 and D2 are now small varicap diodes, such as the BB-809, and the reverse bias is applied via the preset VR2. C4 is a low reactance capacitor to RF ground the bottom of L1.

The basic problem with both of these circuits is making a suitable push-pull transformer to work at the required frequency. For low frequencies this is not too difficult but as the input frequency increases the winding of the transformer to obtain good balance becomes more difficult. Hence, the circuit is really only useful up to about 400MHz. At higher input frequencies a printed 180° 3dB coupler is a better choice, although it will occupy a fair amount of board area even at 1GHz.

ODD HARMONIC MULTIPLIERS

When we need to enhance the odd order harmonics and suppress the even order ones then the *push-push* circuit is a good candidate. The basic circuit is shown in **Fig 7.10**. Here two diodes are connected in anti-parallel so both half cycles of the input waveform are acted on. In the experimental circuit two BAT-85 Schottky diodes were used because they were to hand.

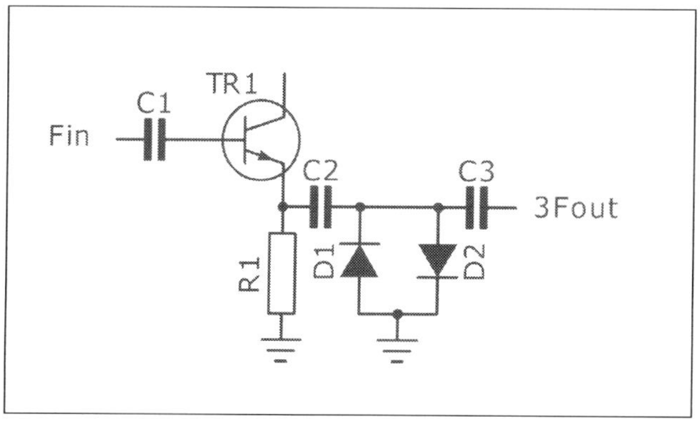

Fig 7.10. Push-push tripler

The input spectrum to this multiplier is shown in **Fig 7.11**. This is the same 50MHz driver as used earlier but the spectrum analyser sweep has been reduced to show 0 - 300MHz. The extreme left-hand spike is zero Hz and 150MHz is in the centre of the display.

The output spectrum of the push-push circuit is shown in **Fig 7.12**, where it can be seen that the even harmonics have been suppressed by approx 35dB, so making the filtering of the unwanted products much easier. The conversion loss of the circuit is approx 6dB at the 3rd harmonic, which yields an efficiency of 25%, a considerable improvement over the simpler circuit.

Fig 7.13 shows the same signal after passing through a simple two-section band-pass filter. The products at 50MHz and 250MHz are now suppressed by

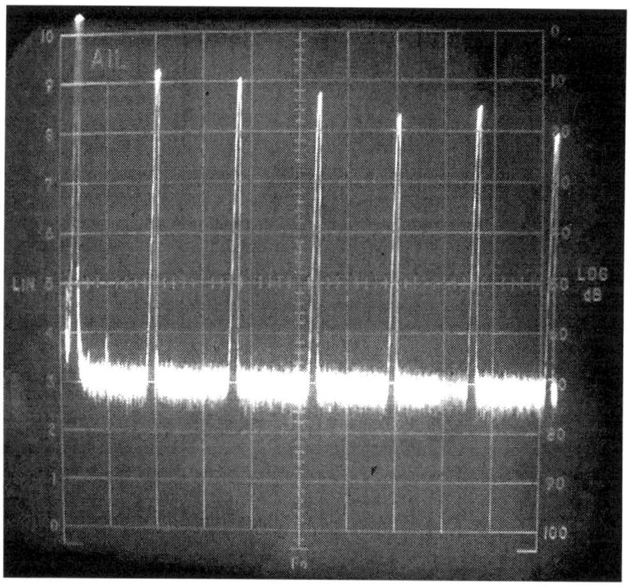

Fig 7.11. Input spectrum to push-push tripler

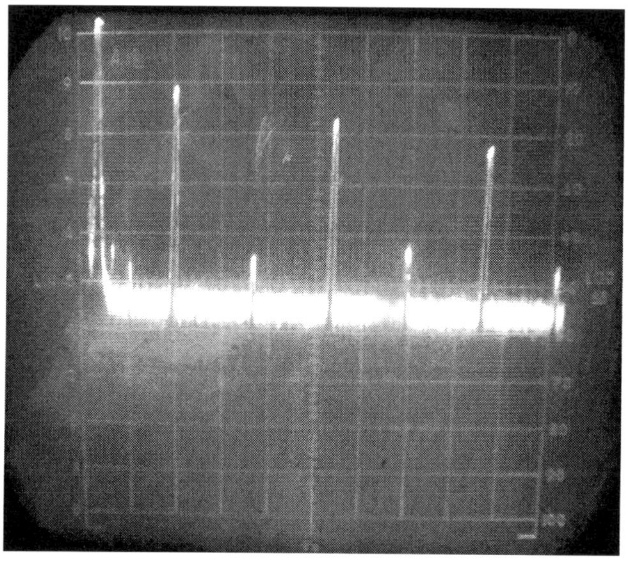

Fig 7.12. Output spectrum of the push-push tripler before filtering

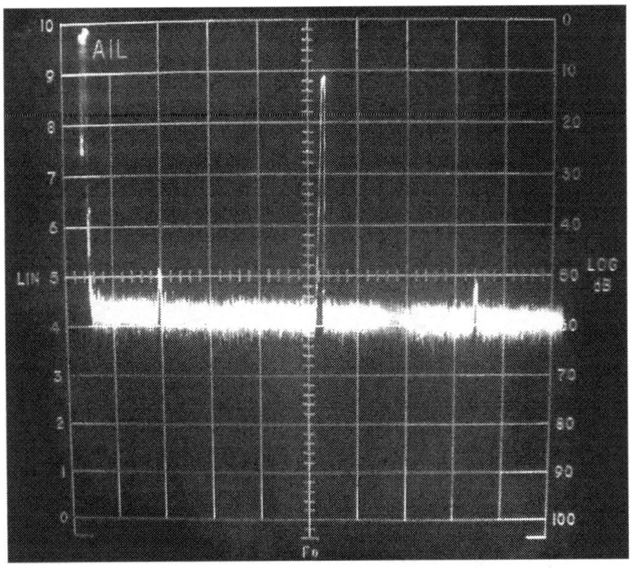

Fig 7.13. Output of diode tripler after filtering

approx 40dB and all others by over 50dB. The following amplifier stage, with further filtering, will provide adequate drive power to the next multiplier with additional suppression of the unwanted products.

ENHANCING MULTIPLIER EFFICIENCY

If the power varactor diode tripler multiplier is examined we see that there are not only tuned networks at the input and output but also an extra tuned network resonant at the 2nd harmonic. This is known as an 'idler circuit' and it increases the 2nd harmonic current flowing in the diode. You can think of the parametric varactor tripler as a type of mixer. The high 2nd harmonic current (2f) is added to the fundamental current (1f) and out pops the sum and difference products. The difference product (2f - 1f) is the fundamental and this reinforces the input current. However, the sum product is 1f + 2f, which is the wanted output frequency. So the clue to efficient multipliers is to add idler networks to the input and output to reinforce the wanted harmonic. This is illustrated in **Fig 7.14** where two idler networks have been added to the basic diode multiplier.

The input network C1 & L1 is series resonant at the fundamental frequency and so appear to be highly reactive to the second, third and higher harmonics, reflecting power back to the diode. Network C2 & L2 is series resonant at the 2nd harmonic generating a high circulating current through the varicap diode at that frequency. Pre-bias is applied to the varicap by the variable resistor and R3. The predominant energy is therefore contained in the 3rd harmonic and is filtered in a band-pass filter to remove unwanted products.

When a circuit such as this is aligned using a spectrum analyser it will be noted that as C2 is adjusted the power at the 2nd harmonic will dip towards zero and the power at the 3rd harmonic will correspondingly peak. This is a much better signal spectrum to try and filter as the main unwanted signal is now at the fundamental and hence further away on the band-pass filter slope so easing the design of the filter.

Additional idlers can be added for the 4th and higher harmonics should they be necessary. Typically this network would be a parallel tuned 'trap' inserted in series with the output and adjusted to null out the harmonic power. With sufficient Q in the idler networks more than 30dB of additional rejection can be obtained. The parallel tuned trap reflects power at the harmonic back into the multiplier diode and this is added to the wanted 3rd harmonic signal. This is demonstrated in **Fig 7.15** where a trap for the 4th harmonic has been added.

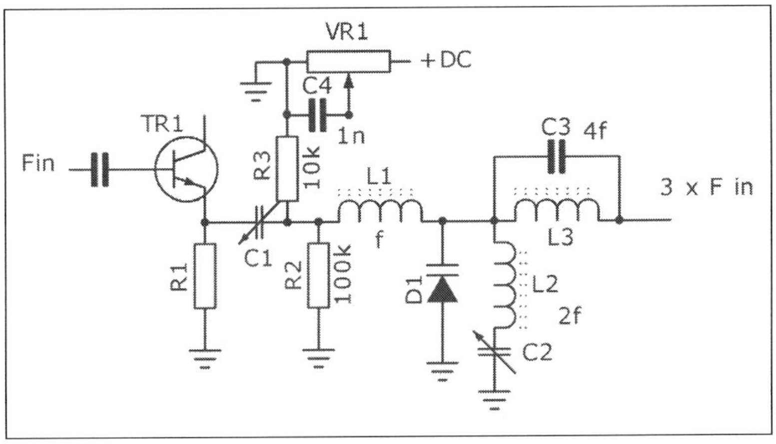

Fig 7.15. 4th harmonic idler added

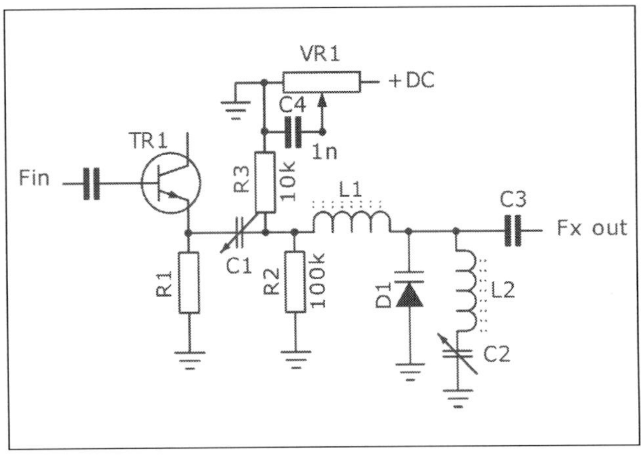

Fig 7.14. Multiplier with idler networks

OTHER DIODE TYPES

Whereas most designs published to date have concentrated on hot carrier, Schottky or varicap diodes there are other less expensive types that can work. As an experiment I substituted the BAT-85 diodes in the push-push tripler circuit shown in **Fig 7.10** with a common switching diode, the 1N4148, to see what difference would result. It was rather a surprise to find that the 1N4148 actually performed far better than the BAT-85. The output power at 150MHz was almost 9dB greater and after changing the band-pass filter component values to provide more rejection to the close in products an overall improvement of nearly 8dB was achieved with better rejection of the unwanted products. It occurred to me that this is probably because the forward voltage of the 1N4148 is 0.7V whereas the Schottky BAT-85 is typically 0.35V. The junction capacity of the 1N4148 is also much greater than the BAT-85 or similar Schottky diodes so enhancing the parametric characteristics. Hence, this bears out the pre-

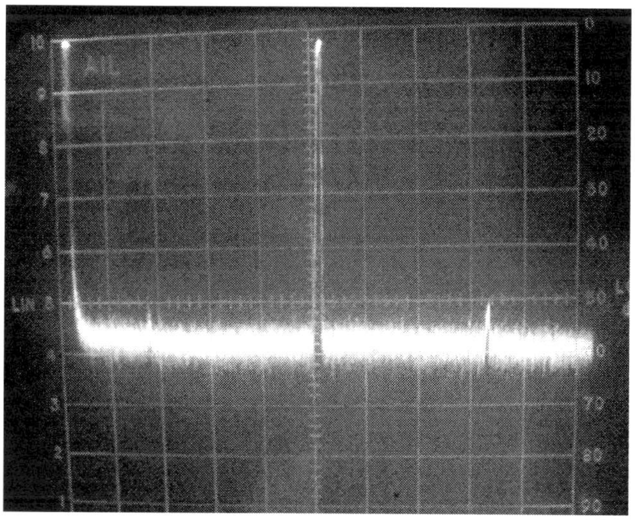

Fig 7.16. Spectrum of push-push tripler fitted with 1N4148 diodes

vious measurements made with reverse biased diodes, where the greater the reverse bias the better the diode acts like a non-linear element. As the cost of the 1N4148 is approx 10% of the BAT-85 this is a worthwhile modification for low frequency applications.

Fig 7.16 is a picture of the spectrum plot after the diodes were changed to 1N4148s and the band-pass filter readjusted. The suppression of the products at 50MHz and 250MHz is improved to about 50dB. The 100MHz and 200MHz products are greater than 60dB below the 150MHz output. A possible choice for the dual-diode is the BAV-99, which contains two closely matched 1N4148-type diodes connected internally in a SOT-23 package. This is shown in **Fig 7.17**.

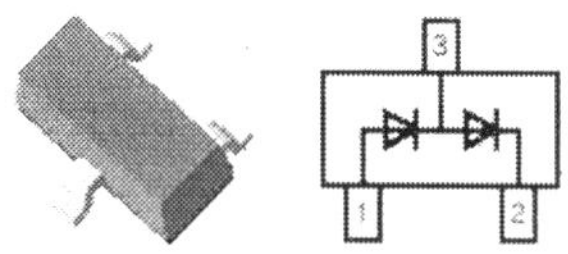

Fig 7.17. BAV-99 dual diode

The configuration is ideal for the push-push tripler and pins 1 and 2 are grounded and pin 3 is the common point.

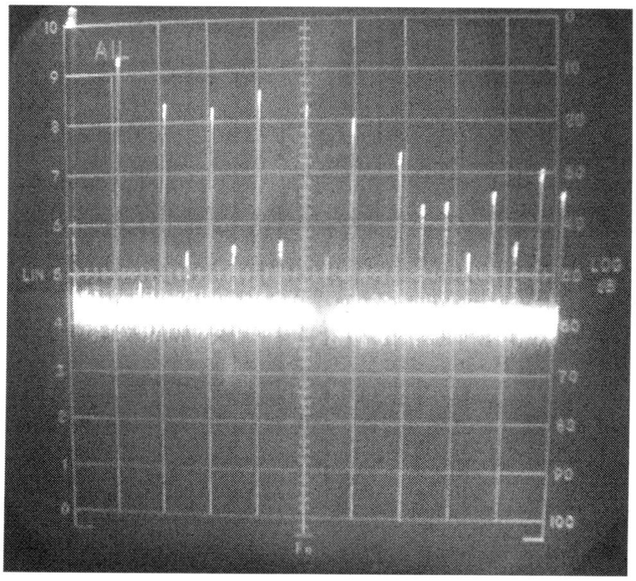

Fig 7.18. 100MHz comb generator output spectrum. The spectrum analyser covers 0 to 1GHz

COMB GENERATOR

For circuits where the push-pull doubler extra components may not be desired a simpler modification is to make a slight change to the circuit shown in Fig 7.1. If a high value resistor is connected across the multiplier diode the circuit effectively becomes self-biasing. A typical resistor value is about 33kΩ. **Fig 7.18** shows the output spectrum of a simple harmonic multiplier driven by a 100MHz low impedance source. This circuit was built as a 'comb generator' to calibrate a spectrum analyser. There are some 50MHz products apparent in the higher portion of the spectrum due to insufficient filtering in the 50MHz to 100MHz doubler stage preceding the final comb generator. This comb generator uses a 1N4148 switching diode to generate the 100MHz comb output, which extends up to at least 4GHz.

The output power from the Comb Generator is adequate for the purpose intended, to calibrate the frequency range of the spectrum analyser. **Fig 7.19** shows the same comb generator covering 1.8 to 2.8GHz.

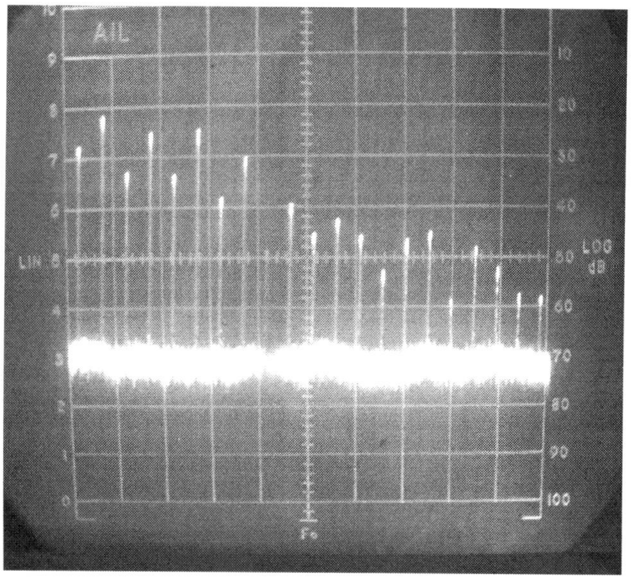

Fig 7.19. Comb generator output on 1.8 to 2.8GHz range

TRANSISTOR MULTIPLIERS

Using an active device such as a transistor can overcome the inherent loss of the simple diode multiplier, because the transistor can have appreciable gain when used as a multiplier, which eases the design of multiplier chains where high multiplication is required. If the transistor is operated in grounded emitter configuration with the input RF drive applied to the base the impedance is medium and a greater RF voltage swing can be obtained from relatively low drive power.

PRINCIPLE OF OPERATION

If a transistor is used as an amplifier for a signal requiring linear amplification it is limited to the output power it can produce because of the maximum voltage swing on the base or collector. If the input voltage swing exceeds the base-emitter forward voltage the junction starts to behave like a rectifier diode and distortion occurs. We can simplistically think of a multiplier transistor as being two separate items. The first is the diode junction contained within the base-emitter and the second is a very linear voltage amplifier between the base and collector. By overdriving the base-emitter junction we generate harmonics of the input frequency and these are then amplified in a highly linear manner by the collector-emitter portion of the transistor.

Unfortunately it isn't as simple as this, but the simplistic idea forms the basis of all transistor frequency multipliers. We could place a multiplier diode across the base-emitter and couple into the transistor base via a capacitor. By generating harmonics in the external diode these are then amplified by the linear biased transistor. Some commercial two-way radio multiplier chains used this technique. Unfortunately the efficiency this achieves is worse than the diode on its own if we include the DC power being drawn by the transistors before and after the diode multiplier.

By utilising the inherent transistor base-emitter diode junction we have a better method of generating high levels of harmonic power. Of course we still have to filter out all the unwanted harmonics at the collector and select the desired one with as little loss as possible. However, by altering the conduction angle of the transistor we have a method of enhancing certain harmonics and hence reducing the unwanted ones.

The most basic circuit is shown in **Fig 7.20**. Although this will work, it has a number of problems, which we will consider later.

The drive signal to be multiplied is applied to the base of TR1 via C1. The inductor L1 across the base to ground holds the base at DC ground and looks like high impedance to the input frequency. The emitter is also directly grounded, so for the transistor to conduct the RF voltage across L1 needs to exceed the base-emitter junction voltage of about 0.6V. This means that when the RF voltage is below the threshold nothing happens, but as soon as the base voltage exceeds about +0.6V on the positive peaks of the RF waveform the transistor will start to turn on. If the RF drive is raised even more the transistor will be conducting for a greater portion of the RF envelope and so more collector current will flow. The transistor is being operated as a Class C amplifier and the waveform at the collector is similar to a square wave, which is rich in harmonic content. All that is required is a suitable band-pass filter to select the wanted harmonic and reject the remainder.

On the face of it this seems to be perfect. But one thing to be aware of is that the multiplier is very critical to the RF drive level, too little and it will snap off and too much and it will saturate and the gain will fall dramatically. Also the threshold is very dependent on the base-emitter forward voltage. If the ambient temperature varies over a wide range the base-emitter voltage does the same. At very low temperatures the voltage required to make the base-emitter junction conduct will be greater than when the ambient temperature is high. So although it works fine in the shack it may not work at all from a contest site on top of a mountain in winter!

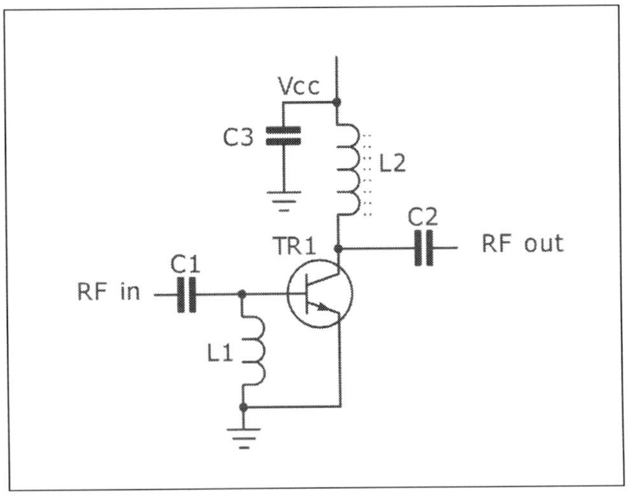

Fig 7.20. Basic multiplier circuit

Another failing with this simple circuit is that the impedance presented to the driving stage varies wildly as the drive level varies. Below the threshold voltage the multiplier appears close to an open circuit, at high drive levels the multiplier appears as low impedance, just like the diode multipliers. The other problem is the fragility of the base-emitter junction. For many small signal RF transistors the maximum reverse bias the junction can safely stand is only a volt or two. If the drive level is too high the junction can fail.

IMPROVING THE BASIC CIRCUIT

The first problem we will address is the temperature sensitivity. In a linear amplifier using the common emitter configuration we normally include a resistor in series with the emitter to ground to provide a small bias voltage. This helps to hold the collector current more stable with variations in temperature.

The second problem is the sensitivity to the RF drive. We can to some extent reduce this tendency by applying a small amount of forward bias to the base-emitter junction with a potential divider of two resistors. By selecting the resistor values such that the base-emitter voltage is just below the conduction point the RF drive voltage swing is greatly reduced. This translates into less drive power being required and higher input impedance.

Fig 7.21 shows the modifications needed.

The value of the emitter resistor is determined by the order of multiplication required. For high multiplica-

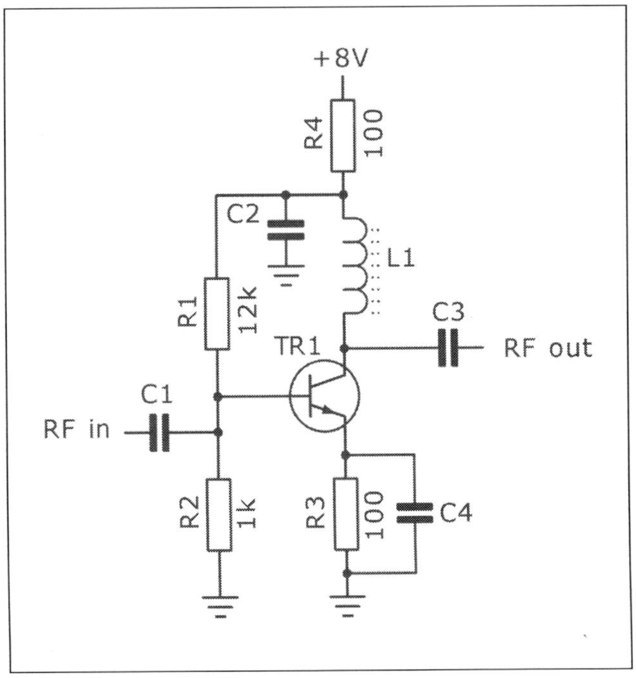

Fig 7.21. Improved frequency multiplier

tion factors the resistor value is higher than for low order factors. This is to reduce the conduction angle to enhance the higher harmonics. In simple terms we need to move the conduction more towards Class C as the harmonic order increases.

The value of the base potential divider inter-reacts with the emitter resistor. Essentially what we need to achieve is the DC voltage on the base with no RF drive to be just below the conduction point or a very small collector current flowing, a few hundred μA typically. With the values shown in Fig 7.21 the base voltage with respect to ground is approx 0.62V. But the collector current flowing in the emitter resistor raises the emitter voltage above ground. For a collector current of 20mA flowing when the base is driven the emitter will be approx 0.2V above ground, making the effective base-emitter voltage about 0.42V. Hence the input RF drive needs to be greater than about 0.3V peak to turn the transistor on.

The disadvantage of this circuit is the emitter decoupling capacitor C4; this needs to be a low reactance to not only the input frequency but also all harmonics. This requires quite a low inductance capacitor and a SMD chip type is normally the best choice. If leaded capacitors are used, for example ceramic disks, the emitter decoupling may not be good enough for the higher harmonics and some loss of gain will occur at the higher harmonic products.

COUPLING THE REQUIRED HARMONIC OUT OF THE COLLECTOR

This is where most published designs get it wrong. Although the multiplier stage appears to work satisfactorily often rearranging the circuit a little can make a distinct improvement.

In Fig 7.20 and Fig 7.21 the collector is connected to a radio frequency choke with a wide bandwidth. The coupling capacitor serves to block the DC supply and also transfers the AC signal to the following band-pass filter. **Fig 7.22** shows the *wrong* way of coupling out the harmonic signal.

The collector of TR1 is tapped down the inductor L1. L1 is resonated to the required frequency by the vari-

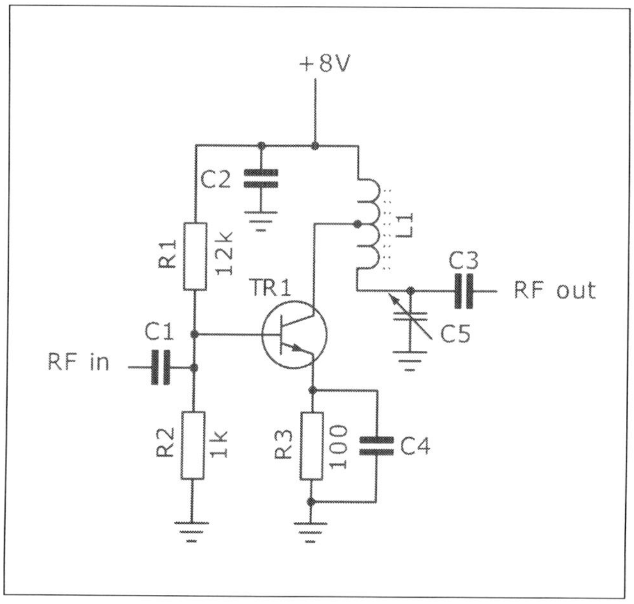

Fig 7.22. Incorrect collector coupling

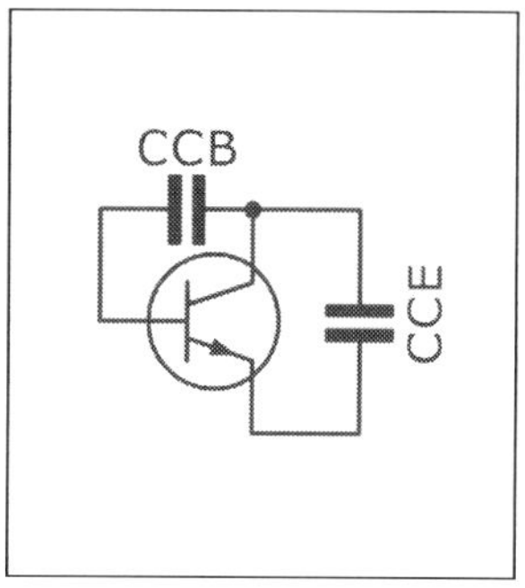

Fig 7.23. Transistor parasitic capacity

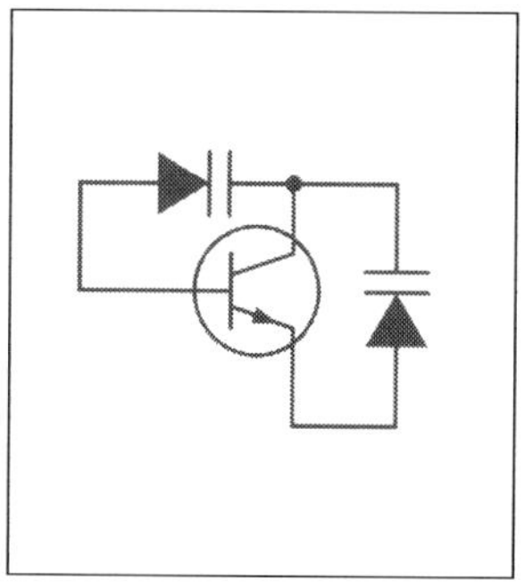

Fig 7.24. Inherent varactor diodes in BJT

able capacitor C5 and this tuned circuit could the first portion of a multi-pole band-pass filter.

This tapped inductor method is commonly used for amplifiers such as intermediate frequency amplifiers where linearity is critical. The tapping of the inductor allows a good impedance match to the collector, the conjugate match. However, in a frequency multiplier this is exactly the opposite of what we require. To understand this logic we need to look more closely at how the bipolar junction transistor (BJT) works and its internal structure.

Fig 7.23 shows the two major parasitic capacities that are inherent in the bipolar junction transistor. These are the collector to base (CCB) and collector to emitter (CCE). There is one other which exists between the base and emitter but this does not affect the operation, only the input matching.

Normally in a small signal linear amplifier application the capacitance is relatively constant and can be tuned out. Because the RF voltages appearing across the junctions are small the capacity is constant. But in a frequency multiplier with correct collector matching the RF voltages are not small, in fact we want to make them as large as possible, without exceeding the collector-emitter secondary breakdown voltage of the device. The reason is that with a large voltage swing the parasitic capacitance is not constant but behaves like a varactor diode. Motorola in an *Engineering Bulletin* [EB-70A: *Frequency multiplication simplified by internal Faraday shields in MRF629*, Motorola Semiconductors] gave details of how the varactor diode characteristic can be exploited to make more efficient frequency multipliers. This used the MRF629 a 12V/2.5W UHF power transistor as a *parametric multiplier*. This exhibited a power gain of over 6dB when driven by a 100mW signal at 150MHz and developed as much as 700mW output at 450MHz when configured as a tripler. I used this technique when building the 1152MHz LO chain for a 2C39A power-mixer SSB transverter for 23cm and obtained nearly 1W output at 384MHz. This used 150mW input at 96MHz with the circuit configured as a 4x multiplier. The use of idler circuits at the input and output is the secret to efficient multiplication.

This technique is not just applicable to the MRF629 (which today is obsolete) but even small signal BJT RF transistor exhibit this effect when a sufficiently large RF voltage is developed at the collector. Hence, we need to match the collector to higher impedance than suggested by the formula for conjugate matching. This also explains why junction field effect transistors (JFETs) are poor fre-

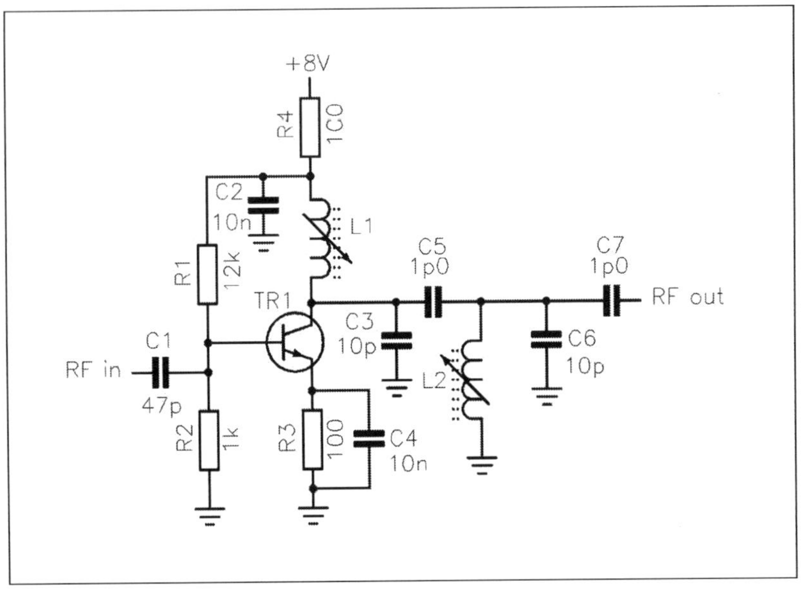

Fig 7.25. Typical frequency multiplier stage

quency multipliers, as they do not exhibit the same varactor effect as BJTs (**Fig 7.24**).

Fig 7.25 shows an example of the correct collector matching for a frequency multiplier. This uses a two-section top-coupled band-pass filter to select the required harmonic. The Q of the tuning components needs to be sufficiently high to develop the required RF voltage swing.

The input to the multiplier is via a capacitor with a medium impedance at the operating frequency. The drive source is a low impedance stage such as an emitter follower and with an output voltage swing of at least 1V p-p. The tuned circuit of L1 and C3 selects the required harmonic. This develops the necessary high impedance, and hence the higher voltage swing, at the collector of TR1. The coupling into the next section of the filter is made via C5, a very low value capacitor to prevent loading of the collector. This capacitor is also the top coupling for the band-pass filter so it needs to be a very low value, which suits the circuit configuration well. The second inductor L2 is resonated by C6, the coupling to the next stage is performed by C7 which is also a low value capacitor because we need to prevent loading on the band-pass filter, which would alter the selectivity response if excessive loading occurs. The values of the inductors and capacitors need to be chosen to suit the frequency. If more selectivity is required it is often better to incorporate this into the following buffer amplifier before the next multiplier stage.

Note that the input impedance to the transistor base is approx 500Ω and if the drive is provided by a 50Ω source (for example a signal generator) a transformation network will be required to develop adequate base-emitter RF signal. This could be a 4:1 bifilar wound transformer. Generally, signal generators develop the correct voltage into the impedance they are designed for, the 'PD condition' but into a very high impedance (approaching an open circuit) they deliver almost twice the voltage (the EMF condition). Into an open circuit the voltage will be twice the PD condition. Also be aware that the output matching in Fig 7.25 assumes it is driving another BJT stage as a buffer amplifier with an input impedance of about 500Ω. If the output needs to drive 50Ω the modified circuit shown in **Fig 7.26** is a better method. This uses a capacitor tapped impedance divider to match to this low impedance.

Fig 7.27 shows the typical spectrum at the output of the band-pass filter. This circuit is the one shown in Fig 7.25 and uses an MPSH-10 transistor as a tripler. The input frequency is 50MHz at a level of 5mW in 50Ω and the wanted output at 150MHz is approx +13dBm (20mW). The worst spurious products are about 50dB down on the 150MHz signal. This will require further filtering in another band-pass filter before it is applied to the next multiplier or used by the equipment.

Note the 5mW refers to the power the signal generator would develop into 50Ω when correctly matched.

MULTIPLIER STABILITY

In general, frequency multipliers using BJTs are inherently very stable because the input and output are tuned to different frequencies. Unlike a tuned amplifier where the input and output are the same, the only thing that can make a frequency multiplier unstable is either poor layout or the internal parametric components of the transistor. If the drive level is marginal the transistor can exhibit erratic operation. If the input drive varies over too great a range the multiplier can snap on or off as the level varies. Hence, it is necessary to have some reserve drive in hand. A good minimum is about 3dB above that required to ensure reliable multiplication.

When tuning up a multiplier stage the power output should vary smoothly as the input and output are aligned. If the multiplier snaps on and off as the tuning is altered it is usually a sign that the input drive

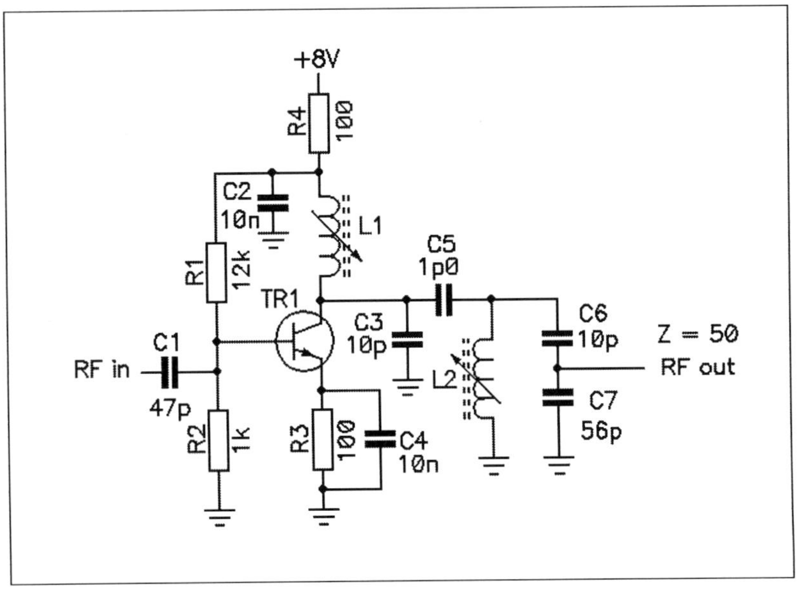

Fig 7.26. Modification to drive 50Ω load

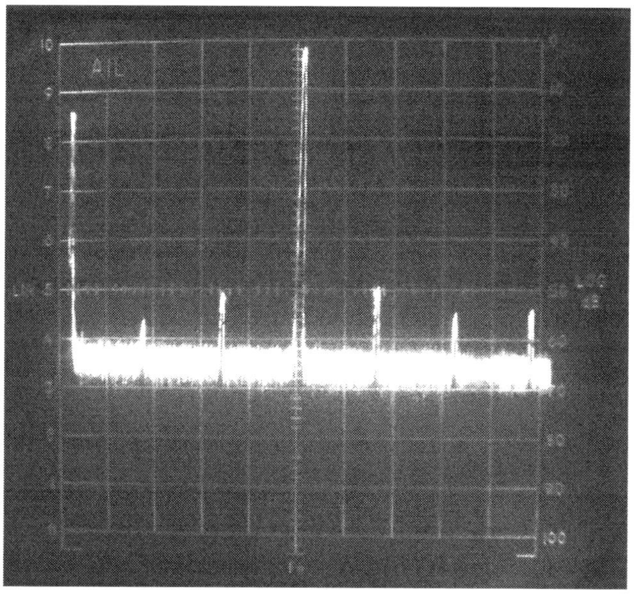

Fig 7.27. Typical spectrum for a BJT multiplier

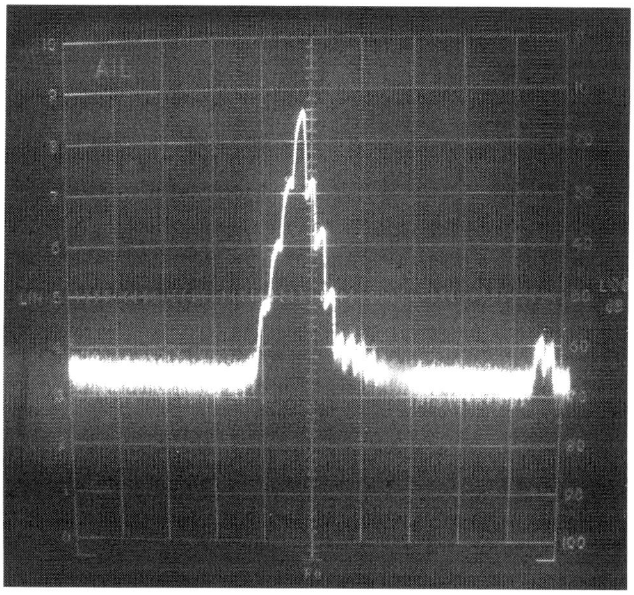

Fig 7.28. Unstable multiplier spectrum

level is marginal. A good check of this is to measure the base voltage of the BJT with full input drive. If the drive is adequate the base voltage should be a voltage slightly negative with respect to ground. A value of -0.25V to -0.75V is normally about right. When the input drive level is more than required the top potential divider resistor in the base bias chain can sometimes be omitted and the drive develops a negative base voltage across the base to ground resistor. In cases such as this it is prudent to connect a reverse biased diode across the base to ground to clamp the base reverse voltage to less than 0.7V to protect the base-emitter junction. **Fig 7.28** shows the spectrum for a multiplier stage where the input drive level has been deliberately reduced to show the effect that instability causes.

OPTIMUM SUPPLY VOLTAGE

In all the years of designing frequency multipliers using BJTs it has been found that there is a certain range of supply voltage where the optimum performance occurs. It is not quite clear why this is so but it is probably to do with the inherent parametric varactor diodes in the BJT. However, from many experiments the optimum supply voltage on the collector-emitter junction is about 7 to 9V, the so-called 'sweet spot'. With voltages far from these limits the efficiency suffers, with too low a voltage the power output falls away and with too high a voltage the efficiency also falls off. Hence, when designing a multiplier chain using BJTs it is often best to use a supply voltage of 8 to 9V. This is convenient as a 3-terminal regulator can be supplied by the nominal 12V-vehicle supply.

If the drive level into the next stage is too high a simple cure is to starve the inter-stage buffer amplifier transistor or MMIC following the multiplier by increasing the collector feed resistor value to limit the collector voltage.

CHOICE OF TRANSISTOR TYPE

As strange as it may seem we do not need to use a 'super transistor' for an efficient multiplier. Often the choice of a transistor is dictated by the frequency of operation and hence the f_t as an amplifier needs to be quite a bit higher than the operating frequency. This, however, is for a transistor operated as a linear amplifier, when used as a highly non-linear multiplier we can often get by with less exotic devices.

From experiments made on a wide selection of transistor types it has become apparent that even lowly devices such as the 2N2222 with an f_t of only 100MHz work perfectly well up to a few hundred MHz when operated as a doubler or tripler stage. For the frequencies higher than this a move to a popular type such as the MPSH-10 or the BFR92A in the SMD SOT-23 package work well even as high as 2GHz, which the manufacturers data does not suggest. Of far more importance is the drive level and the filtering components used to select the wanted harmonic. If the filter has excessive loss then the advantage of using an exotic transistor is wasted.

TEST POINTS

When aligning a multiplier chain it is helpful to be able to break into the chain at critical places to attach test equipment to see what is going on. Some designers make all the multiplier stages to terminate in 50Ω so that normal test equipment can be connected. This not only introduces extra loss but also add unnecessary components to the circuit. We can safely assume that the base input impedance of a typical BJT when driven at the optimum level is about 200 to 500Ω. Therefore the drive power is not going to be too high; perhaps +10 to +13dBm in 50Ω when transformed up will give sufficient base-emitter voltage swing.

Probing a high impedance portion of the circuit with a RF AC voltmeter often causes detuning because of the probe capacity. A simpler method is to incorporate some test points where a 50Ω instrument can be attached without breaking the chain. Often two resistors can achieve this and the 50Ω equipment can be attached as and when required without upsetting the system. This concept is illustrated in **Fig 7.29**. The two resistors are a high value one to tap into the high impedance point and a second termination resistor to suit the test equipment. A coaxial cable can be connected and taken to a spectrum analyser to see the spectrum. The amplitude will not be correct but the spectrum should still be the same.

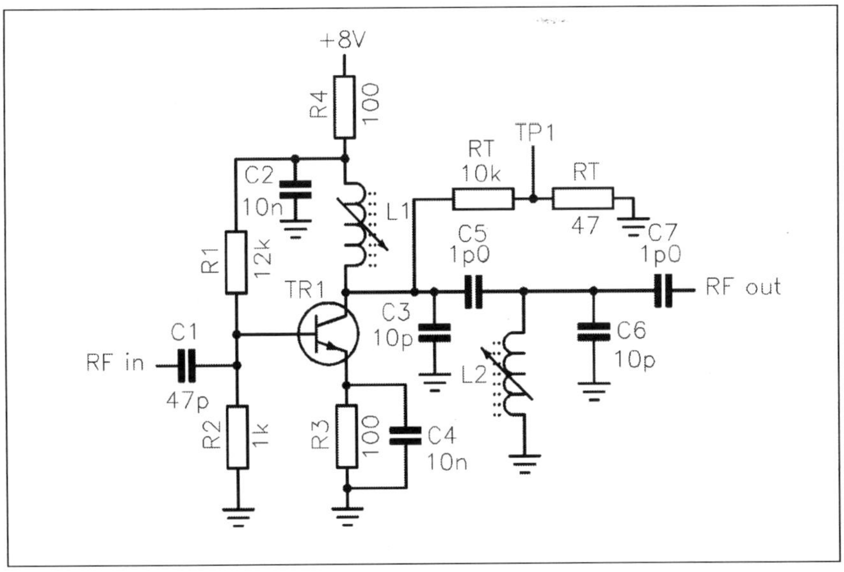

Fig 7.29. Test point added to multiplier stage

USING MMIC AMPLIFIERS AS MULTIPLIERS

There have been several application notes and other articles published concerning the use of wide band *microwave monolithic integrated circuit* (MMIC) amplifiers as frequency multipliers. The choice of devices is somewhat limited, but several writers have reported good results. However, because of the internal biasing components of typical MMICs these devices do not really lend themselves to this application. Whereas it is true that any transistor can be over-driven to force it into a non-linear operating portion this seems not so easy with the typical MMIC. As the typical MMIC is designed to interface with 50Ω terminations intuitively we can see that we cannot run the collector into the high impedance necessary to induce the varactor diode effect.

Where MMICs can be made to work is with an external shunt diode across the input, with DC blocking capacitors to prevent the bias being upset, to generate the harmonic energy and then to use the MMIC as a conventional amplifier to raise the level. However when you run the calculations of the total power required the efficiency still ends up being quite low and a cheaper BJT will probably out perform it at far less cost. MMICs are good as inter-stage buffer amplifiers but not so good as multipliers.

SPECIAL MULTIPLIERS

Where we need additional suppression of certain harmonics we can utilise some of the ideas found in diode multipliers. **Fig 7.30** shows a push-pull doubler using two BJTs. The circuit suppresses the odd order harmonics by about 30dB and leaves just the even order harmonics. The variable resistor VR1 allows the circuit to be balanced by adjusting the emitter currents of TR1 and TR2. This provides the best null to the unwanted products. For optimum performance a dual-transistor in a single package offers the best results. The wideband transformers T1 and T2 need to be wound for the best balance and this limits the higher frequency performance.

This circuit can also be modified to form a push-push tripler where the predominant harmonics are the odd order ones. This is shown in **Fig 7.31**.

MULTIPLICATION ORDER

In many cases we need to multiply by a fixed number and this is clearly defined by the starting and finishing frequencies. For example a receive-converter for 23cm may require a 24x-multiplier chain to arrive at the final LO injection frequency. A 24x-multiplier can be arranged in several different ways. We need to multiply by 24, which can be broken down into doublers, triplers or quadruplers (2x, 3x or 4x). Choosing which multiplier factor to use and where in the chain is often not simple. As a general rule it is often better to make the first multiplier stage with the greatest multiplication factor, so in this case we might

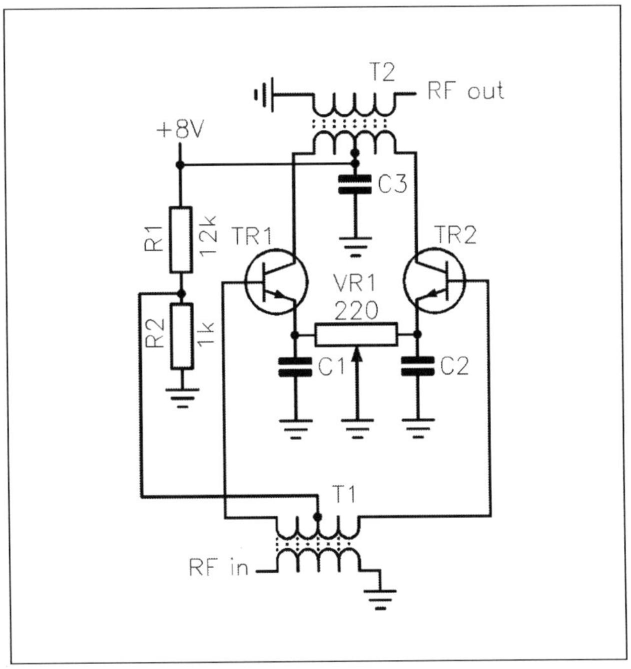

Fig 7.30. Push-pull doubler

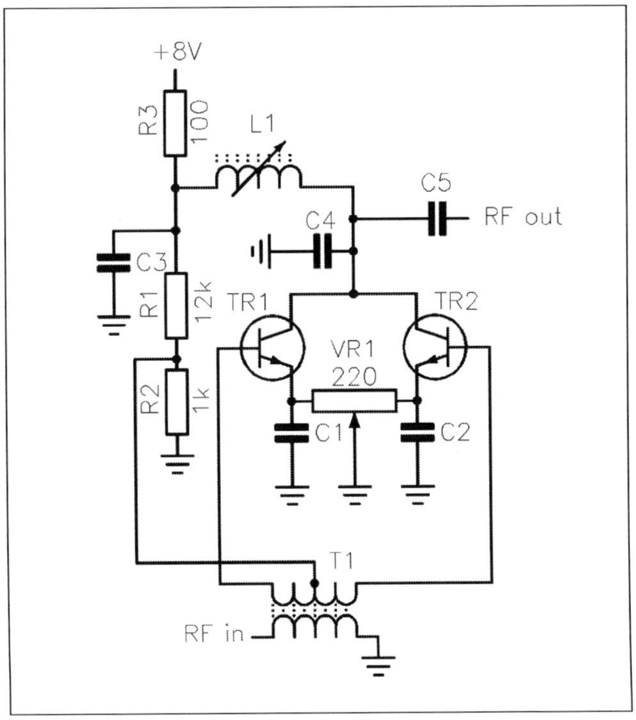

Fig 7.31. Push-push tripler

choose the first multiplier stage to be a 4x. Against this must be weighed the difficulty of filtering out the harmonics close to the wanted multiplier product.

As an example, consider the 23cm receive-converter for an IF of 144MHz. The starting frequency is 24x less than the LO injection, which in this case is 1152MHz. This determines the crystal frequency as being 48MHz. This is a good choice as the 3x product is 144MHz and if a test port is provided on the first multiplier then a small amount of the 144MHz signal can be used to set the band edge on the 144MHz tunable IF. If the 144MHz receiver is accurately calibrated in frequency readout the receiver can be used to zero beat the crystal oscillator exactly onto 144MHz. Hence, in this case the first multiplier needs to be a tripler.

However, consider the case where the filtering of the first multiplier is insufficient to reject all the 48MHz spaced products. If these are allowed to pass through the subsequent multipliers then we will end up with 48MHz products spaced either side of the final LO frequency. The 1152MHz product is not of concern but the image at 1008MHz (288MHz away) will cause degradation in the receiver performance.

Hence, it is better to first double the 48MHz to 96MHz, band-pass filter the result as well as possible to reduce the 48MHz products to negligible levels, and then proceed from there. No integer multiplication factor using 96MHz can produce a signal at 1008MHz so this is a better plan.

Assuming we choose the first multiplier as a doubler, this leaves a total of 12x further multiplication to arrive at the final LO injection. Filtering at 1152MHz will require a fairly good filter and to ease the design of the filter we need to place the unwanted harmonic products as far away as possible. We could choose the final multiplier to be a doubler and hence the injection frequency into the final multiplier would then be 576MHz. If this choice is made then we need to assess if it is possible to arrive at 576MHz from 96MHz. The answer is we can by multiplying by 6, this would be a 2x and a 3x multiplier in either order.

If we chose the more traditional route of multiplying by 3 from 384MHz to arrive at 1152MHz then we need a total multiplication of 4x from 96MHz; this could be 2x and 2x in two separate stages or a 4x in one stage. Again we need to assess the difficulty of making a band-pass filter at 384MHz that can reject the 96MHz spaced products with sufficient attenuation. The safest method would be the dual 2x approach as the filtering of a 192MHz spaced product with a 384MHz band-pass filter is likely to be easier. This also simplifies the filter design and it is likely to have less insertion loss than the narrower filter previously required. In some cases commercial helical filters can be obtained aligned for the required frequency which allows a more compact design. Companies such as Toko and Temwell list suitable filters in their standard product range that occupy very little board area and these have adequate attenuation to the unwanted products.

Unfortunately not all amateurs have access to a spectrum analyser to ascertain if the multiplier chain filtering is adequate. Many constructors have little more than a RF diode probe and perhaps an absorption wavemeter or milliwatt meter to align the multiplier chain. Although it may appear to work satisfactorily it is only when it is examined on a spectrum analyser that the truth emerges, sometimes causing a big surprise!

Unwanted products at the final LO of a receive-converter only cause the operator a problem, but if the same dirty LO is also used to mix up in a transmit converter then we are potentially causing spectral pollution to other users of the spectrum. This is not only irresponsible behaviour but probably in contravention of our licence conditions. These spurious signals may interfere with sensitive receivers used by other services, perhaps radio astronomy, civil or military communication services and should be avoided for obvious reasons!

One factor which is misunderstood by most amateurs is how clean the LO signal needs to be. If the signal is being used to drive a mixer to down-convert or up-convert then the harmonic content of the final signal is not important, in fact given the correct phase relationship a bit of 2nd and 3rd harmonic content can be beneficial to the following mixer. The following mixer will regenerate the harmonics of the input signal so there is no need to suppress the 2nd and higher harmonics of the LO signal. What is of far greater importance is to eliminate as far as possible the lower multiplier products especially the ones closest to the wanted LO frequency.

To illustrate the problem take a look at **Fig 7.32**. This is the tripler circuit shown in Fig 7.27 but with the spectrum analyser sweep increased to cover up to 1GHz. Although the products close to the wanted 150MHz are acceptable the far out products around 700MHz are very bad. They are barely 30dB down and are due to a poor layout where the harmonic energy was able to hop over the filter and get picked up in the output amplifier stage. A contributory factor is the topology of the band-pass filter; it is a top-coupled design and hence tends to act like a high pass filter to the higher harmonics. If the spectrum analyser sweep had not been opened up we would be blissfully unaware of how bad the spectrum was. With my spectrum analyser covering from 10kHz to 12.4GHz it is possible to see significant levels of harmonic energy all the way up to nearly 3GHz. This is with a transistor officially 'spec'd' only to an f_t of 650MHz.

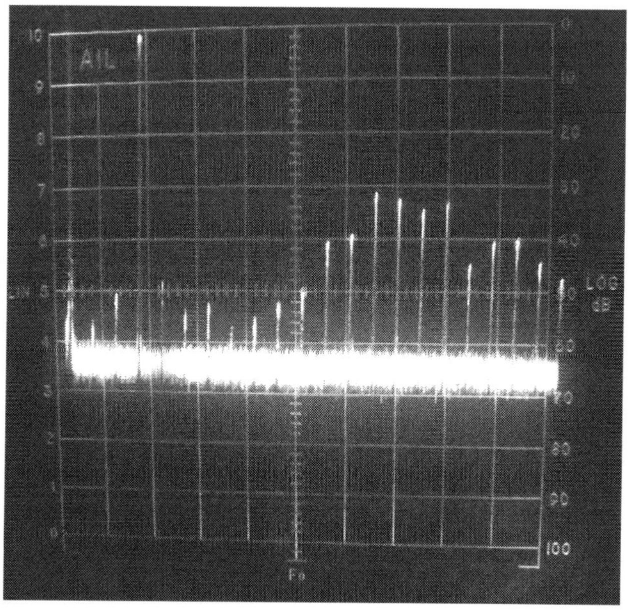

Fig 7.32. Wide band sweep of LO signal

If significant levels of spurious signals at the LO exist these are potential interference generators for frequencies far removed from the wanted channel frequency. Most microwave receiver front ends do not have very high selectivity because of the nature of LNA design; few LNAs have anything like the required rejection characteristics because this would destroy the low noise figure required. If the unwanted signal is of sufficient amplitude to get past the front end filtering and into the mixer then these will appear as 'birdies' or 'sprogs' when the receiver is tuned across the band. If you are really unlucky you may find that your weak signal DX channel is completely obliterated by a strong signal far from the wanted frequency. Running some simple mixing spurious calculations can highlight potential problem areas.

CHOICE OF LO AND IF

The choice of the 1st IF and hence the LO frequency can have a dramatic effect on the image and other spurious response characteristics. If the 23cm band is used as an example, the choice of 28MHz as the 1st IF is not good. For this we need a LO at 1268MHz and the image will then be at 1268 - 28 = 1240MHz which is in the aeronautical DME (distance measuring equipment) and radar aeronautical surveillance band world-wide! Radar surveillance systems are operated to keep track of aircraft approaching airports and other in-transit sites. These radiate up to 10MW of pulsed power. Surveillance radars operate in the 1215 to 1240MHz band. As the amount of image rejection it is possible to obtain at 23cm with a 28MHz IF is not likely to exceed 50dB this means the radar signal can still be a very high level.

I recently had to do a significant system revision for a radio telescope receiver for the Hydrogen Line at 1420MHz that is under development. It was originally proposed to use a 1280MHz 1st LO and the 1st IF was at 140MHz. Only recently it became apparent that a significant amount of strong signals from aircraft DME transponders on or near the image frequency of 1140MHz existed and these caused severe interference, forcing a change in 1st LO and 1st IF.

The 1420MHz receiver has only 40dB of image rejection because of the nature of the LNA and front-end filter design to obtain the required sensitivity. The minimum discernible signal of this receiver with the LNA noise figure of 0.18dB is less than -146dBm, which is about 20dB better than a good 23cm receive converter. In this case it was found that changing from a low-side injection scheme to a high-side injection placed the image frequency at about 1700MHz which is a protected portion of spectrum reserved for SETI research. The 1st IF then occurred at 152MHz, which is another protected piece of spectrum for Radio Astronomy. The 1st LO is then at approx 1572MHz. This allowed a commercial dielectric GPS receiver front-end filter (1575.4MHz) to be used for the final LO filtering. The GPS receiver filter is approx 24MHz wide at the 3dB points because the GPS signal is a spread-spectrum type with a ±12MHz signal. Hence, shifting the LO signal 3MHz from the filter centre does not significantly attenuate the signal level.

> DME systems are used for distance measuring and the aircraft DME transponder interrogates a land-based transponder, this then sends a reply pulse delayed by a known amount, normally 50μs. By measuring the time delay the aircraft can calculate the exact distance to the ground transponder. A typical aircraft DME transmitter is about 300W and the ground station anything up to 10kW. Of course the distance measured is the 'slant range' but with input from the aircraft altimeter the computer calculates the flat earth distance to the aircraft. DME transponder channels are paired with VHF Omni-directional Ranging (VOR) channels and the two systems work in tandem.
>
> In the DME allocation the aircraft transmit between 1025 and 1150MHz and the ground stations from 962 to 1024MHz and 1151 to 1213MHz. There is a fixed 63MHz split between the aircraft and ground station channel. The DME channels are 1MHz apart. Sometimes the frequencies used are reversed. The normal system as detailed above is known as 'Mode X' when the frequency plan is reversed the transponders operate under 'Mode Y'. Hence, DME transmissions can be found anywhere between 962 and 1213MHz.

SPECIAL LOCAL OSCILLATOR TECHNIQUES

During the upgrade to my 23cm transverter an unusual approach was used. The original design used a 5th overtone crystal at 96MHz, but this suffered from excessive warm-up drift and other problems. Because the desire was to make as little change as possible it was decided to make a new 96MHz-oscillator module that then fed into the existing modules. This new module used a 12MHz TCXO, which was multiplied up to 96MHz, a factor of 8x. The first multiplier used a 2x and the second a 4x. After filtering and amplification the 96MHz signal was fed to the original oscillator multiplier chain. A test port on the second multiplier provided a small amount of 144MHz harmonic energy to calibrate the 144MHz transceiver.

Because the first multiplier output spectrum contained products spaced 12MHz and 24MHz either side of the wanted frequency the filtering required was quite severe. If these products are not removed at the source it is practically impossible to eliminate them further down the multiplier chain. The aim was to attenuate these close-in products by at least 70dB, in the end a figure of over 80dB was achieved by careful attention to the early band-pass filters. The 96MHz final output was clean; the worst spurious product was -86dBc, which was considered sufficient. Filtering, shielding and supply line bypassing achieved a compact and clean module suitable to drive the existing multiplier chain. One of the other problems with the 96MHz crystal oscillator was its susceptibility to strong RF fields when transmitting 250W on 23cm. This caused FM when using SSB or a chirp when CW was used. With adequate shielding and supply line bypassing these problems disappeared with the new oscillator module.

It must be appreciated that the filtering inductors need to be well shielded; the best option is pre-wound canned coils or helical filters such as those made by Toko. If air-wound inductors are used the coils radiate spurious energy and this can hop over shielding plates and effectively bypass the filter. The RF voltage at the collector of a multiplier stage can be many volts, and these make good 'mini transmitters' unless the RF field is contained in shielded compartments. Grounding, layout and shielding are all vital parts of any oscillator chain design and construction.

Chapter 8

Oscillators

OSCILLATORS are used in a variety of communication systems. For radio frequency items such as receivers and transmitters several types may be required. A simple Morse code transmitter would require one to generate the RF carrier that can be switched on and off to send the coded signal.

BASIC PRINCIPLES

All oscillators work in a similar fashion and there are many different types of circuit employed. However, they all work by feeding a portion of the output signal back to the input via a frequency-determining network. In order for oscillation to commence two criteria need to be met. The first is the amplitude of the feed back signal. If the 'open loop' gain is less than one oscillation cannot commence. Therefore the device used to amplify the signal needs to have a gain greater than one.

The second criterion is the phase of the feedback signal. This needs to be in phase with the input signal. This is performed by a *phase-shift network* (PSN). Hence, the total phase shift from the output to the input must be 0° or 360° or some multiple of 360°. The amplifier stage will often have a 180° phase shift between the input and output, so the network needs to have a 180° phase shift to get the two signals in phase.

Oscillation starts due to the inherent broad-band noise all practical amplifiers possess. If the amplifier were noiseless (noise figure of 0dB) the oscillator could not start. The amplified noise, fed back to the input again amplifies the small amount of noise, and it builds up in amplitude until oscillation is sustained. This is a form of *regeneration* and is known as the 'Barkhausen effect' after the scientist who discovered it. The lower the noise output from the device the longer it takes the oscillations to build up to a maximum. For an RF crystal oscillator operating at 1MHz this may be as long as 1ms before the oscillator is running at full output. Hence, keying an oscillator by switching the supply on and off is not a very good method. It can also cause small shifts in the frequency as the oscillation builds up: this we call 'chirp'. The time it takes for the output amplitude to reach maximum is also determined by the selectivity of the frequency-determining network. If the network has a very narrow bandwidth (i.e. high Q) the time required is longer than one with a wider bandwidth (lower Q). Hence, a crystal oscillator can be slow to start if the open loop gain is less than about 3. There are methods to get around this problem. One method is to apply a very narrow pulse to the crystal at switch on to 'kick start' it. (This method was developed by me and was the subject of a patent to get high Q crystal oscillators running in less than 5μs in a frequency hopping jammer developed in the 1980s.) This method is also commonly used with large klystrons and magnetrons used for radar transmitters. Normally the transient caused by switching on the DC supply to the crystal oscillator is sufficient to 'jolt' the oscillator into activity. If the DC supply voltage is slowly increased from zero the oscillator may not start if the loop-gain is too low.

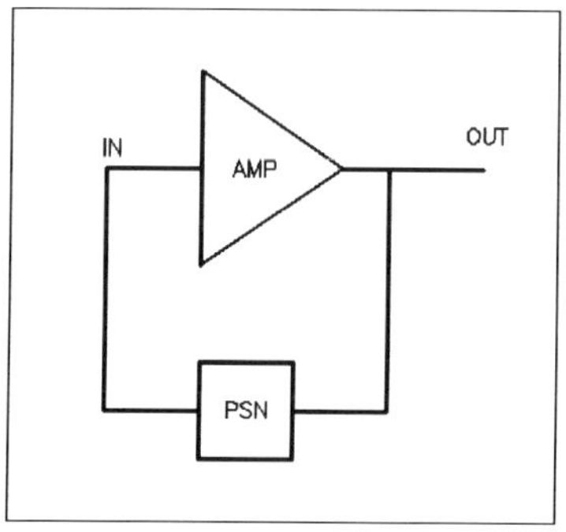

Fig 8.1. Block diagram of an oscillator

When the amplitude of the oscillations builds up, the amplifier stage is driven into saturation and the gain falls slightly. This causes the output amplitude to stabilise and is only affected by changes in the supply voltage. Hence, it is normal to use a very well stabilised supply to power an oscillator. Changes in the supply voltage also cause small variation in the parasitics associated with the amplifier and this causes a small variation in frequency. So for maximum stability the DC supply voltage needs to be very stable and very low noise.

Because of the inherent self-limiting action of the amplifier we can expect that the output spectrum will not be a pure sinusoidal waveform and it will contain harmonics of the fundamental carrier. These can be easily filtered out in subsequent stages if needed. If the output must be a pure sine wave then often the output can be taken from the input with very loose coupling and fed to a high impedance buffer amplifier.

Unfortunately analysing a complex item like a closed loop oscillator is not easy. Although for small signal linear amplifiers we can use readily available software with the S parameters of the amplifier device this is not the case for limiting amplifiers that form the basis of all oscillators. However, with the correct non-linear analysis software now available the task is simpler than in days gone by. One of the problems is that the amplifier is not in full saturation for the whole RF cycle. Over about 10 to 20% of the waveform the amplifier is driven into severe limiting, over the remainder of the waveform it behaves more like a linear amplifier. The modern form of software uses a technique called 'harmonic balance' to assess the performance and a good correlation between simulation and actual results is possible. This especially is the case with oscillators with fundamental outputs, those not used as multipliers, and the level of the harmonics can be predicted with a fair degree of certainty.

PULLING & PUSHING

Most amateurs are aware that the output of a high stability oscillator needs to be buffered with an amplifier stage that presents a constant load to it. If the load applied to the output of the oscillator varies so will the carrier frequency. This is known as 'pulling' and often high impedance buffer amplifiers are used with very little load to the oscillator. This isolates the oscillator from the following circuitry.

The phenomenon of 'pushing' we have already touched on. Pushing is the frequency variation due to the supply voltage changing.

NOISE PERFORMANCE

Although some internal noise needs to be present to make the oscillation start, using an inherently noisy amplifier is not a good idea. If the amplifier has a poor noise figure it will modulate the carrier and cause spurious side bands. The random noise inherent in the amplifier device causes two types of noise, one being random variations in the amplitude and the second being random variations in the phase. To counter this problem oscillator designers use two tricks. Firstly the amplifier gain is made very high so little noise is required to start the oscillator. This of course forces the amplifier into its non-linear region harder and this has an effect of lowering the noise power generated internally. For the best noise performance the amplifier should be running in a saturated manner, that is delivering a high output power. This means that it develops much more power output and the phase shift network needs to have considerable loss. This is normally an advantage as crystals used for oscillators can have quite high insertion loss. In an amplifier running in a limited mode the amplitude variations are largely smoothed out by compression but these are then converted into phase variations. This is known as AM to PM conversion. Hence, the close-in noise product is referred to as 'phase noise'.

Another common technique where a very low noise oscillator is required is to place a second identical crystal in series with the output acting as a very narrow band filter. Using this technique a very low noise oscillator can be constructed. However, this only works for a single channel oscillator, as each channel requires two crystals, one for the oscillator and one for the filter. This makes the oscillator more expensive. Ensuring the two crystals have identical centre frequencies with temperature variations is not simple and these types of oscillator are normally reserved for reference oscillators for high quality PLL synthesisers or multiplier chains.

> Synthesised oscillators used with PLL synthesisers have other spurious products close to the carrier. The frequency dividers cause these and these generate spurious modulation in the voltage-controlled oscillator (VCO). If the channel spacing is 25kHz we can expect to see small carriers spaced either side of the main carrier by 25kHz. The design and layout determine how well these are suppressed. Generally synthesised oscillators are poorer than crystal oscillators and need more care in the design and construction to reduce the reference side bands to acceptable levels. If these are excessive it severely degrades the potential adjacent channel rejection in a receiver.

If the oscillator is noisy the noise sidebands only fall away to a certain point and remain flat for a considerable bandwidth. Above this bandwidth the noise then falls off to low levels. This is known as a 'noise pedestal' and it looks like an up-turned wineglass. This is shown in **Fig 8.2**. The values shown are fairly typical for a poor oscillator. The zero noise level point isn't in fact zero noise; this is what the spectrum analyser is able to measure to. In practice with a good spectrum analyser we would be able to see about 30dB lower. The bandwidth of the pedestal for a good crystal oscillator at 100MHz would be about 500Hz and for a LC oscillator about 20kHz.

The shape of the noise spectrum is determined by several factors. The slope of the noise further away from the carrier changes shape. The frequencies very close to the carrier, perhaps within 200Hz, are predominantly dictated by the low frequency noise performance of the amplifier device. This is known as '1/f' noise or 'flicker noise' and it can be heard in audio amplifiers as a random hissing noise. This low frequency noise falls in amplitude as the frequency increases until at about 15kHz other noise types generated in the amplifier device mask it. At higher frequency offsets from the carrier the predominant noise is broadband (white noise) and this extends up to several MHz. Hence if the noise spectrum is examined the point where the 1/f noise ceases to be dominant and the broadband noise takes over shows a change in the slope of the noise. Measuring the absolute value of the noise is not simple and a very good spectrum analyser is required often with some sort of additional filtering to reduce the carrier amplitude to prevent over loading on the analyser. Often a very narrow band crystal filter is used to explore the noise skirt without allowing the carrier to pass to the spectrum analyser.

In **Fig 8.3** three distinct types of noise are evident. The portion denoted A is due to the Q of the phase shift

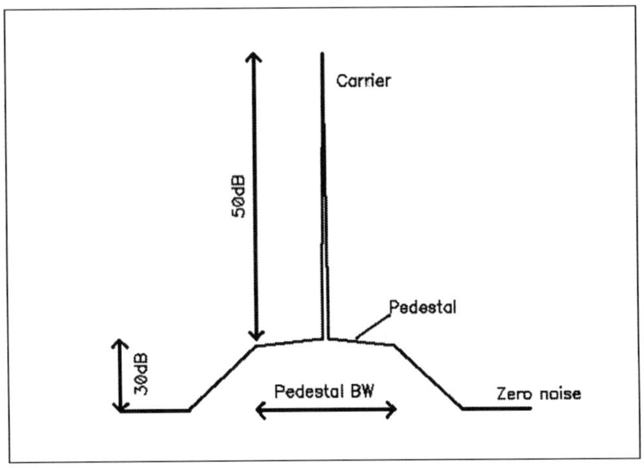

Fig 8.2. Typical noise pedestal

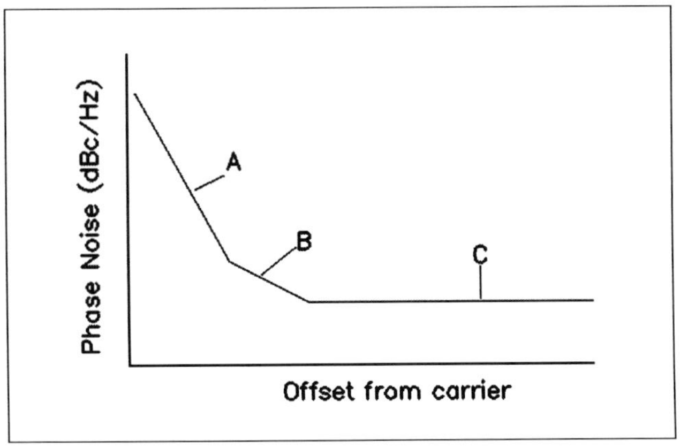

Fig 8.3. Break points in the close-in noise

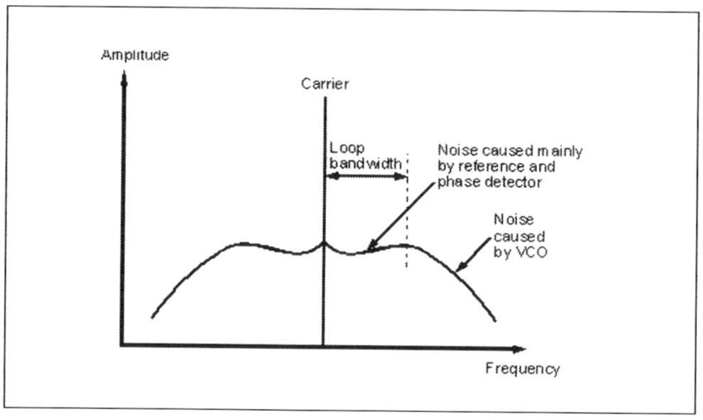

Fig 8.4. Phase noise for synthesised oscillator

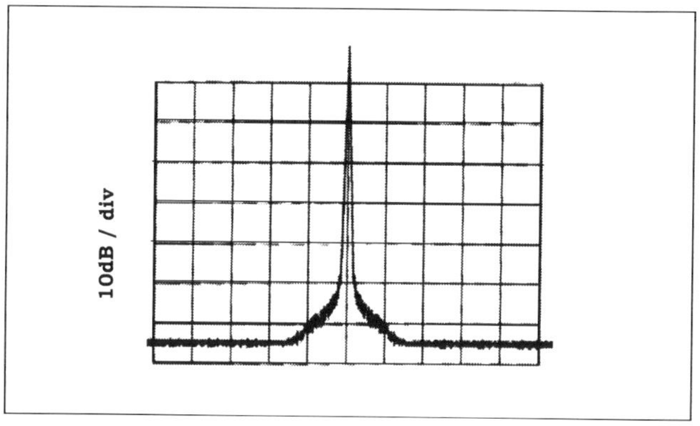

Fig 8.5. Typical spectrum analyser plot of close-in phase noise

(feedback) network, if the Q is low the amplitude is high close to the carrier. This portion is often referred to as 'flicker FM noise'. The portion denoted B is due to the 1/f flicker noise in the amplifier. The portion denoted C is due to the broadband noise performance of the amplifier.

For the best noise performance the amplifier needs to have low 1/f and broadband noise. For most applications a junction FET (JFET) or a low-noise bipolar junction transistor (BJT) is the best choice. Because of the high closed loop gain required this means that the device selected needs to have a relatively high cut off frequency (f_t). Generally if the f_t is about five times the oscillator frequency this is normally adequate.

In synthesised oscillators (PLL) the noise spectrum close to the carrier is modified by the 'loop bandwidth' of the control circuits. With a wide loop bandwidth the noise is suppressed over a wider portion of the spectrum than one having a narrow loop bandwidth. Outside of the loop bandwidth the noise is predominantly broadband and this will decrease in amplitude further away from the carrier. This effect causes the noise pedestal to be lower in amplitude within the loop bandwidth. Generally the loop bandwidth is made as wide as the reference frequency to suppress the reference side bands as far as possible.

PHASING

Many amateurs have difficulty understanding how the phasing in the feedback network works. A simple analogy to a pendulum serves to explain the effect. Imagine a pendulum with a long arm and it is swinging slowly. View the pendulum so that it swings left to right and then back again. Now imagine what happens when we push on the bob weight on the end of the pendulum. If the pendulum is swinging from the right to the left and we are standing at the left, the pendulum is swinging towards us at one time and away a little later. If we hit the bob weight with a small hammer as it approaches us it retards the pendulum. If we do this often enough the pendulum will eventually stop swinging. If we are slow and only hit it when it has started to swing away from us we do not impart as much effort as possible and it also eventually slow down and ceases swinging. The ideal case is when the pendulum bob weight has just comes to a stop before it reverses in direction. This is the effect of phasing or timing the input of energy to maintain the pendulum swinging. If two people stand at the two extremes of swing each can impart a little energy input to keep the pendulum swinging. However, if one gets their timing wrong it can cause the pendulum to lose energy and it will slow down and eventually stop.

SERIES MODE COLPITTS OSCILLATOR

One of the most common circuits used is the Colpitts oscillator. This can use a bipolar transistor, FET or a valve. There are several variants of the basic Colpitts circuit. The one important point to note is that the collector (drain or anode) must be well decoupled to RF signals for the oscillator to work in the way it is intended. Some designs only partially decouple the collector and the output signal is tapped off a low impedance point. These versions are not as low noise as a proper Colpitts circuit or provide as high an output signal. A basic Colpitts oscillator is shown in **Fig 8.6** using a bipolar transistor.

The oscillator is formed around TR1 which is a high gain, low noise bipolar transistor such as a BFR-92 (SMD) or MPSH-10 in a TO92 package. The base is biased to half the supply voltage by two equal high value resistors, in the example circuit 47kΩ. The emitter is raised above ground by a medium value resistor for biasing. Hence, the emitter voltage will be approx 0.6V below the half rail voltage, in the circuit this is about 3.4V due to the 8V supply voltage. The quiescent current flowing in TR1 is

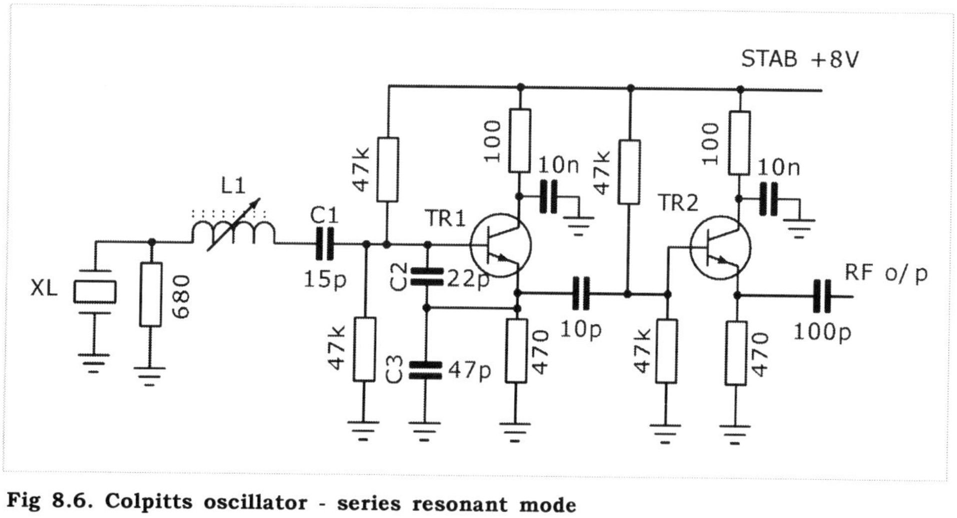

Fig 8.6. Colpitts oscillator - series resonant mode

determined by the value of the emitter resistor and for best noise performance (high power output) the collector current needs to be relatively high. With the value shown the collector current is approx 7mA.

The collector is decoupled with a low reactance capacitor to hold it at zero RF potential. A capacitive divider between the emitter provides the feedback to the base formed from two low value capacitors, C2 and C3. The transistor operates as an emitter follower with essentially zero phase shift between the base and emitter. The output from TR1 is fed to a second bipolar transistor TR2 via a low value coupling capacitor and this stage operates as a unity voltage gain buffer amplifier (emitter follower) to reduce the loading on the oscillator.

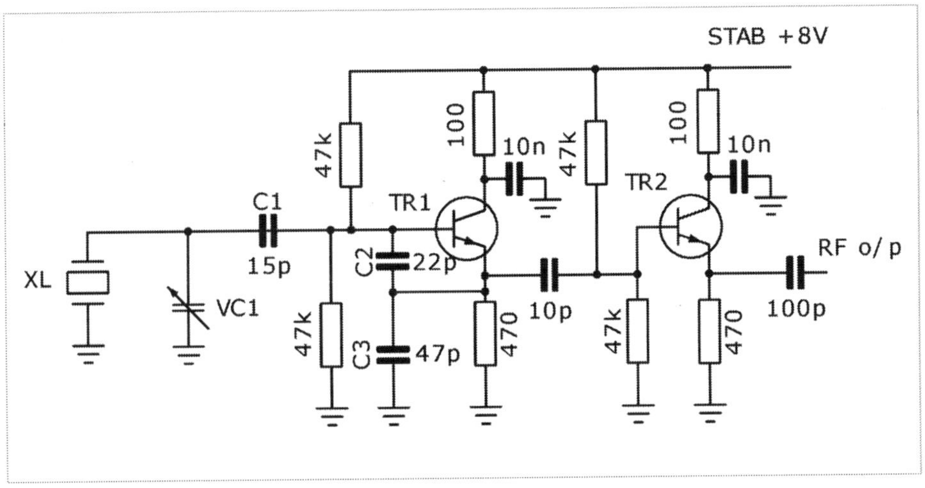

Fig 8.7. Colpitts oscillator for fundamental crystals

Note that the base-emitter capacitance (gate-source for a FET) of a small signal transistor is of the order of 4pF so this needs to be taken into consideration when selecting the value of C2. For oscillators below about 50MHz this can usually be ignored.

The frequency determining components are the crystal XL, the series inductor L1 and capacitor C1. These are chosen so L1 and C1 are series resonant at the crystal frequency. The crystal is a series mode type that exhibits a low resistance at the centre frequency but a high resistance at other frequencies. To prevent spurious modes being excited the crystal is shunted with a low value resistor. This resistor should be about ten-times the ESR of the crystal in use. For most series mode crystals a value of 680Ω is normally adequate.

The resonant frequency of the crystal is determined by the load presented to it by the oscillator. Since L1 and C1 can be adjusted to the crystal frequency a small variation in crystal frequency (about 0.001% maximum) can be achieved by varying either L1 or C1. For convenience L1 is the adjustable component in the circuit and is a screened inductor with a ferrite slug to alter the inductance. The circuit shown suits 3rd and 5th overtone crystals between approx 20MHz and 125MHz.

The value of the feedback capacitors C2 and C3 depends on the frequency and the gain of the amplifier device. If C2 is made smaller and C3 increased the feedback is increased. Too much feedback can cause spurious modes and make the crystal jump from the correct series mode to another mode; hence the feedback should be adjusted to provide just enough feedback for stable operation and reliable starting when the supply is switched on. Similarly C1 needs to be chosen to be series resonant with L1 at the crystal frequency.

FUNDAMENTAL MODE COLPITTS OSCILLATOR

For crystals that are cut to operate in fundamental mode (parallel resonant) another version is used. This is shown in **Fig 8.7**.

The difference between the series resonant design is the way the crystal is fed. It is now connected across the base to ground of TR1 with a low value capacitor in series. This circuit is rather trickier to get working correctly due to the way a parallel mode crystal operates. The manufacturer of the crystal will specify a particular shunt load capacitance for the crystal to be on frequency. A typical value is 30pF.

Before we continue it is as well to look at how the crystal behaves and then we will see why the value of the capacitor divider, the series feed capacitor and shunt trimmer are important.

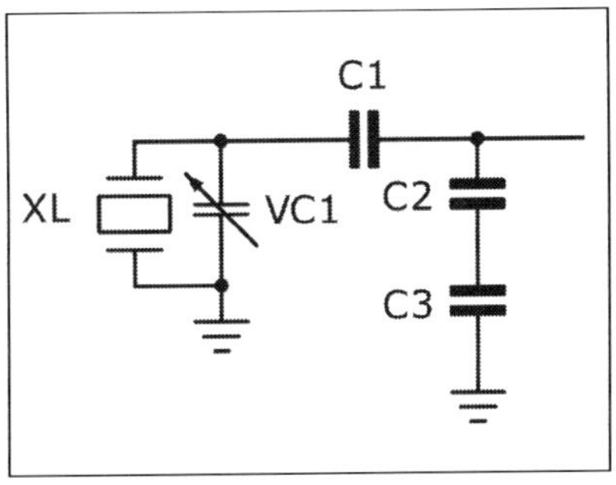

Fig 8.8. Equivalent crystal load network

The crystal is effectively connected across a network as shown in **Fig 8.8**. At resonance the impedance the crystal presents to the network is approaching an open circuit and is typically a few 100kΩ to as much as several megohm depending on the type and frequency.

The equivalent circuit of a crystal at parallel resonance is the same as a parallel L-C network with very high Q. Unloaded Q values for crystals run between 10,000 and 500,000 so the *dynamic resistance* (R_D) at resonance is very high, approaching an open circuit.

For low frequency fundamental crystals (approx 1 to 4MHz) the feedback capacitors C2 and C3 will need to be fairly large values, perhaps as much as 470pF. As these are connected in series the equivalent shunt capacity is 235pF. Because the crystal needs to 'see' a load capacity of 30pF to be on the correct frequency we can't simply connect the crystal directly between C2 and ground. This would pull the crystal too low in frequency and it may not oscillate with such a high load capacity. To get around this problem we need to connect a lower value capacitor in series with the crystal and C2. Also we need to be able to 'tweak' the crystal on to the correct frequency and we normally use a small value variable capacitor to achieve this. Placing a capacitor in shunt with the crystal and adjusting it so the total capacity the crystal sees is 30pF should bring the crystal exactly on frequency. If the shunt capacity is less than 30pF the crystal will be higher in frequency and if the capacity is more than 30pF the crystal will be lower in frequency. Some circuits show the variable capacitor in series with the crystal ground, this works in a similar manner and either is acceptable.

So we need to calculate the effective capacitance the feedback capacitors provide and then select a suitable series capacitor for C1 so that the shunt capacity is less than 30pF. The variable trimmer makes up the difference. A typical series capacitor for C1 might be 33pF and the variable would be about 20pF maximum, so when the crystal is on frequency the trimmer is approx half meshed at 10pF effective capacitance.

If the feedback capacitors are two 120pF connected in series between the base and ground the effective capacity is 60pF. To reduce the crystal shunt capacity to about 20pF requires a series capacitor of about 47pF, which with the shunt trimmer set to 10pF will adjust the crystal to the correct frequency. However, the parallel resonant crystal appears as close to an open circuit when on frequency (very high resistance) and the base biasing resistors will load this. So it is preferable to select a smaller value for C1 to counter this effect and to increase the value of the variable capacitor across the crystal to bring the crystal on frequency. In most case a value for C1 of about 22pF to 33pF will suit fundamental oscillator with 30pF crystal load. The smaller the loading on the crystal the greater the amplitude will appear across the crystal and hence across the base of TR1.

QUARTZ CRYSTAL RESONATORS

Before we continue, it is as well to look at crystals in a little more detail to understand the parameters. Quartz exhibits the piezo-electric effect whereby electrical charges are produced across the opposite faces of the crystal when it is subjected to mechanical stress. Conversely the crystal is distorted by the application of an electrical charge across it. Acoustic waves are able to propagate through the material with very little attenuation and the transit time for the acoustic wave to pass through the material is stable.

Consider a rod made of quartz. An acoustic wave propagating along it eventually strikes an end where there is a large and abrupt change in acoustic impedance. A strong reflection occurs, and the wave propagates back to the other end where it is similarly reflected. (Just like a transmission line with an open or short circuit termination reflects all the power back to the transmitter.) At an applied frequency, where the thickness of the quartz material is one half-wavelength, waves from successive round trips coincide in phase and reinforce each other, thus increasing the wave amplitude. This is the fundamental resonant frequency. The RF analogy is the half-wave dipole.

The crystal will also resonate at odd multiples of the fundamental frequency known as overtones. Third, fifth, seventh and ninth overtone crystal resonators are available. The overtone frequencies are not exact odd multiples of the fundamental, although they are very close. This is due to a small phase shift error that is introduced during reflection of the acoustic waves at the crystal ends. It should be noted that a crystal cannot resonate at even numbered overtones, since this would imply that opposite faces of the crystal would be at the same polarity. This would prevent the crystal from distorting or resonating.

The most common crystal type is known as an AT cut where crystals are manufactured between about 800kHz to 40MHz in fundamental mode, and from 10MHz to 360MHz in overtone mode. AT cut resonators are *thickness shear* vibrators, i.e. they vibrate perpendicular to the plane of the electrodes. The crystal is normally a round disc. The diameter of the disc does not directly affect the resonant frequency. Diameters between 5mm and 15mm are commonly used.

The following points should be noted with regard to crystal manufacture:

All crystals have all modes. The crystal manufacturer can adjust the disc faces to make the wanted resonance (overtone) a little more active and the other modes a little less active, but this does not guarantee oscillation at the wanted overtone. Some round disc crystals have a small flat edge ground to reduce the spurious modes. This method is also used to move the operating frequency slightly during manufacture to get the crystal exactly on frequency in a test oscillator.

All crystals have spurious or unwanted resonant responses ('spurs'), besides the fundamental and overtone resonant responses. In the AT cut these spurs exist only above the fundamental resonance frequency and appear close to the desired resonance point. They are not well controlled and may vary substantially between crystals made to the same specification. In a good crystal the magnitude of the spurs is much reduced when compared with the response at the wanted resonance and the spurious modes may be ignored.

The equivalent electrical model for a quartz crystal is shown in **Fig 8.9**.

The motional capacitance CM, the motional inductance LM and the motional loss resistance RM model the electro-mechanical behaviour of the crystal. In addition the metal electrodes that are plated onto the two opposite faces of the quartz disc to allow connection to an external circuit gives rise to the parallel capacitance denoted by CO. This is known as the static capacitance and includes any stray capacitance between the disc and the metal can that enclose the crystal disc.

The value of CM is very small, often as little as 20fF (femtofarads) and consequently LM is very large. For a 10MHz crystal LM is about 15mH. RM will be low and values of between 5Ω and 150Ω are typical for a fundamental and a 9th overtone type respectively. Some typical maximum values for different overtones (OT) are:

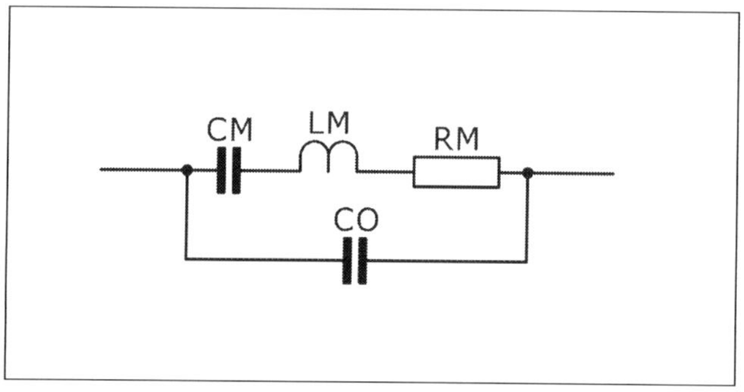

Fig 8.9. Equivalent circuit of a crystal

3rd OT ≈ 40Ω
5th OT ≈ 60Ω
7th OT ≈ 90Ω
9th OT ≈ 150Ω

The crystal is essentially a series resonant circuit, but because of CO and other external shunt capacitance it does exhibit a parallel resonant property where an impedance peak is found above the series resonance frequency. The shunt capacitance resonating with the motional inductance LM causes this.

There is no difference between the construction of a series resonant crystal and a parallel resonant crystal, which are manufactured exactly alike. The only difference between them is that the series resonant frequency of a parallel resonant crystal is set approx 100ppm lower than the desired operating frequency. Parallel resonance also requires the external application of a small capacitance across the crystal, often about 30pF, to obtain the correct oscillation frequency.

> It should be noted that there is no such thing as a *parallel resonant crystal oscillator*. All crystal oscillators operate either at or near (above) series resonance. What is usually meant by parallel resonance is that the crystal has a high load impedance across its terminals. Series resonance physically exists in the crystal, but parallel resonance exists only as a crystal measurement phenomenon when external shunt capacitance exists.

The major advantage of the quartz crystal resonator is the very high unloaded Q. Because of this the frequency of oscillation is very stable and only affected by external influences such as ambient temperature and loading of the oscillator components.

Designations used by most crystal manufacturers identify the crystal cut and the overtone number. For example, 3AT denote an AT cut with 3rd overtone characteristics. Note that if specifying a parallel resonant crystal the shunt capacitance is required to be specified. Most crystal manufacturers prefer to use a value of between 20pF and 40pF. If the crystal is required to be a series resonant type then the shunt capacitance is not required but the maximum value for RM (known as the crystal ESR) is required. The value of this (or what is economical to produce) will define the type of oscillator to use. As a rough indication, a 3rd overtone crystal for about 50MHz will have ESR values between 10Ω and 39Ω. The lower the ESR the higher grade (purity) of quartz required and the more expensive it is to manufacture. The lower the ESR the more active the crystal is and the lower noise performance it can produce in a practical oscillator. Economics normally dictate a somewhat less than perfect crystal is used, except in special circumstances where cost is no object.

Quartz is the same chemical composition as glass but considerably harder. They can be used to cut glass.

To summarise, the following points should be noted when using a quartz crystal as a resonator:

It is recommended that quartz crystals be operated free from any DC voltage by connecting an appropriate blocking capacitor in series with the crystal. This is especially important if a valve is used as the amplifier where high voltage is present.

When specifying a crystal for parallel resonance, the value of external parallel load capacitance must be specified when ordering a crystal. Standard values of load capacitance are 10, 12, 15, 20, 30 and 50pF.

Overtone crystals are generally operated in series resonance mode and are difficult to pull. Their pulling range is reduced by a factor of n^2 where n is the overtone number. Hence a 5th overtone crystal has a pulling range of 1/25 of a fundamental at the same frequency.

> Quartz crystals are grown in large autoclaves under very high pressure (≈100bar) and temperature (≈2000°C). The 'seed' crystal is a slice of quartz carefully selected from previously grown quartz to be as perfect as possible. This is hung on a copper wire on a frame in the autoclave. Raw, impure mined quartz is packed into the bottom of the autoclave and a caustic liquid poured in to a depth of about 30% of the height of the chamber. The seeds are hung above the raw quartz and liquid and the autoclave is shut. A large heating coil is powered to raise the temperature and pressure and maintained for about 10 days. During this time the raw, impure quartz is dissolved in the caustic liquid and it is deposited onto the seeds to form an exact 'clone'. (At one crystal manufacturer the autoclaves were made from WWI 18' naval gun barrels with the breech serving to seal the top of the autoclave.) The crystal slowly grows and becomes a large block, known as a 'boule'. The angle of 'cut' of the seed determines the finished crystal angle of cut. When the growing process is complete the new boules are taken out and sliced up into thin wafers using diamond saws (the only method possible, as quartz is almost as hard as a diamond) to form blanks that the new crystals will be made from. Part of the process is the accurate grinding and lapping to suit the frequency of oscillation. For very high frequency crystals the disc thickness may be as little as 0.5mm. A typical raw boule can furnish several hundred new crystals with careful cutting and grinding. The off cuts are used for the next batch of crystal growing.

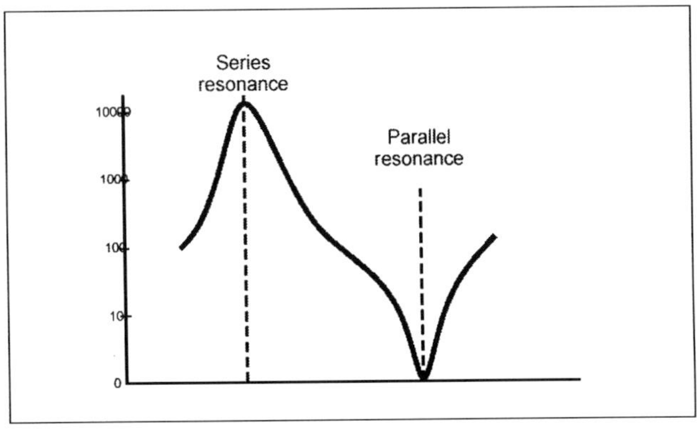

Fig 8.10. Series and parallel resonance modes

Fig 8.11. A typical small quartz boule. This is a reject piece that has been cut in half and made into a paperweight

> The reason the crystal ageing slows down over an extended period is due to the grinding and lapping processes during manufacture. This leaves minute particles of quartz 'dust' in the pores of the crystal, which over time gets ejected as the crystal vibrates. Heating the quartz disc to a high temperature and running at power accelerates the ejection process. The criteria of when to terminate the aging is the rate of drift, which is measured automatically during the aging process. Even as little as 24 hours accelerated ageing gives a large improvement in the drift characteristics.

Over excitation of the crystal causes a long-term change in the crystal parameters due to internal heating i.e. ageing. Severe over excitation will destroy the crystal due to mechanical stresses causing fracturing of the brittle material. To avoid damage the crystal dissipation should be less than 2mW. To minimise crystal ageing the excitation level should be less than 20μW. A drive level of about 0.1mW is normally a good choice.

For a crystal to control the frequency of an oscillator circuit, the crystal must maximise the oscillators gain at the oscillation frequency and minimise it at all other frequencies.

Very high quality crystals are 'accelerated aged' by immersing them in oil at about +80°C for a period of 10 days to reduce the drift to close to zero. This of course delays the delivery. If a non-aged crystal is used, the user will need to adjust the crystal periodically for about the first three months (it will tend to drift higher in frequency) until the drifting reduces to zero. Hence, if a crystal is needed in a hurry and the ageing process is omitted, it is a good idea to order two crystals, one non-aged and the other correctly aged and to change to the aged crystal when it is delivered sometime later.

Another important factor is the ambient temperature. If the temperature varies over a wide range the frequency changes slightly. If this is unacceptable, enclosing the crystal in a temperature-regulated oven will give very high stability. Also, if the crystal is expected to encounter sub-zero temperatures, it is prudent to specify a 'nitrogen-purged' envelope. Water molecules trapped in the envelope will freeze on the quartz disc and cause a loss of activity and a lowering of the frequency.

Older metal can enclosed crystals used a small hole in the can for purging with inert gas or dry air. These were then plugged with solder. Less expensive types had the metal can soldered to the base. The free solder vapour trapped in the envelope causes a slow decrease in frequency and activity as the particles get deposited on the quartz disc. Today the metal envelopes are either resistance-welded or compression welded in a nitrogen atmosphere to eliminate the tin-lead in the solder causing this effect. Very high stability crystals are normally enclosed in an evacuated glass envelope (see **Fig 8.12**).

Of interest is that we can often use a crystal on one of its overtones although it may not be specifically made for this application. For example an 8MHz fundamental crystal can be made to oscillate at 24MHz by designing the oscillator to excite the crystal on the 3rd overtone. This can save a multiplier stage and reduce the filtering required. Similarly a 3rd overtone crystal can be made to oscillate on its fundamental with the correct oscillator design, for example a 33MHz 3rd overtone crystal can be used to produce an 11MHz signal. It is for this reason that in overtone oscillators it is imperative to suppress the fundamental mode with a shunting network to prevent this mode. Some designers prefer to use a low value inductance across the crystal for this, but it can be tricky to get the correct value and often a shunt resistor works just as well. When the overtone number is high, for example 7th and higher, suppressing the lower overtones can be quite a challenge and most designers prefer to limit the overtone number to 5 for this reason.

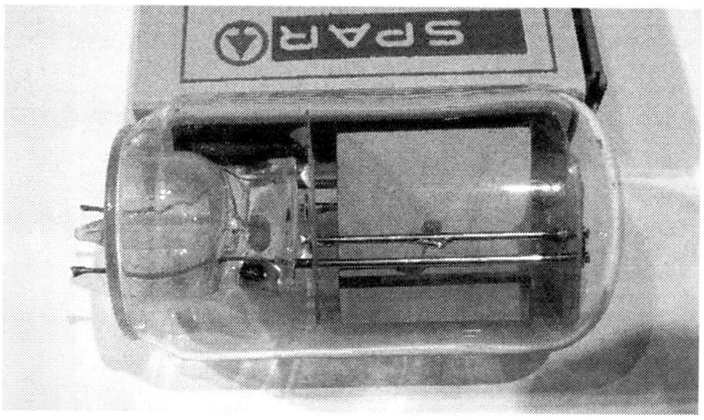

Fig 8.12. 100kHz crystal in a glass envelope. The quartz wafer is approx 25mm square and suspended on two stiff wires from the thick wires inside the envelope. The quartz wafer is gold plated for the connection points.

The crystal when manufactured is cut for a particular 'inversion' or 'inflection temperature'. The inversion temperature is the centre of a curve; often the temperature chosen is +25°C when normal ambient temperatures are encountered. As the ambient temperature deviates from the inversion temperature the error in frequency gets progressively worse. At the two extreme ends of the temperature range the frequency error will be greatest. Over the centre portion of the curve the slope of the curve is quite flat and so very little error occurs. So if the inversion temperature was chosen as +25°C and the crystal temperature varies over ±15°C the frequency error falls to very low values.

Some amateurs have experimented with changing the frequency of surplus crystals. After WWII a large number of FT-243 packaged crystals were available on the surplus market. These differ from modern crystals by the mounting method. In the FT-243 crystal the quartz is

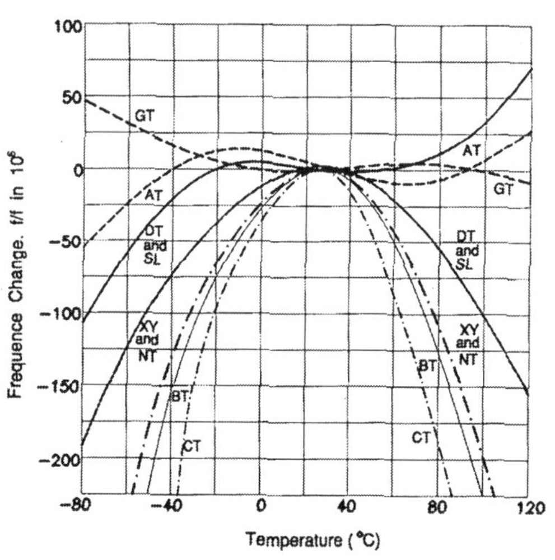

Fig 8.13. Frequency variation versus temperature for various crystal types

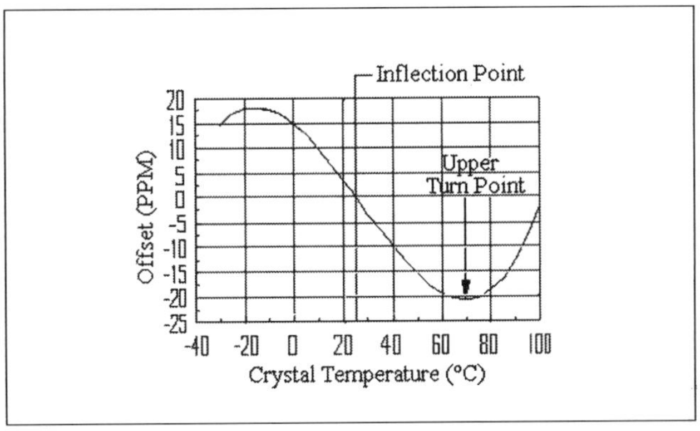

Fig 8.14. Frequency error versus temperature for AT-cut crystal

a square or rectangular slab clamped between two metal pieces and retained in the holder with small screws. It is a simple matter to take one to pieces and then to reduce the thickness by dipping in a solution of hydrofluoric acid, which etches away the crystal. However, the frequency can only be raised by this method because it thins the crystal. To lower the frequency requires a different method, ideally the addition of new quartz onto the slab. Some amateurs experimented and found that rubbing the faces with a graphite pencil or a piece of solder could lower the frequency by a small amount. However, this lowers the activity and if taken too far can cause the crystal to no longer oscillate.

WORKED EXAMPLE

We can now examine some of the parameters when designing a crystal oscillator from scratch. We will choose the popular Colpitts type, although there are better types for some applications. There is a minimum crystal load resistance required below which the circuit will not oscillate. This is approximately equal to:

$$R_{shunt} \approx \frac{100 \times 10^3}{f}$$

where f is in MHz.

Ideally the crystal shunt load resistance should be 5 to 10 times this value. This gives a useful starting point for the design of the oscillator. For example, assuming the DC current gain of the transistor is 100, the value of emitter resistor to a first approximation is:

$$R_e \approx \frac{10 \times 10^3}{f}$$

where f is in MHz.

There is no special relationship between the capacitor divider values. For most Colpitts oscillators between about 1MHz and 20MHz a value for the capacitor across the emitter resistor of about 100pF to 220pF is appropriate. Where the crystal shunt capacitance needs to be a specific value, to ensure it being on the correct frequency, the value of the top capacitor needs to be calculated with the value of the emitter shunt capacitor to get the correct load capacity.

Note: Only the RF components are shown in the equivalent circuit (**Fig 8.16**). Similarly the collector is shown as being connected directly to ground, this is correct for the Colpitts because the de-coupling capacitor holds the collector at RF ground potential. The base resistors are assumed to be high values and do not significantly alter the network performance. A worked example serves to illustrate the method.

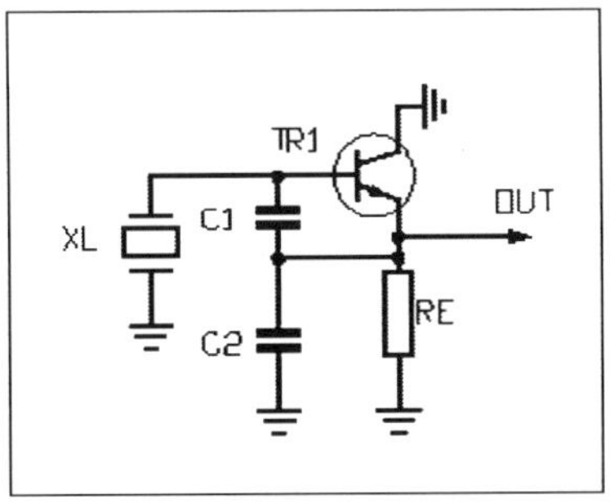

Fig 8.16. Equivalent circuit of Colpitts fundamental oscillator

We will choose a crystal with a required shunt capacity of 50pF and a bipolar transistor with a DC current gain of 100. The frequency is ≈10MHz. For this the value of the emitter resistor is 1kΩ. If we select C2 to be 120pF the value of the top capacitor in the divider network (C1) is given by:

$$C1 = \frac{C2 \times C_{shunt}}{C2 - C_{shunt}}$$

where C2 is the value of the capacitor across the emitter resistor, C1 is the top capacitor in the divider and C_{shunt} is the required crystal capacitance for the correct frequency.

$$\text{Hence } C1 = \frac{120 \times 50}{120 - 50} = 86\text{pF}$$

A standard value of 82pF will suffice. As we have already mentioned, the value of C1 can be modified by reducing it slightly and connecting a variable capacitor across the crystal to tweak it on to the correct frequency. Alternatively C1 can be made variable to achieve the correct crystal load capacitance.

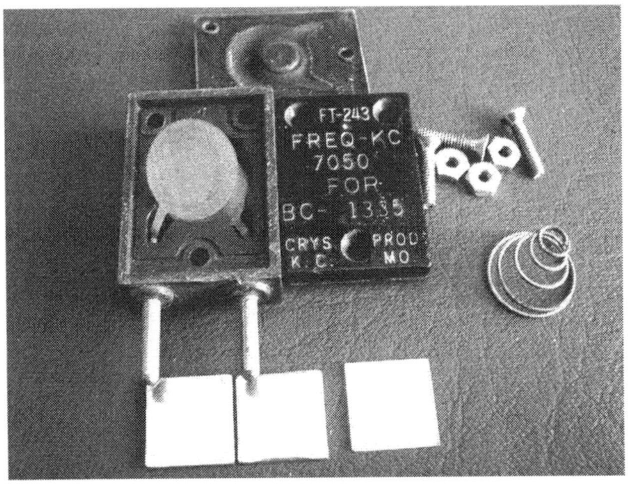

Fig 8.15. FT-243 crystal dismantled. The quartz slice is on the extreme right at the bottom (next to the metal plates that form the connections). The spring provides pressure to force the metal plates in contact with the quartz slice.

L-C COLPITTS VARIABLE FREQUENCY OSCILLATOR

A version using an inductor and capacitor is easy to design but the inductor Q needs to be as high as possible

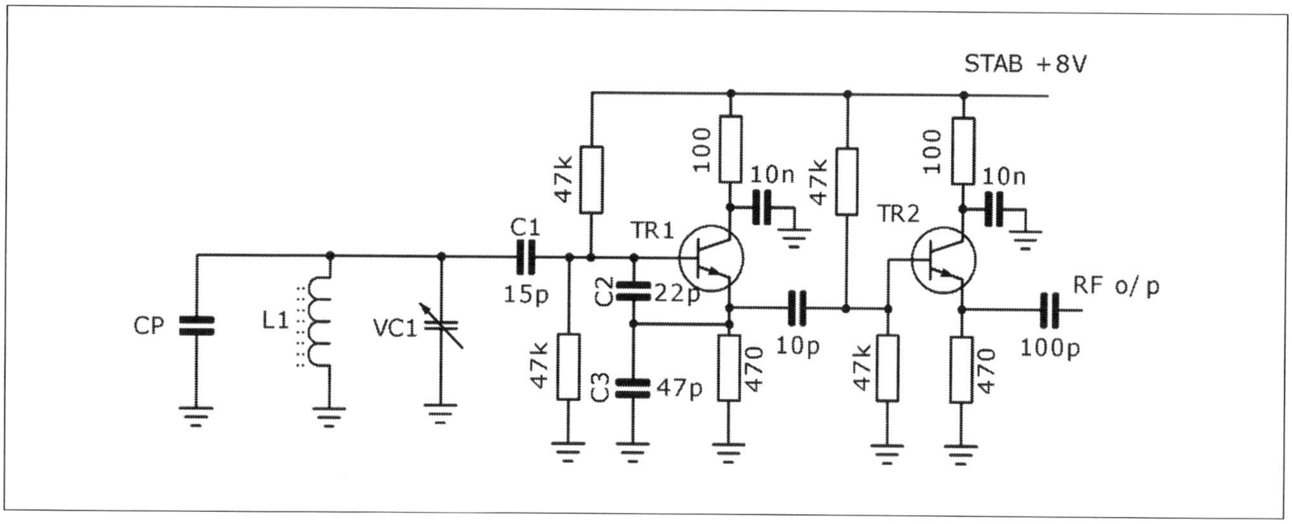

Fig 8.17. Variable Frequency Oscillator

to obtain good noise and frequency stability performance. If an L-C network is substituted for the crystal we can construct a variable frequency oscillator that can be varied over quite a large range by varying the capacitor value. A typical variable frequency oscillator (VFO) is shown in **Fig 8.17**.

The parallel resonant network consists of CP, L1 and VC1. CP (padding capacitor) makes up the bulk of the tuning capacitance and VC1 allows a small variation about the centre frequency. Dimensioning CP and VC1 allows more or less band to be covered. In many cases CP will be made up from several different temperature coefficient capacitors to compensate for the positive temperature coefficient of L1. Often a mixture of negative and positive temperature coefficient capacitors (e.g. N150 and P150 materials) will be needed to eliminate drift with varying ambient temperature and these normally have to be ascertained empirically under test. The VFO range of frequencies will determine the value of the feedback capacitors, the value of L1, CP and VC1 etc.

The frequency stability versus temperature will be worse than a crystal oscillator because of the limited Q possible from a wound inductor, but with care in construction and temperature compensation this can be reduced to acceptable levels. Mechanical rigidity of the components will assist in making a low microphonic VFO. For the best frequency stability with temperature the inductor should be tightly wound on a zero temperature coefficient former (ceramic or PTFE coil former). Although ferrite cores can be used to reduce the number of turns required ferrite has temperature dependence and this degrades the ultimate stability, although for less critical applications they may be suitable. In most cases some temperature compensation will be required. Ferrite wound inductors normally have a strong positive temperature coefficient, placing a negative temperature coefficient capacitor in parallel can often trim this out.

BUTLER OSCILLATOR

This oscillator was originally designed for valves, as were most of the classic oscillator circuits, and normally a dual-triode such as an ECC82 (12AT7) was selected. Today with the availability of semiconductors the Butler is a good choice where good noise performance and high output level is required. The basic Butler oscillator is shown in **Fig 8.18**. This uses a JFET and a BJT as the gain stages. A slightly lower noise version would use two JFETs. Often a dual-JFET can be used where the two transistors are closely matched because they are formed on one die.

The JFET TR1 operates as a grounded gate amplifier with the crystal connected to the source. This is an amplifier with essentially zero phase shift. The drain of TR1 has a wide band RF choke. The signal on the drain is connected to the base of TR2 that also operates as a zero phase shift amplifier. The feedback loop is effectively via an emitter follower whereas the output signal at the collector of TR2 is an amplifier with ≈180° phase shift. The crystal is a series resonant type, either a fundamental or overtone. This oscillator is capable of very low phase noise and can also serve as a multiplier. If the tuned circuit in the collector of TR2 is set to 3-times the crystal frequency it acts as a tripler, although the output amplitude will be lower than a 'straight-through' type. The crystal frequency is trimmed by VC1 in series with the two ground resistors for TR1 and TR2. For high overtone crystals a shunt resistor may be required across XL to suppress lower overtone activity and a series tuned L-C network placed in series with the crystal to enhance the required overtone. Generally up to 5th overtone crystals

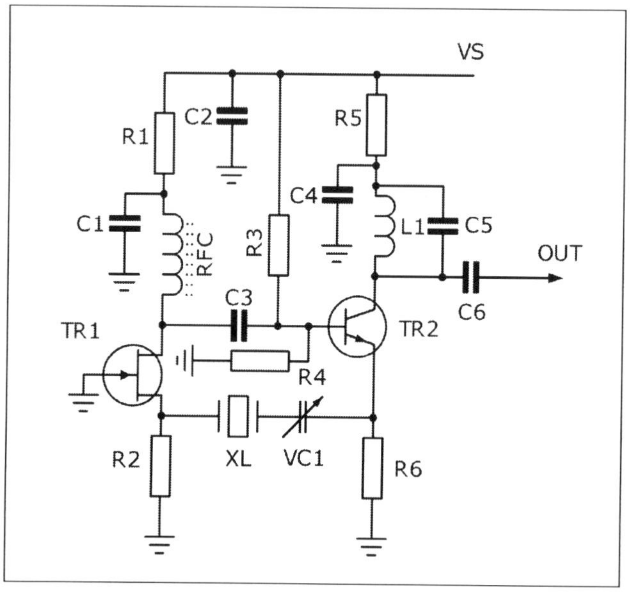

Fig 8.18. Butler oscillator

the source and emitter resistors serve to suppress the unwanted overtones as they are effectively in series across the crystal.

The values for R2 and R6 will normally be of the order of 330 to 680Ω for R2 and 470 to 1kΩ for TR2 assuming a supply voltage of about 8 to 12V. R1 and R5 can normally be chosen to be about 100Ω. Decoupling capacitors C1, C2 and C4 will be determined by the frequency, often 10nF will be satisfactory for use up to about 30MHz. Coupling capacitor C3 should be a fairly low value to eliminate loading on the drain of TR1; often a value of 47pF will suffice.

Note that for use as a frequency multiplier TR2 needs to be a BJT and not a JFET. JFETs have poor multiplier characteristics compared to BJTs and will give lower power output at the multiplied signal (harmonic). If the oscillator is a fundamental type, (no frequency multiplication), the values of the base biasing resistors for TR2 need to be selected to provide approx half the supply rail voltage at the base. If the oscillator is a *harmonic type* the base bias divider will need optimising for the harmonic required, higher harmonics require less base bias voltage to maximise the multiplier efficiency. As with all oscillators the output signal should be fed to a constant load buffer amplifier to eliminate pulling. An emitter or source follower is a good choice for this application.

PIERCE OSCILLATOR

Another popular type for fundamental or overtone operation is the Pierce oscillator. Normally the amplifier stage of the Pierce oscillator generates a phase shift of -180° and so the phase shift network needs to generate a similar shift to get the two signals in phase. The best solution is to use a type of π network. The impedance the series resonant crystal provides is low, but the impedance at the collector / drain of the amplifier is higher, perhaps as high as 5kΩ. This allows a larger output amplitude swing to be developed into a high impedance load.

The main disadvantage of the Pierce oscillator is that the crystal is now being used to simulate an inductor with a very high Q. This means that the frequency of oscillation will be slightly higher than the marked crystal frequency. Ordering the crystal with a slightly lower nominal frequency can compensate this, once the exact factor has been determined. Adding an inductor or capacitor in series with the crystal will not significantly move the frequency and will lead to poorer noise performance.

A practical Pierce oscillator is shown in **Fig 8.20** for crystals of about 20MHz. The output impedance is high and will require buffering in a suitable amplifier. The advantage of the Pierce oscillator is the higher loaded Q

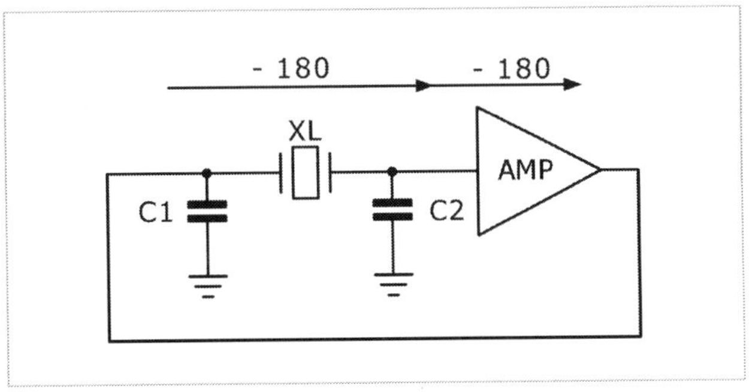

Fig 8.19. Pierce oscillator equivalent circuit

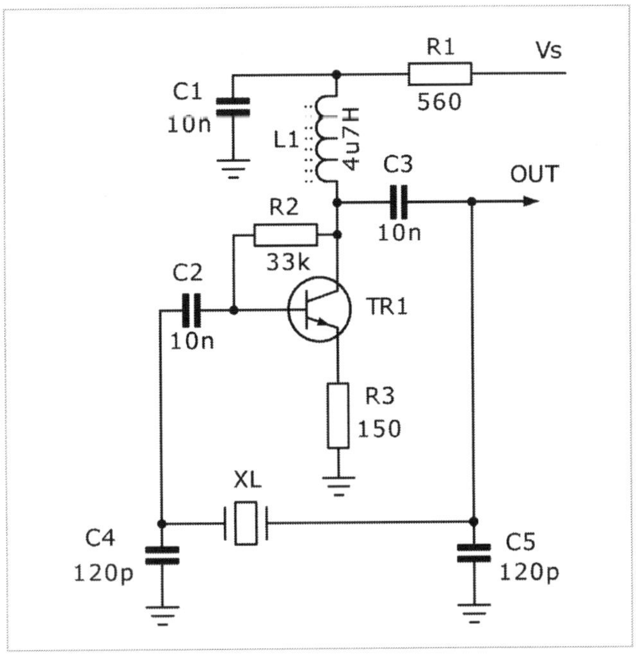

Fig 8.20. Practical Pierce oscillator circuit

the crystal can attain. Often this is about two to three times that of a comparable Colpitts and so the close-in noise performance and stability is somewhat better. However, at large frequency offsets the broad-band noise is determined by the amplifier device noise performance.

HIGH STABILITY CRYSTAL OSCILLATORS

Where the requirement is for exceptional stability, for example in a frequency counter or other measuring instrument, the best option is the use of an oven-controlled crystal oscillator (OCXO). The crystal is cut for a high ambient temperature and maintained within close tolerance by a heating element at about +70°C. The downside of this technique is the higher power consumption and the delay from switch on before the oscillator is on frequency. (During the warm up period the frequency drifts from being above the nominal until the drift is zero). When first powered up the oscillator frequency will be approx 25 to 50ppm above the nominal. Because the heating element consumes several watts of power and it may take as long as 20 minutes to bring the crystal up to the stable temperature these factors often mean an alternative solution needs to be found. However, if exceptional short and long-term frequency stability is required the OCXO is the best option. It is not too difficult to arrange for the oven to be permanently powered even when the equipment is switched off. A typical oven-controlled oscillator can hold the frequency within 0.1ppm over the ambient

Fig 8.21. Cathodeon oven for HC-6/U crystals

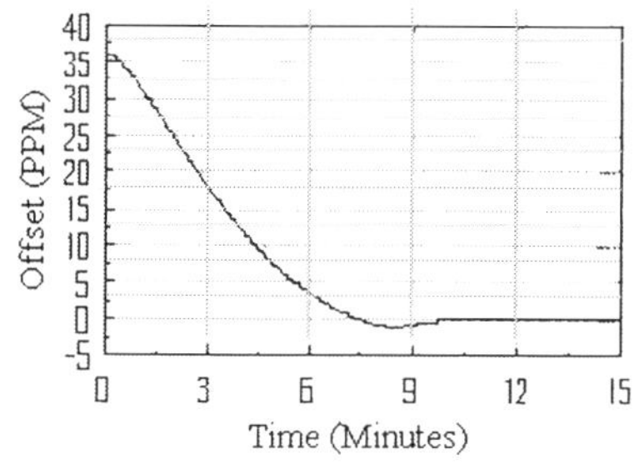

Fig 8.22. Warm up characteristics of an OCXO

temperature range of as much as -50°C to about +70°C. At nominal ambient temperatures the short-term stability is about 1 x 10^{-11} per °C. At very low ambient temperature the heater is on for a large portion of time and hence the energy consumed is high. When the ambient temperature is raised the current consumption falls because the oven needs to supply less energy to maintain the stable temperature.

A typical oven oscillator heater that accepts two HC-6/U crystals and runs at +75°C is shown in **Fig 8.21**. This type can often be found on the surplus market and was used in a number of two-way radio base stations and operates on 6V or 12V AC or DC and made by several manufacturers. I used this oven to stabilise the crystal oscillator for a 4m (70MHz) beacon. Over the ambient temperature range of -10°C to +40°C the maximum drift was 4Hz.

Fig 8.23. Typical surface mount TCXO for 12.8MHz (12mm x 12mm)

An alternative to the OCXO is the *temperature compensated crystal oscillator* (TCXO). These are normally packaged oscillators that have built-in temperature compensating components. These contain a temperature sensor that develops a variable DC voltage proportional to the ambient temperature. This is applied to a variable capacitance diode (varicap) across the crystal to adjust its frequency. These consume little power compared with an OCXO heater but suffer from poorer phase noise due to the varicap diode lowering the Q of the crystal. TCXOs do not suffer from the warm up drift of an OCXO but have inferior stability. However, a frequency stability of 2.5ppm is normally acceptable for most applications, which most TCXOs can achieve at a reasonable price. At 1MHz the total drift will be of the order of 2.5Hz over the ambient temperature range of -30°C to +75°C. At 100MHz it is 250Hz and at 1GHz it is 2.5kHz for the same conditions, which is normally adequate for amateur purposes.

Today there are some newer types that use a Peltier junction to control the temperature. A Peltier junction is a semiconductor that acts somewhat like a diode. When connected one way round across a DC supply they generate heat. When connected in the reverse polarity they act as heat absorbers and cause the temperature to be lowered. By switching the polarity between the two states they can act as either a heater or a cooler and so can hold a very stable temperature over a wide ambient range. However, like the OCXO heater, they consume a fair amount of power and are quite expensive.

LOW NOISE OSCILLATOR SUPPLY

The stabilised supply feeding the oscillator and buffer stage needs special attention. This needs to be as stable and as low noise as possible, do not even consider using zener diodes for this application they are simply too poor. Zener diodes generate an incredible amount of wide band noise, which if allowed to enter the oscillator will cause very poor phase noise performance because the noise riding on the DC supply amplitude modulates the carrier. Although the carrier is only amplitude modulated when the signal then passes through a non-linear multiplier stage the process of *AM to PM conversion* occurs which causes the phase noise performance to degrade.

The best choice for a stable supply is a three-terminal regulator with adequate decoupling. Some modern three-terminal regulators offer noise performance figures as little as 1µV over a broad bandwidth. In some cases an additional noise filter is required to clean up the DC supply rail. The use of an emitter follower with a large value capacitor connected in shunt with the base to ground can reduce the noise voltage by 100 times or more.

With a 2N2222 transistor, which has a current gain of about 400 at small currents, the effective capacitance appearing across the emitter to ground is the product of the current gain multiplied by the value of C5.

In **Fig 8.24** the effective shunt output capacitance is nearly 19,000µF. Of course we lose approx 0.6V across the pass transistor TR1 but this is not of too much concern as the pass transistor is only powering the crystal oscillator and first buffer amplifier and the total current will only be about 20mA or so. (If you are concerned about this then simply add a silicon diode in series with the common lead of the LM78L08 regulator to raise the output by 0.6V or so). The remaining stages are fed directly from the output of another 8V regulator, perhaps the 1A version such as the LM7808.

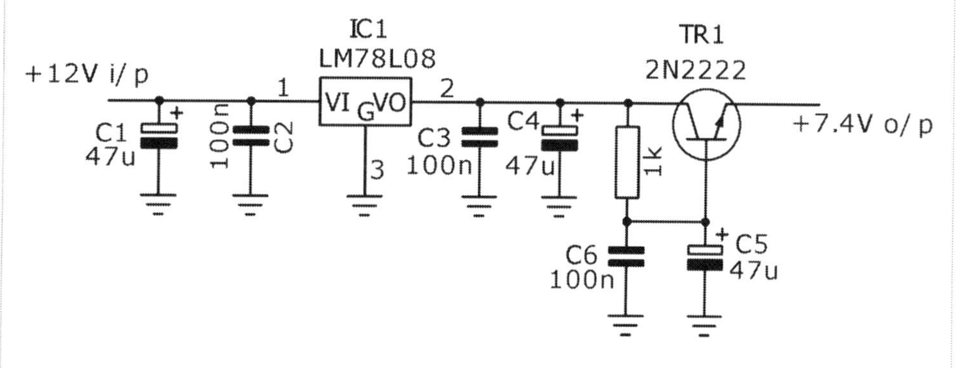

Fig 8.24. Additional noise filtering for oscillator supply line

Chapter 9

Understanding dual gate mosfets

MANY amateurs use the popular dual gate mosfets as RF amplifiers for the VHF bands. They are a good choice as they offer substantial gain and low noise figures and are relatively inexpensive. However, they are not optimum for many published circuits that I have seen. This is because the designer does not fully understand the various parameters and how they affect the gain and noise figure.

Very often when used as a masthead LNA the gain is too much and this causes problems with IMD generation in the following receiver. In order to reduce the IMD generation it is necessary to reduce the gain of the LNA. Many amateurs do this by altering the gate-2 bias voltage. Although this lowers the gain the required amount what they do not realise is this also severely degrades the noise figure. So in fact they could get better results by removing the LNA and replacing it with a lower gain part with a worse noise figure.

To understand how a Dual Gate MosFet works we need to study the data presented by the manufacturer. Unfortunately not all manufactures give the complete data. For example companies such as Vishay (originally Telefunken) and Philips omit some vital data, whereas Siemens / Infineon normally supplies the full data required. I recently had some correspondence with a semiconductor manufacturer on their range of Dual Gate MosFets. The noise figure data showed the devices noise figure increasing with decreasing frequency which is certainly not the normal trend. It turned out that the test circuits they used for frequencies below 100MHz had such high resistive components that the input circuit Q was severely degraded. In fact the noise figure they quoted at 45MHz was 1dB worse than that measured at 200MHz. I measured the same device in a narrow band high Q circuit and the correct figure was some 2dB lower at 30MHz than 200MHz.

Let us start by looking at a typical Dual Gate MosFet data. I have selected a modern device, the BF998, which is an SMD part and capable of working up to 800MHz. I have extracted the data from the Siemens / Infineon data sheet (used with their permission). Firstly, we will look at the short-form data, which is normally in tabular format for easy inspection. From this we can make several informed assumptions as to whether the device will perform the function we desire.

There are normally two sets of data, the DC characteristics and the AC characteristics. For designing at RF we need both, but the AC set is the more important.

The AC characteristics of the BF998 is reproduced in **Fig 9.1**. The first few items are not of much interest, but from halfway down the gain and noise figure are given at two test frequencies of 200MHz and 800MHz, along with the AGC range. These are the factors that concern us. If 144MHz were our design frequency we could estimate that the gain would be a little higher than the 200MHz data and probably over 30dB should be possible. The same estimate can be made for the noise figure although this is very dependent on the circuit Q values attained. We will see a little later that the test circuit used to derive the measured values is not optimum and better results can be achieved in a narrow band circuit.

Electrical Characteristics

Parameter	Symbol	Values			Unit
		min.	typ.	max.	
AC characteristics					
Forward transconductance V_{DS} = 8 V, I_D = 10 mA, V_{G2S} = 4 V	g_{fs}	-	24	-	mS
Gate 1 input capacitance V_{DS} = 8 V, I_D = 10 mA, V_{G2S} = 4 V, f = 1 MHz	C_{g1ss}	-	2.1	2.5	pF
Gate 2 input capacitance V_{DS} = 8 V, I_D = 10 mA, V_{G2S} = 4 V, f = 1 MHz	C_{g2ss}	-	1.2	-	
Feedback capacitance V_{DS} = 8 V, I_D = 10 mA, V_{G2S} = 4 V, f = 1 MHz	C_{dg1}	-	25	-	fF
Output capacitance V_{DS} = 8 V, I_D = 10 mA, V_{G2S} = 4 V, f = 1 MHz	C_{dss}	-	1.1	-	pF
Power gain V_{DS} = 8 V, I_D = 10 mA, V_{G2S} = 4 V, f = 200 MHz	G_{ps}	-	28	-	dB
Power gain V_{DS} = 8 V, I_D = 10 mA, V_{G2S} = 4 V, f = 800 MHz	G_{ps}	-	20	-	
Noise figure V_{DS} = 8 V, I_D = 10 mA, V_{G2S} = 4 V, f = 200 MHz	F	-	0.6	-	
Noise figure V_{DS} = 8 V, I_D = 10 mA, V_{G2S} = 4 V, f = 800 MHz	F	-	1	-	
Gain control range V_{DS} = 8 V, VG2S = 4 ... -2V, f = 800 MHz	ΔG_{ps}	40	-	-	

Fig 9.1. Data for the BF998 Dual Gate MosFet. ***(Reproduced with permission of Infineon Technologies AG)***

We should firstly note that the manufacturer chose a supply voltage of 8V and a drain current of 10mA for the gain and noise figure measurements. The BF998 has a limited safe supply voltage; Infineon recommends it not to exceed 15V maximum. 8V is a nice voltage to work with as for a mobile radio we can easily obtain it from 12V using a 3-terminal IC regulator such as a LM7808 or via a dropper resistor from a higher voltage.

The other thing to note is that the gate-2 bias voltage is specified as +4V. We will see the significance of this a little later. But one thing we can assume from this short form data is that +4V gate-2 bias voltage, 8V drain-source voltage and 10mA-drain current are the optimum for gain and noise figure. Deviating away from these values will lead to reduced performance. The BF998 is quite

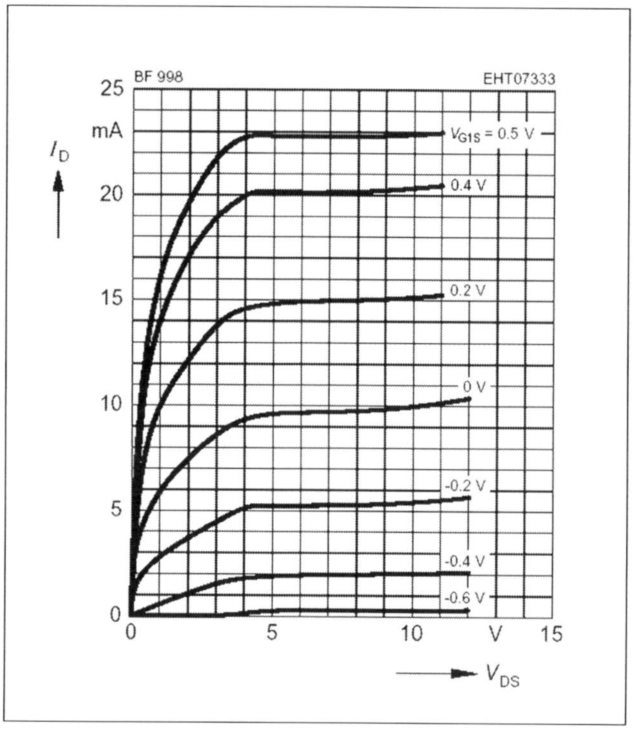

Fig 9.2. Drain current versus gate 1 voltage. ***(Repro-duced with permission of Infineon Technologies AG)***

a good device for 2m as it can attain a noise figure comparable with more expensive devices and sub 1dB is very simple to achieve.

Now we will look at a graph of drain current versus drain-source voltage to see how it effects performance. As we saw the optimum drain current for gain and noise figure is 10mA. **Fig 9.2** shows how the drain current varies as the drain-source voltage is varied with the bias voltage on gate-1. We need to aim for a drain current of ≈10mA when the drain-source voltage is 8V. Inspection of Fig 9.2 shows this occurs when the gate-1 voltage is approx 0V. This means that gate-1 can be connected DC-wise to the circuit ground and no source resistor is required. This eliminates the potentially lossy source decoupling capacitor, which can cause instability problems at high frequency. Hence, the source terminal can be grounded directly. This also saves the cost of two components.

If we now look at **Fig 9.3** we can see how varying the gate-2 bias voltage (V_{G2S}) alters the gain. This shows the case where gate-1 is tied to ground (V_{G1S} = 0V).

With gate-2 at +4V the gain is approx 28dB, a power gain ratio of about 700. When gate-2 voltage is reduced the gain falls as expected. When gate-2 voltage is reduced below +1V the gain starts to fall very rapidly. Between +1V and +4V the variation is only a few dB. But when gate-2 voltage falls to 0V the gain has fallen to approx 15dB. This is a gain reduction of about 13dB (20x). To achieve the full AGC range the voltage on gate-2 needs to be brought negative. When gate-2 is at -0.5V the gain has fallen by 58dB (≈900x) from the full gain figure. Hence only 0.5V change caused a gain reduction of approx 45dB. So we can see that for maximum AGC range we need to use a control circuit that operates from a split supply rail and can swing positive up to +4V and negative to at least -0.5V.

This sounds excellent but we haven't looked at how the AGC alters the noise figure. For this we need to look at **Fig 9.4** for the same DC conditions.

Fig 9.4 shows how the noise figure varies as gate-2 bias is varied. For minimum noise figure gate-2 needs to be greater than approx +2V and varying the gate-2 voltage between +2V and +4V has virtually no effect. But when gate-2 is reduced to +1V the noise figure has started to degrade, it has risen from 0.6dB to almost 1dB. When gate-2 is lowered to +0.5V the noise figure has severely degraded. Plotting the same bias point on Fig 9.3 shows this is a gain of 23dB, or a gain reduction of 5dB. If we need to reduce the LNA gain to, say, 15dB to stop IMD products this requires a gate-2 bias voltage of 0V and the noise figure is now approx10dB by extrapolating the curve in Fig 9.4. Hence, for optimum noise figure we cannot

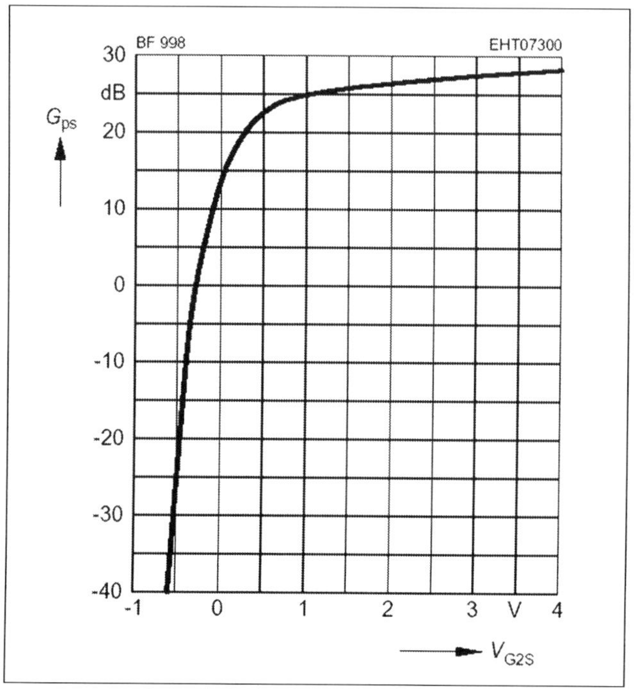

Fig 9.3. Gain versus gate 2 voltage. ***(Reproduced with permission of Infineon Technologies AG)***

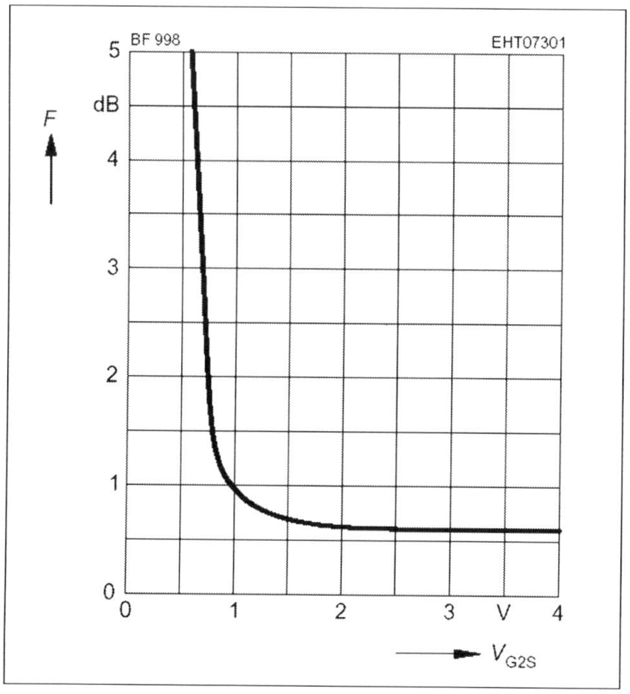

Fig 9.4. Noise figure variation versus gate 2 voltage. ***(Reproduced with permission of Infineon Technologies AG)***

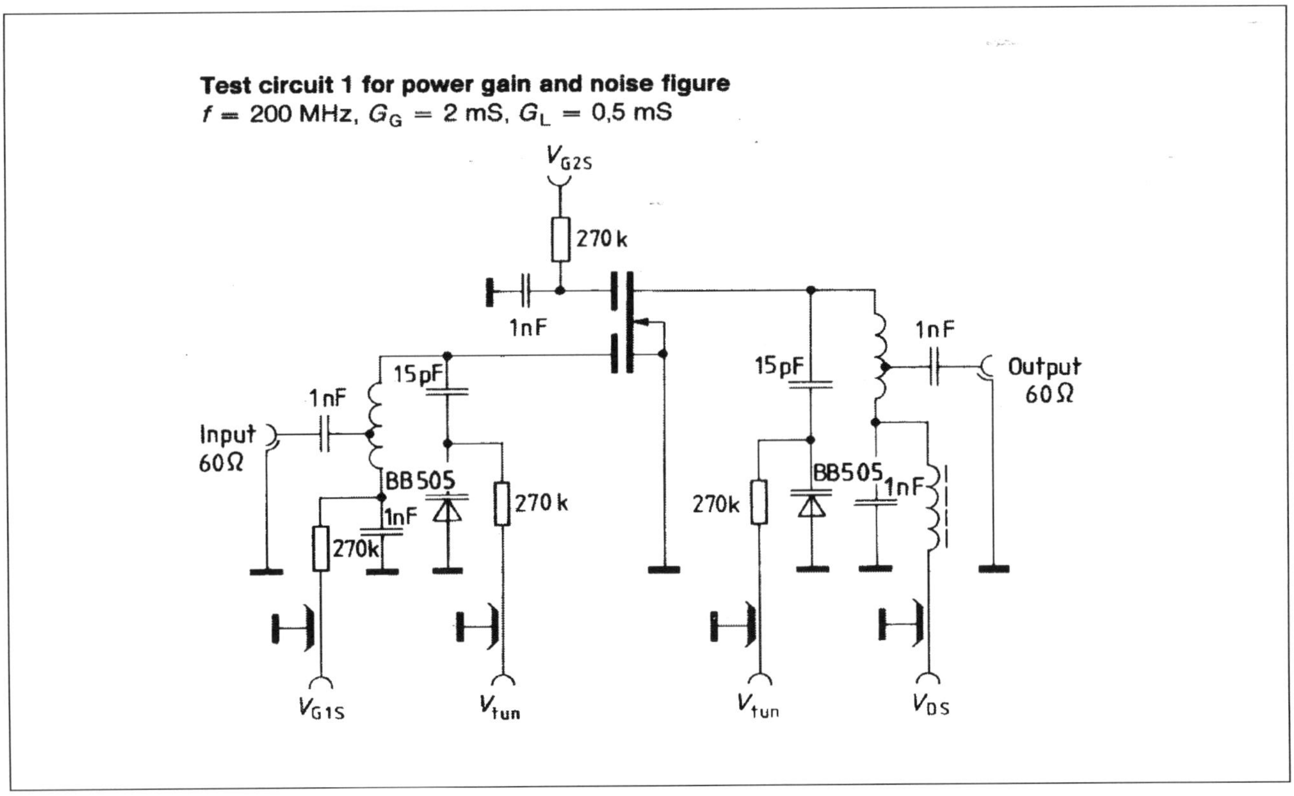

Fig 9.5. 200MHz test circuit. ***(Reproduced with permission of Infineon Technologies AG)***

lower gate-2 bias to below +1V. At +1V the gain is 25dB or only 3dB less than the possible maximum. The BF998 is not the only Dual Gate MosFet to exhibit this effect; all other similar MosFets show the same trend.

The only way to get around this problem is to place an attenuator after the LNA and before the main receiver to lower the overall gain. This will not degrade the system noise figure by any significant amount unless the gain is reduced so that the overall gain, including the attenuator loss, falls to below 6dB. The way the writer achieves this is to use very lossy coax cable between the masthead LNA and the receiver. As the LNA is likely to be some distance from the main receiver this is a low cost solution. By using cable such as RG-58/U rather than lower loss types this solves the problem.

Finally let us look at the test circuit that Siemens/Infineon used to measure the gain and noise figure at 200MHz. This is shown in **Fig 9.5**.

The 200MHz test circuit uses the BF998 with the source connected directly to ground and the bias voltage for gate-1 and gate-2 fed in via high value resistors acting as RF chokes. The input signal is tapped down an inductor and connected to gate-1. The circuit centre frequency is varied by two varicap diodes, one across the gate-1 inductor and the other tuning the drain circuit. The problem with this arrangement is the serious 'de-Q-ing' of the input inductor by two effects.

Firstly, any resistor placed across the gate-1 to ground tuned circuit will lower the Q and introduce additional thermal noise because of the kTB effect. The value of the resistor has almost no effect on the kTB-generated noise; a 0.1Ω resistor will generate as much thermal noise as one of 1MΩ at the same temperature. However, what the resistor does affect is the Q of the inductor. For a low value of shunt resistance the Q will be lower than a high value resistance. The Q of the input inductor likewise needs to be high to reduce the series resistive losses. Some designs show a resistor in shunt with the input inductor across gate-1 to ground, this is a serious mistake and will severely degrade the noise figure because of the kTB effect. The varicap likewise has a limited Q; it can be much lower than the potential inductor Q when the reverse bias is low for tuning to a lower frequency. (In effect the two shunt resistances are in parallel and the Q of the input network is reduced.) These two effects add to degrade the overall noise figure. (The effective shunt resistor values can be found by using the standard series to parallel conversion formulae).

To illustrate the problem we can examine the manufacturer's data on the BB505 varicap diode used in the test circuit input and output networks. The series resistance (Rs) is given as 0.7Ω worst case for the BB505B version and 1Ω for the BB505G version. At the test frequency of 200MHz the Q values can be ascertained. The capacitance range of the BB505 is approx 18pF to 2.5pF, with an applied bias voltage of 1V and 28V respectively.

At 1V bias the capacity is 17.5pF and Xc is 45.5Ω at 200MHz. With Rs at the maximum value of 0.7Ω the Q is only 65. At mid range of 10pF the value of Xc is 79.5Ω at 200MHz and the Q is 113. At the minimum capacity of 2.5pF the value of Xc is 318Ω and the resulting Q is now 454. Even a Q of ≈450 is poor compared with a conventional fixed capacitor, which can be expected to be at least 1000. The Q of the inductor can be expected to be of the order of 150 to 250, if correctly wound.

From this we can ascertain that when the amplifier is tuned to a low frequency, that is approaching maximum capacity of the BB505 varicap diode, the network Q will be worse than when tuned to a higher frequency. This degrades the potential noise figure. In a television tuner (the application for which the BF998 is primarily designed), the lower Q translates to a wider bandwidth. As the composite vision and sound transmission bandwidth is wide, this is not a problem and in fact is an advantage. But where the ultimate noise figure is required in a narrow band circuit, such as an amateur application for an LNA, the degradation of the network Q limits the potential noise figure.

In order to attain the lowest noise figure the impedance presented by the parallel tuned network to gate-1 needs to be in excess of 10kΩ, using a poor Q inductor or tuning capacitor will lower the impedance too much. Hence, the type of inductor and capacitor need to be carefully selected. Generally enameled copper wire or silver copper plated wire will be the best choice at 144MHz. For 50MHz an inductor wound on a ferrite core can attain Q values of approx 150 to 200 depending on the ferrite type used. At 144MHz with an enameled copper wire inductor a Q value of 200 to 300 is possible if wound correctly.

The reason Siemens / Infineon chose this test circuit is because it was also used by the earlier generation devices (BF992/4/6 etc). The BF998 is primarily intended for VHF and UHF TV tuners where the circuit bandwidth needs to be at least 6MHz to fully encompass the composite vision and sound signal. From measurement I made at 50MHz and 144MHz using the BF998 in a narrow-band LNA (approx 2MHz BW at 50MHz and 4MHz BW at 144MHz) the noise figure stated by Siemens / Infineon of 0.6dB typical at 200MHz can be improved by about 0.1 to 0.2dB. The ultimate limitation is of course the circuit Q attained. These sorts of noise figure one normally associates with expensive GaAsFet types. For 144MHz it simple to get below 1dB noise figure and unless you intend using the LNA for EME there is no need to strive for less than about 1.5dB because of the high sky noise at low frequencies. It is only at 432MHz and above where a noise figure of less than 1dB can be useful, even here 0.5dB is the lowest required for 432MHz for EME. The BF998 in a suitable circuit can achieve close to this at low cost.

V_D = 8V I_D = 10mA V_{G2S} = 4V

Freq	S_{11}		S_{21}		S_{12}		S_{22}	
MHz	Lin mag	Angle	Lin mag	Angle	Lin mag	Angle	Lin mag	Angle
100	-0.02	-8.3	7.84	168.5	-55.67	83.0	-0.04	-3.7
200	-0.11	-16.1	7.70	156.6	-50.01	76.4	-0.09	-7.4
300	-0.35	-24.0	7.49	144.8	-47.20	70.3	-0.16	-10.8
400	-0.62	-31.6	7.21	133.6	-45.60	65.1	-0.23	-14.3
500	-0.97	-39.2	6.93	122.5	-44.88	60.0	-0.31	-17.9
600	-1.39	-46.4	6.59	111.9	-45.25	54.5	-0.42	-20.9
700	-1.76	-53.2	6.27	101.9	-46.51	57.4	-0.48	-24.1
800	-2.25	-60.3	5.97	92.1	-47.19	61.4	-0.55	-27.3
900	-2.67	-67.1	5.71	82.8	-49.28	76.0	-0.58	-30.6
1000	-3.16	-74.1	5.46	73.3	-48.99	107.1	-0.60	-33.8
1100	-3.72	-81.1	5.07	63.3	-48.03	123.3	-0.73	-37.2
1200	-4.30	-88.0	4.85	54.6	-45.15	147.6	-0.73	-40.6
1300	-4.87	-94.4	4.63	45.4	-42.46	157.0	-0.73	-44.3

Fig 9.6. S parameter data for the BF998 dual gate mosfet. ***(Reproduced with permission of Infineon Technologies AG)***

Chapter 10

Solid-state power amplifiers

ALTHOUGH the traditional method of generating high levels of RF power in earlier days was with valves, with the advent of affordable devices in the 1970s many manufacturers now use semiconductors. The advantage is that they are able to operate from quite low voltages and this suits equipment designed for fitting into vehicles. Prior to this the equipment needed to include a DC-DC converter to generate the high voltage for the valves. Today it is possible to construct quite high power amplifiers for HF, VHF and UHF to operate on 12V DC supplies. However, the power output and linearity is more easily achieved if a higher voltage, such as 28V or higher, is used.

For the best linearity, and hence lowest IMD products, when operated as an amplifier for SSB, some manufacturers make devices for a supply voltage of 50V DC or even 100V DC. This means that operation in a vehicle is not possible unless a DC-DC converter is used. This somewhat precludes their general acceptance, but these types find wide use in fixed station equipment operating from the 230V AC mains supply.

The general advantage of using a high supply voltage is the higher impedance the collector or drain offers to the matching network. At low DC supply voltage the impedance becomes very low and hence high values of RF current are circulating in the networks which, with the component associated series resistive loss, dissipate a fair bit of power. This makes it difficult to obtain a good match, especially when wide band operation is required.

One of the advantages often put forward in favour of solid state power amplifiers is that they are fixed tuned and therefore the operator does not have to perform any 'tuning-up'. Also they can be made to cover a wide bandwidth, for example 2 - 30MHz, so suiting an HF power amplifier. However, this can be a distinct *dis*advantage if the antenna presents a reactive load to the amplifier and often requires the use of an antenna tuning unit (ATU) to allow a good transfer of power.

LOAD LINE

As with the valve amplifier the transistor presents a resistive part to the matching network required to transform from the low impedance at the collector (or drain if the device is a mosfet) up to the antenna impedance of 50Ω. The supply voltage and current flowing in the collector or drain determine the value of the load. Transistors have a parameter known as the 'saturation voltage'. Valves also have a saturation voltage, but this is typically a small percentage of the supply voltage, and can generally be ignored if the anode voltage is greater than about 400V. But unlike valves the transistor saturation voltage is a relatively large percentage of the operating voltage.

A typical 12V bipolar transistor of NPN structure has a collector-emitter saturation voltage of approx 1 to 2V. This means that when the collector current is driven to maximum the collector-emitter has a forward voltage drop, just like a diode. This voltage does no useful work and it needs to be subtracted from the supply voltage to obtain the effective voltage the collector can swing over.

The formula to derive the optimum load resistance is:

$$R_L = \frac{(V_s - V_{ce\text{-}sat})^2}{2 \times P_o}$$

where:
R_L is the optimum load in ohms
V_s is the supply voltage in volts
$V_{ce\text{-}sat}$ is the device saturation voltage
P_o is the required output power in watts
An amplifier for 50W output using a NPN bipolar transistor operating on 12V which has a saturation voltage of 1.5V requires a load resistance of:

$$R_L = \frac{(12 - 1.5)^2}{2 \times 50} = 1.1\Omega$$

By comparison an amplifier using a 28V transistor of similar parameters to obtain the same 50W output requires ≈7Ω. Because we have to convert from the collector impedance to 50Ω the 28V device is much easier to work with and will have lower network losses, which equates to more power into the antenna.

The saturation voltage is determined by the manufacturer with a test fixture that pulses the collector current. From the DC supply is a resistor which sets the peak collector or drain current value. A low frequency and low duty cycle square wave signal is used to drive the base or gate and a cathode ray oscilloscope monitors the collector or drain voltage.

WIDE-BAND AND NARROW-BAND MATCHING

If the amplifier needs to cover a wide range of frequencies a type of wide-band matching is required. However, if the amplifier only needs to cater for one band, such as 144MHz, a narrow-band matching scheme is easier to design and will have fewer losses associated with it. This equates to more power output. The second advantage of narrow band matching is that we can obtain more gain than when the amplifier needs to cover a wide band. This is the effect of the 'gain-bandwidth' product of all devices.

If the device bandwidth is stretched over a wide band the average gain we can achieve falls quite a bit. For example, manufacturer's when they give details in data sheets normally list gain and power output for one or more frequencies, such as 175MHz and 50MHz if the device is intended to operate up to 175MHz. The measurement is made in a narrow band test circuit, normally not more than 15% bandwidth and is given to indicate the device potential.

A typical device rated at 50W output might offer a gain of 12dB at 175MHz and as much as 16dB at 50MHz. But if we needed to cover between 50MHz to 150MHz the average gain would be approx 10dB across this range. Part of the reason is due to mismatch losses in a wide band network because in some part of the band the collector is matched better than at other parts. In general the matching will be closer to a narrow band match at the frequency where the device has least gain, that is the higher frequencies. The majority is due to mismatch losses in the amplifier-input network. To lower the gain at 50MHz means we have to deliberately mismatch the input so that we do not achieve its ultimate gain, this is what 'broad-banding' an amplifier does. Effectively the input matching takes the form of a high pass filter to reduce the drive level at low frequencies. The disadvantage with wide band networks is maintaining a low input VSWR to the driver across the full band. Often the VSWR will vary up to as much as 3:1 in some parts of the band.

A transistor wide-band amplifier to cover the HF spectrum of 2 to 30MHz is a very wide-band device and hence the device chosen needs to possess much more gain to obtain a flat frequency response across this range. At 2MHz the device might be capable of 23dB of gain but at 30MHz only 13dB, a difference of 10dB. Using resistive loading on the base (or gate) can flatten the gain response and give a reasonably low VSWR across the band. These resistors, however, dissipate some of the drive power.

CHARACTERISTICS OF TRANSISTORS

It is as well to realise that a typical power transistor for linear RF amplification is not as linear as a similarly rated valve. The power out versus power in curves often show a distinct 'humped-back' trace. This is due to the parasitic components affecting the transistor. These are predominantly the junction capacity and the lead inductance. The junction capacity acts like a varactor diode (**Fig 10.1**) at a large RF voltage and causes the impedance to alter as the power level varies. This degrades the matching to the circuit.

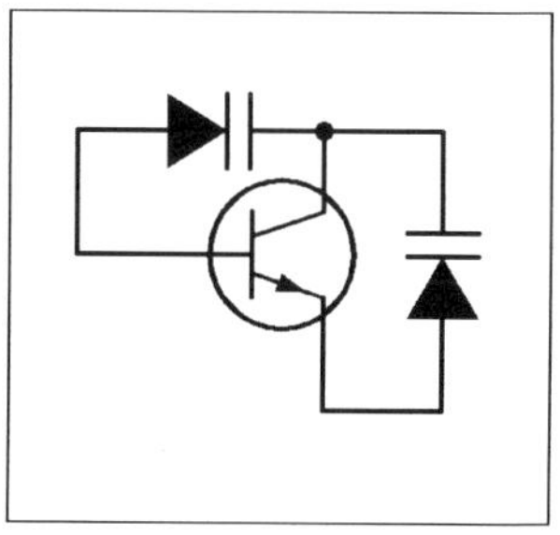

Fig 10.1. Inherent Varactor diodes in an RF power transistor

As an example of how the gain curve varies we can take a look at a typical 12V RF transistor to see the effect. **Fig 10.2** shows the curve for a 12V device rated at a nominal 5W output.

As we can see the curve never really approaches

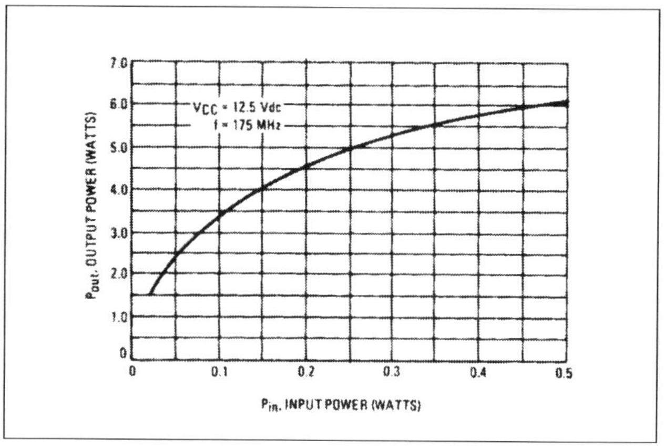

Fig 10.2. Power out versus power in for 12V 5W transistor

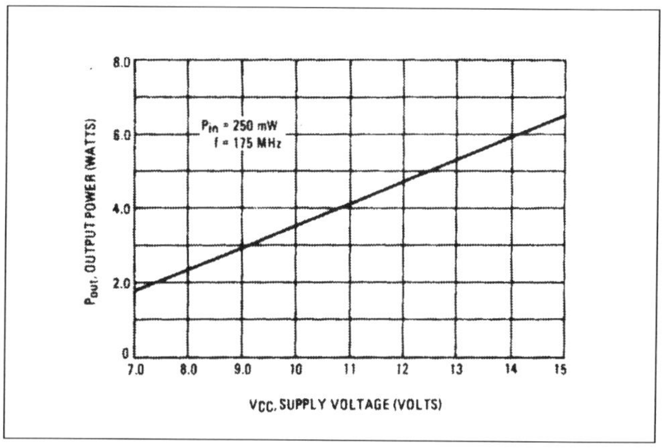

Fig 10.3. Power output versus supply voltage with constant drive level

a straight-line condition. At the rated output power of 5W it requires 0.25W of input drive, this is a gain of 5/0.25 = 20 which in logarithmic terms is 13dB. At an input power of 0.5W the transistor delivers 6.1W, this is a gain of 6.1/0.5 = 12.2 or 10.8dB. At the lower end an output power of 3.4W requires an input drive of 0.1W, a gain of 34 or 15.3dB. A look at the power curve for different operating supply voltage is also enlightening. This is shown in **Fig 10.3**.

The gain is very dependent on the supply voltage. With a constant input drive power of 250mW the power output varies between 1.9W when the supply voltage is 7V and up to 6.5W when the supply voltage is 15V. Hence, the supply voltage will need to be well stabilised so as not to affect the linearity. A technique used in very

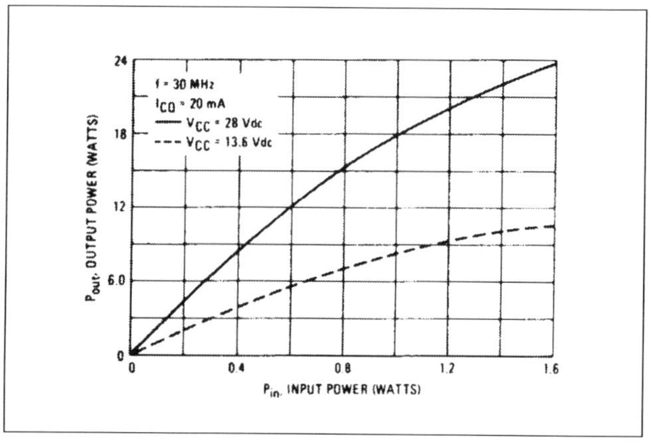

Fig 10.4. Power curve for a 28V transistor

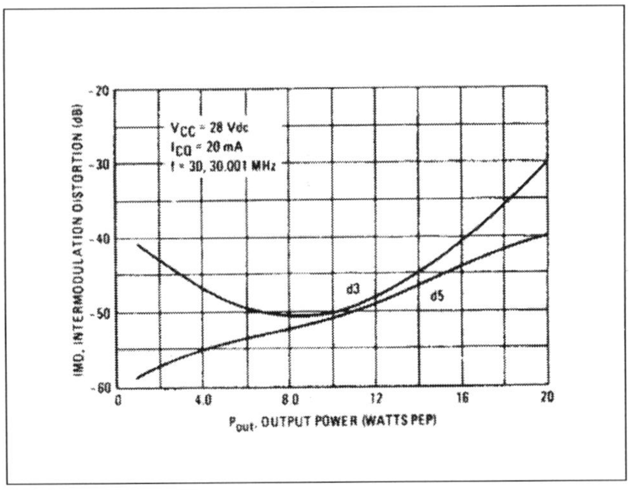

Fig 10.5. IMD versus output power

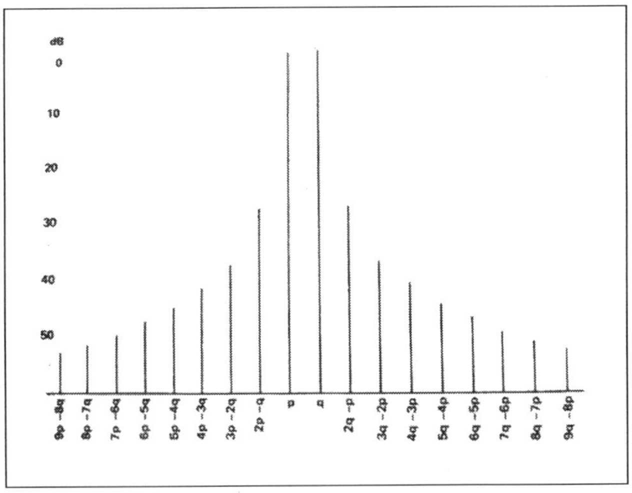

Fig 10.6. Typical IMD performance of a linear transistor

high linearity amplifiers is deliberately to modulate the DC supply voltage in sympathy with the input drive power to counter this non-linearity. This allows lower cost devices to be used. If the curve in **Fig 10.2** where used with a higher supply voltage for the higher power output conditions the curve can be pulled more into a straight line. Using this technique improvement in the IMD figures as much as 20dB is possible. By comparison compare this to a 28V device shown in **Fig 10.4**.

Here we can see a much straighter line for the 28V transistor. Of interest is that the manufacturer has also plotted the power curve when the device is operated at 13.6V. Of particular interest is how the transistor will amplify a composite two-tone signal when IMD measurements are being made. This can be seen in **Fig 10.5** for a 10W PEP linear transistor designed for use as a driver in a mobile HF SSB transmitter operating from 12V.

The curve of 3rd order IMD (d3) shows that at very low output power it is higher than at the rated power of 10W. As the power output increases the IMD fall to a lower level and then increase again with rising output power beyond the device rating. This is in direct contrast to a valve where the IMD level starts out very low and gradually increases with increasing output power. The 5th order curve is similar to that which a valve would achieve. The cause of the IMD variation is due to mismatch of the collector for varying power output. The collector is only correctly matched at the 10W output level.

So it should now be fairly obvious that if we require low IMD we need to be using 28V devices and not 12V devices. A further benefit is the higher gain the 28V devices offer and the reduction in supply current. An amplifier operating on 12V and drawing an input power of 160W requires a DC supply of 13.3A. An amplifier using 28V will only require 6A. Because the efficiency of 28V transistors is normally a bit higher than 12V types the current required for a certain output power will be less.

A typical IMD spectrum for a linear transistor is shown in **Fig 10.6**.

> The old Kenwood TS-700 2m transceiver series made in the 1970s used a 20W PEP 28V linear transistor in the power amplifier stage and a type of switching supply to generate the supply from the raw 13.8V input. This amplifier exhibits very low IMD products at the 15W output level. Unfortunately the TS-700 when manufactured only used a 20V supply for the power amplifier, but by changing the switch-mode supply to a modern type allows up to 28V. This improves the gain and IMD by several dB.
>
> Most 28V linear transistors give more power and better linearity at voltages above 28V; some are best at about 35V. The vehicle battery voltage range for a 24V vehicle range between 21V and 32V in normal operation.

LINEAR AND NON-LINEAR

The difference between a linear and non-linear transistor is the *die area*. Linear transistors have approximately double the die area to distribute the current flowing over more transistor cells. When a linear power transistor is designed, the manufacturers normally use a generic small signal die, such as a BFY90 type which has good high frequency characteristics, and then connect several dies in parallel to obtain the required collector current. To ensure equal current flowing in each cell the emitters have a low value resistor inserted in series to the common emitter pin.

The emitter current sharing resistors are made from the bond wires used to attach each cell to the common emitter tie point in the package. Normally the bond wire is a low resistance material such as gold or platinum, but in the linear transistor a high resistance nichrome wire is used. The value of the *emitter ballasting* resistance cannot be too high so the values we find are in the order of a few milliohms. A typical 10W linear transistor might consist of 64 die-cells connected in parallel. This also helps to spread the heat dissipating parts over a larger area and so lowers the junction temperature. A 100W transistor can use as many as 1000 cells. The way to tell how many cells are being used is to measure the capacitance.

The disadvantage of this technique is that the overall gain is lower than a Class C transistor because the emitter resistors are a form of negative feedback, lowering the gain, but widening the bandwidth. A further step is the so-called *controlled Q* transistor, often known as the *j-zero* type. In this the manufacturers use additional bond wires on the base input with an integrated MOS capacitor to form a T network which brings the base impedance closer to a resistive part.

A further disadvantage of the linear transistor is the efficiency suffers because of the power dissipated in the emitter ballasting resistors. For a Class C transistor efficiency figures of 55 to 70% are quite practical, a linear transistor is typically only 45 to 60% at best. The efficiency is also frequency dependant, low frequencies have better efficiencies than higher frequencies. The downside of the lower efficiency is the greater amount of heat generated that needs to be dissipated in a heatsink. This often means a larger heatsink and/or cooling fans for high power amplifiers. For very high power often water-cooling is required.

NON-LINEAR TRANSISTORS

All the transistors we have looked at so far are ones especially designed for linear operation. However, most amateurs and some commercial manufacturers of so-called 'linear amplifiers' do not use these types. They use transistors designed for Class C operation in FM transmitters where linearity is not important. If the amplifier is used only for FM or CW this is of no concern. But some manufacturers specify the amplifier for SSB operation as well. The construction of a transistor intended for FM operation is quite different to a transistor designed for linear operation and there is no effective way to use the non-linear transistor in a truly linear mode. The reason the manufacturers use Class C devices is purely cost; linear transistors cost much more. One of the methods used to try to linearise the transistor is to apply an abnormally high amount of idle current. This is wasted power and this generally causes a loss of gain to occur and a high dissipation in the transistor.

INPUT NETWORK CIRCULATING CURRENT

Because transistors are low impedance devices the value of circulating current in the base (or gate) of a typical large transistor can reach high values. If we take a NPN bipolar rated at 40W output and examine the base input impedance we will see figures as low as 0.6Ω for the resistive part. This device has a nominal gain of 10dB so it requires an input power into the base of 4W to attain full output. Assuming an input network loss of 1dB this means we need about 5W of drive. Power is I^2R so with a base resistive part of 0.6Ω the current flowing in the base-emitter junction is approx 2.6A.

If we now look at a larger device of 100W output power we find the gain is 7dB and the base resistive part is now only 0.4Ω. A gain of 7dB is approx 5 as a power ratio and so requires a power of 20W into the base-emitter junction to deliver the full output. This is a base current of approx 7A.

SWITCH-MODE MOSFETS

Today the most popular devices for linear operation using 28V are mosfets. These possess certain advantages over the earlier bipolar types; typically having a little more gain and slightly higher gate impedance, which makes the matching easier. They are also inherently self-protecting. In a bipolar transistor if the junction becomes too hot the collector current creeps up and the device can enter a runaway condition, which increases the collector current more and becomes a spiraling effect. A mosfet does the opposite; it reduces the drain current if the junction becomes too hot. Because the individual die-cells are self-protecting the need for source *ballasting resistors* is reduced and lower values can be used, which equates to more gain.

There are other types of mosfets manufactured that are intended for switch-mode power supplies. These are much less expensive and seem attractive due to the lower cost. The lower cost is largely due to the special packages that the RF transistors use. The switch-mode mosfet is normally packaged in the cost-effective TO-220 package, which is a fraction of the cost of a RF package. However, their use as a RF amplifier is hardly ever satisfactory. The reason is the subtle difference between the two types. Large mosfets are comprised of many individual cells all connected in parallel. A 50W device might contain as many as 1000 cells each of which need to carry similar current to avoid localised heating.

In a switch-mode mosfet there is an inherent drain-source Schottky diode that is designed to clamp the drain-source junction to protect it from transient negative voltage spikes. The drain-source reverse polarised diode is quite a big device and relatively fast when normal switching speeds are used, typically from about 25kHz to about 2MHz. However, when used for RF amplifier duties the diode is quite slow and hence tends to absorb a lot of the power. The RF mosfet also has a similar diode but in this case it is a relatively weak diode and has little effect on the signal. (During the manufacture of a RF mosfet the inherent P-N junction diode made in the process needs to be removed, this makes the RF mosfet more expensive to manufacture and hence another reason they cost more. In actual fact the diode cannot be completely removed but by strapping the parasitic base-emitter junction of the body diode together it is largely suppressed.)

The major difference between a true RF mosfet and a switching mosfet is the gate-source junction. In an RF mosfet the gate-source path is very short, typically only about 100μm, this allows the RF signal current to propagate through the material quickly and hence all cells on the junction are driven in phase. By contrast a switching mosfet has a very long gate path and the different material used has a much slower propagation, about 1000-times slower. So the gate area is not evenly excited and the area near the extreme edges may not even get any exciting signal because of the propagation delays. This means that the individual dies are not driven in phase, some could be 180° out of phase and this robs power and drive energy. What effectively happens is that cells closest to the gate bonding wire connection point are carrying the bulk of the gate charge current and the extreme edges receive very little or none at all. This causes hot spots in the die and is likely to lead to failure. In severe cases the mosfet has been known to literally explode due to the stress.

Another factor concerns the parasitic inductance within the package. RF mosfets have very low parasitic inductance between the gate, drain and source connection pins because of the way the bonding wires are applied. Switching mosfets have quite high inductance and this sets up large transient voltages within the die at RF, as much as 80V in some cases, and this can lead to other localised heating and subsequent failure. Although mosfets are inherently self-protecting because of the positive temperature coefficient of the drain-source region there is a limit to how hot the material can safely withstand.

Like all semiconductors that use silicon the maximum safe junction temperature is around +200°C.

Paralleling up many cells to get the required dissipation leads to high junction capacitance. In a typical RF mosfet the gate and drain capacitance is quite low, but a large switching mosfet can have gate capacitance as high as 10nF and this makes it very difficult to drive it from a 50Ω source.

The only RF amplifier using switching mosfets that works with good efficiency is the Class D or E type, which are limited to about 2MHz switching frequency by the high junction capacitance found in switching mosfets. These can achieve high efficiency (90% or so) but are limited to approx 1MHz RF carrier frequency due to the method used. The aliasing products limit the switching frequency; the minimum switching frequency is twice the carrier frequency. Siliconix in the 1980s described a 1kW medium-wave AM broadcast transmitter that occupied the volume of a briefcase and operated off a 115V AC supply. This used Class D mosfets in a high-efficiency design.

Another cause of poor efficiency is the stray capacity between the drain and ground. For the TO-220 package the surface area the drain tab occupies approx 0.2 square inches (1.2cm^2). With a mica washer as the insulator this causes a high drain to ground capacity. Mica has a dielectric constant of about 7:1 compared to air. This for a 1mil thick mica sheet is a capacity of approx 300pF. At VHF this equates to a very low reactance and considerable RF current flows between the drain and ground. This is 'lost power'. A proper RF transistor has a very low drain to ground capacitance, perhaps 50pF at maximum. This perhaps explains why the various amplifier designs published using switching mosfets have such poor efficiency and high IMD figures.

A little later we will explore a new type of RF mosfet that can generate substantial power and gain up into the VHF region, these are new technology and becoming very popular for high power applications as they can be used on high voltage supplies of up to 300V DC. Output powers of as much as 1.5kW at 30MHz can be attained with just two devices and the linearity is quite acceptable for SSB operation.

MATCHING NETWORKS

In order to use a transistor as a power amplifier we need to design input and output networks to transform from the input and output to the system impedance, often 50Ω. For devices where the input and output impedance is not too low the simple L-match is usually the best option. A single section L-match can transform over a ratio of about 10:1 with minimal loss.

Let us begin by designing the input network for a low power bipolar operating as a Class-C amplifier. First we need to get the manufacturers data for the frequency of interest. Assume we are designing for 145MHz and data is only available for 150MHz, this is normally close enough to design a suitable network that can be adjusted under test. By including some variable capacitors we can normally 'tweak' the input for optimum matching.

We will choose a 15W transistor designed for 12V operation, the 2N6081. This is intended for FM amplifiers operating in Class-C. The manufacturer gives the following data for the input impedance:

Frequency MHz	Zin ohms
130	1.64 +j0.15
140	1.73 +j0.44
150	1.80 +j0.76
160	1.92 +j1.00

Because 145MHz falls halfway between the 140 and 150MHz data points we can interpolate that Zin will be approx 1.765 +j0.6 at 145MHz. The 1.765 is the resistive portion in ohms and the +j0.6 means the device is inductive with a reactance of 0.6Ω. This is equivalent to an inductor of 0.635nH at 145MHz. These values are the *series equivalent impedance* and before we can start designing we need to convert them to the equivalent parallel components. This is performed with formula for series to parallel conversion. This can also be done either with a Smith chart, but to make life simpler we have included the necessary calculations with the software supplied with this book. Running this and plugging in the numbers we obtain the value for X_p which is the equivalent parallel reactance, this value is 5.792Ω. We also obtain the parallel resistance R_p, which is 1.969Ω. By entering the operating frequency of 145MHz the software computes the required value of capacitance required.

We need to cancel the inductive reactance with a capacitive reactance of 5.792Ω. This requires a capacitor of 189.5pF at 145MHz. Hence, we need to connect 189.5pF between the base and ground to cancel out the reactance. Practical capacitors have some inherent self-inductance internally plus the lead lengths add to this. A typical ceramic plate capacitor has approx 2nH of internal inductance and the lead inductance is about 1nH per millimetre of lead length. With 1mm for each lead this amounts at approx 4nH. This means that the inductance cancels out part of the capacity and we need a lower value capacitor to obtain the correct reactance. The formula to find the equivalent capacity is:

$$C_{equiv} = \frac{C}{1 + \{(2\pi f)^2 LC\} 10^{-9}}$$

where:
C = effective capacitance required in pF
L = parasitic inductance in nH
f = frequency in MHz

It is normally best to split the capacitor into several smaller values to distribute the base current flowing, so here a parallel connection of two ceramic plate capacitors is a good choice. This reduces the dissipation in the capacitors, which can cause the capacitor to overheat and change its value. This also lowers the self-inductance to half the normal value. Similarly if the capacitance is made up from four paralleled capacitors the self-inductance falls to a quarter. We will select four parallel capacitors, that is two on each side of the base-terminating track to ground. This gives an even distribution of the RF current flowing. Hence we use the value of 1nH in the

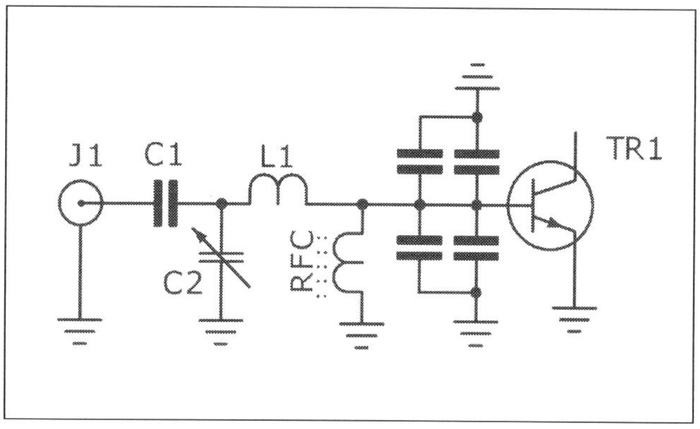

Fig 10.8. L-match for base

formula.

Assuming an inductance of 1nH for the capacitor requires a value of 120pF total. This can be made up from two 33pF and two 27pF capacitors.

Now the base looks like a pure resistance of 1.969Ω, which we need to match to 50Ω. The general circuit we require is shown in **Fig 10.8**. The inductor between the base and ground is required to carry the circulating DC flowing in the base-emitter junction. This needs to present a reactance of at least ten-times the impedance at the base, so a value higher than ≈18Ω is required. It is preferable for this to be a small molded RF choke and a value of 0.1μH (100nH) will be suitable for 145MHz. The input drive signal causes rectification to occur in the inherent base-emitter diode and this voltage is the base bias. The base-emitter diode produces a negative voltage that forces the device harder into Class-C, in order to counteract this the resistance between the base and ground needs to be very low and a RFC is the best option. In some designs the RFC can be replaced with a few turns of enamel copper wire with a ferrite bead slipped over the bottom end to suppress low frequency oscillations. As long as the reactance is greater than 10 times the base impedance it will have an insignificant effect on the operation.

The capacitor in series with the input connector (C1) is a low value, and serves as part of the input matching and also as a DC block to the base bias voltage. It could be another trimmer capacitor but often a fixed value can be used.

An L-match can only match over a narrow range of impedance, typically only about 10:1 so here we need to use another approach. If the step is too large we often need to split this into two or more smaller steps. This we do by cascading two or more L-match sections. The first section might bring the 50Ω down to, say, 8Ω and the second section then transforms the 8Ω to the base impedance. An example is shown in **Fig 10.9**.

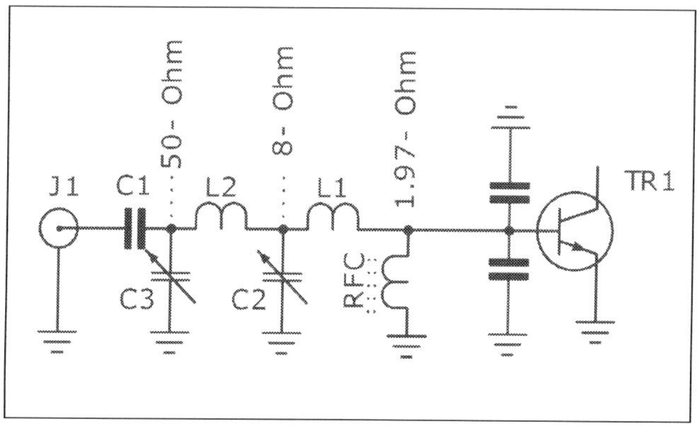

Fig 10.9. Two section L-match network

Note that the input network is a low pass filter topology and so it will only work up to a certain frequency. If we try and drive the amplifier with, say 200MHz it will attenuate the signal and we will get little output power.

If we calculate the required values for Fig 10.8 the shunt tuning C2 and series inductance, L1 requires a piece of 16SWG wire about 16mm long, C1 is approx 19pF and C2 is approx 62pF. The exact values required are L1 = 16.1nH, C1 = 19.84pF and C2 = 62.6pF. C1 could be made up from two paralleled 10pF capacitors to handle the current flowing and C2 from a 30pF trimmer and a fixed capacitor of 47pF. L1 can be a single turn coil about 8mm ID.

COLLECTOR NETWORK

We will now turn our attention to the output-matching network. Looking at the manufacturers data again we also need to interpolate between the 140MHz and 150MHz values to obtain an approx value at 145MHz. We obtain a value of 4.10 -j2.02. The minus j term tells us the collector is capacitive at this frequency with a reactance of 2.02Ω for the series equivalent impedance part. Again we need to convert from the series to the parallel part before we can begin designing the matching network. This gives a value of 5.095Ω for R_p and 10.34Ω for X_p. A reactance of 10.34Ω is equivalent to a capacitance of 106pF or an inductance of 11nH at 145MHz. Hence, to cancel the capacitive reactance we need an inductance of 11nH.

However, we first need to consider how we are going to feed in the DC supply current to the collector. If we use an RFC this needs to have a reactance of at least ten-times the collector load of 5.095Ω. It therefore needs to be at least 51Ω at 145MHz. This requires an inductance of 56nH, which is a very small value. One thing we could do is to make the RFC the value required to eliminate the capacitive reactance, so the inductor would be 11nH. This is a piece of 16SWG wire about 11mm long, so it is a bit impractical. So we will select a collector feed RFC with a much larger value. Again we could choose a molded RFC because the DC current flowing is not too high, only about 1.5A average. It is a good idea to use a RFC with a low Q to suppress low frequency oscillations. It must be remembered that the 2N6081 will have considerably more gain at lower frequencies and we need to prevent the amplifier becoming an oscillator! A popular choice is a ready wound 6-hole ferrite bead manufactured by Siemens and Philips. This is rated at 5A and a nominal inductance of about 6μH. This is far more inductance than we require but it is effective to quite low frequencies where oscillation can be a serious problem. Amplifiers generally oscillate when the output is severely mismatched, which will be the case for low frequencies if the collector matching is designed for 145MHz.

So having chosen a very large inductance for the collector feed we can ignore it for the time being and simply focus on the collector matching. We still need to cancel the collector capacitive reactance with an inductor of 11nH. This is a very small value and difficult to construct.

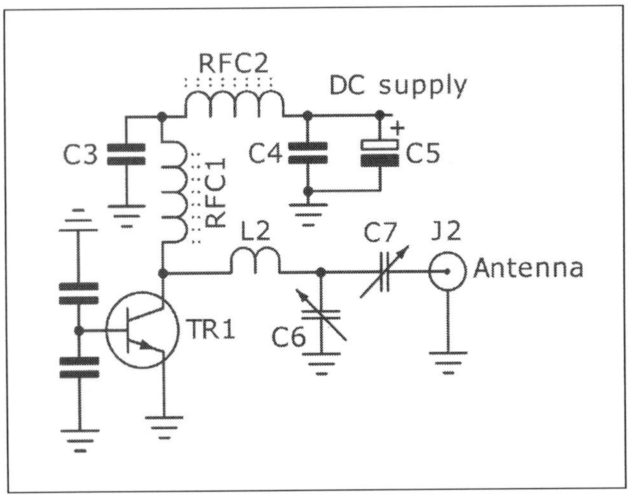

Fig 10.10. Collector matching network

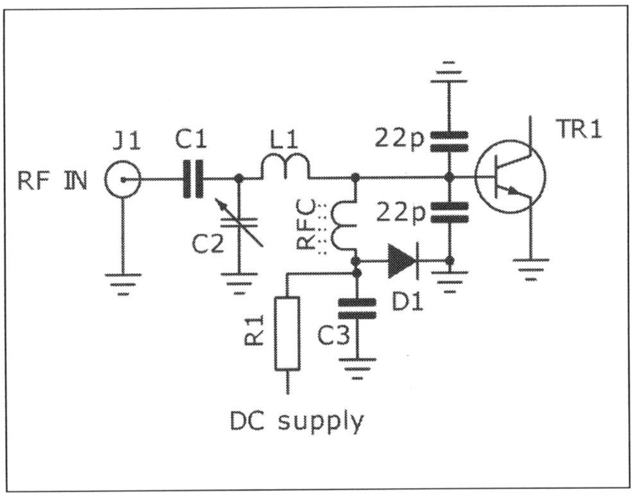

Fig 10.11. Simple biasing method for linear operation

So we will over-compensate and choose a higher value. This will mean we then have to add some extra capacitance to compensate. We need to end up with a practical circuit that also takes care of DC blocking of the supply voltage and this means we need a series capacitor between the collector and the 50Ω antenna. Remembering the rule with L-match networks that the shunt capacitor needs to be across the higher impedance part of the network, so the inductor is the first portion in series. The general circuit is shown in **Fig 10.10**.

Capacitor C3 is chosen to present a low reactance at the operating frequency, for 145MHz a value of 470pF to 1nF is generally adequate. The supply is fed via another RFC designed to suppress low frequency parasitic oscillations. This is decoupled by C4 and C5, which will typically be 10nF and 10µF. In some circuits it is necessary to add a *parasitic-stopper* network to tame a very high gain transistor. This is a series connected low value resistor (typically 10Ω / 0.5W) and a capacitor of approx 10nF from the junction of C3 to ground. This acts like a low impedance to the low frequencies and damps the collector at these frequencies.

The calculated values for L2, C6 and C7 are 26nH, 38pF and 19pF. L2 needs to be wound with substantial gauge wire and 16SWG would be appropriate for 145MHz. C6 can be a 10 - 60pF trimmer and C7 a 5 - 30pF trimmer.

CLASS-C PROBLEMS

The problem with Class-C is that the transistor needs a certain minimum drive power before it will start to turn on. If the drive power is adequate the power out versus power input curve is reasonably linear over about 50% of the drive power range. If used for SSB this means it requires a minimum power input but it will be very non-linear at low drive and high drive levels. For SSB operation we need to bias the transistor on to a small collector current, typically about 5 to 10% of the peak collector current when fully driven.

BIASING FOR LINEAR OPERATION

For linear operation we need to have a small collector current flowing when no drive is applied. This requires a voltage on the base of approx +0.6V. Often this is provided by a 1A-rectifier diode used as a shunt regulator similar to a zener diode. The diode has a current of about 200mA flowing from a dropping resistor fed from the DC supply rail. This diode needs to track the base-emitter forward voltage with temperature. Often this diode is thermally coupled to the device heatsink so the two forward voltages are approx the same. The bottom of the base RFC is decoupled with a large value capacitor and the diode is connected to this point and ground. This is shown in **Fig 10.11**.

A problem with this simple method is that the transistor and diode often do not track accurately and with variations in the ambient temperature the idle current may vary over too wide a range. There are special bias circuits that overcome this problem, some use NTC thermistors and an Op-Amp to control an external pass transistor feeding the base bias. Others are self-contained diodes and compensation components that are designed to bolt to the heatsink. These however are quite expensive. The base-emitter voltage of a typical silicon bipolar transistor has a temperature coefficient of approx -20mV/°C. As the device heats up the forward bias voltage required decreases to maintain a constant collector current. If we applied a fixed base voltage of +0.6V, at very low temperatures the collector current falls to zero and at high ambient temperatures the collector can reach unsafe values.

A secondary problem is that as the drive varies it generates a base current that opposes the DC-generated bias current. This shifts the operating point away from the optimum linear condition. The bias current needs to be supplied from a very low impedance source to prevent bias current variations and the simple diode circuit is often not acceptable for high power transistors where considerable base current is flowing under drive. If we analyse the base-emitter of a typical NPN RF transistor the simplified series equivalent circuit is as shown in **Fig 10.12**.

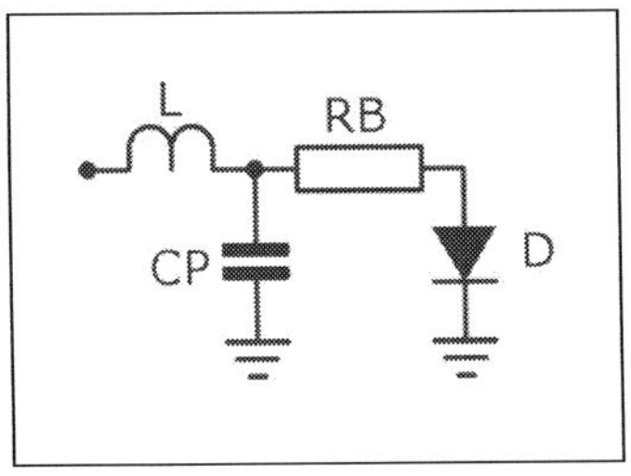

Fig 10.12. Equivalent circuit of the base-emitter junction

The inductor L is the inherent bond wires and package inductance plus the base-emitter inductance. RB is the real part of the impedance and the diode is the base-emitter junction. CP is the stray capacitance the package introduces which is normally quite small.

The other problem is obtaining sufficient base current for linear operation. The base current required can be derived from the formula:

$$I_b = \frac{I_c}{h_{FE}}$$

where:

I_b is the base current required

I_c is the peak collector current

h_{FE} is the current gain (given in the data sheet)

For the 2N6081 transistor delivering 15W and a 12V supply the peak collector current will be approx (15/12) / expected efficiency. For a Class C amplifier we can expect the efficiency to be approx 60%. At an efficiency of 60% the DC input power is 15 / 0.6 = 25W. So the peak collector current is approx 2A. The minimum data sheet value for h_{FE} is about 5. So the base current required is 400mA. The h_{FE} is likely to be somewhat higher than the minimum data sheet value, let us assume it is 10, therefore the base current needs to be approx 200mA to ensure semi-linear operation. It must be remembered that the 2N6080 series are not linear transistors, the range is intended for FM amplifiers and some experimentation will be needed to find the optimum bias conditions for lowest IMD. However, for an output power of approx 10W PEP it can be expected to be fairly linear.

MORE EFFICIENCY CALCULATIONS

Using the amplifier we have just designed let us now run some more accurate efficiency calculations. We saw that we may need as much as 400mA of base bias current to achieve the required quiescent collector current for semi-linear operation. In order for the shunt diode to regulate correctly we need at least 50% more current flowing in it to ground. So the bias current being drawn is 400 + 200mA = 600mA.

This current has to come from somewhere and that is normally the raw 12V supply. Assume we end up with a DC to RF collector efficiency of 50% in the amplifier, which is fairly typical for a linear amplifier. The DC input to the collector at 15W output will be 30W. So our collector current at peak power will be (30 / 12) = 2.5A

But we also need to include the 600mA base regulator current, so the total DC input is 2.5 + 0.6 = 3.1A. This is a total DC input power of 3.1 x 12 = 37.2W. So our real efficiency figure is 15 / 37.2 = 40.3%.

This is why we can only expect an efficiency of about 45% at best with a linear amplifier. The semiconductor manufacturers normally 'cheat' when they give efficiency figures and often do not include the bias current and the other associated losses, quoting only the collector current. This makes the transistor appear to be a little better than we would measure in a real amplifier.

Efficiency by definition is the '*RF power out obtained divided by the DC power in*'. The DC power input includes all the other currents drawn to make the amplifier work, this includes the base bias, relays, cooling fans etc. If you treat an amplifier as a 'black box' and measure the input current drawn and compare it to the RF output obtained it is quite possible to see efficiency figures as low as 25% because of all the other losses involved.

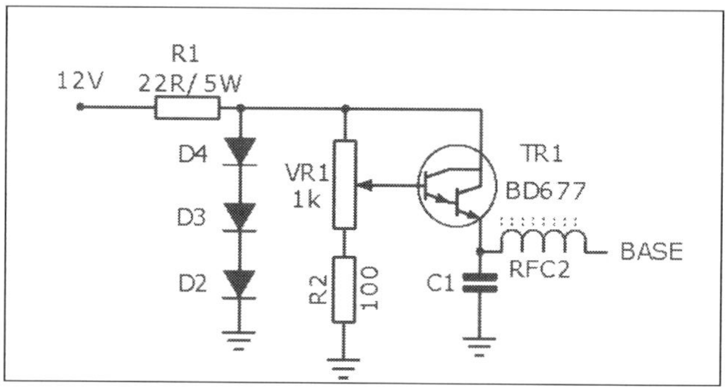

Fig 10.13. Improved base bias system

IMPROVED BIAS STABILISER

Where the base bias needs to be more than a few hundred milliamps a better scheme is to use a power transistor to control the base bias. The method shown in **Fig 10.13** is a good system.

The base current is set by VR1, which controls the voltage on the base of TR1, a NPN Darlington transistor with a current gain of 1000. The type shown is rated at 4A-collector current. The diodes connected in series generate a stable reference voltage of approx 2V. D2 is thermally attached to the heatsink to vary the bias voltage with temperature. RFC2 connects to the base of the RF transistor and is decoupled to RF by C1. This circuit allows a finer control of the quiescent current with some adjustment by VR1. Resistor R1 needs to be adequately rated for the dissipation and the supply voltage. With the value shown the bias generator can deliver up to about 300mA base current to the RF transistor. If we require more current then we need to lower the value of R1 and insert another series diode in the voltage reference string.

RF MOSFETS

We will now look at the mosfet and see how it differs from a bipolar transistor. The gate appears as a very high resistance to DC, typically 1MΩ, and a high value of capacitance to RF. At high frequency the capacitive reactance makes it difficult to match to. The Drain-Source junction also appears to be a large capacitance. The difference between the mosfet and the bipolar is not as obvious with the drain-source behaving in a similar way so the matching networks required are quite similar.

A bipolar transistor has a parameter $V_{ce\text{-}sat}$, which is the voltage appearing across the collector-emitter junction at full collector current and is measured in Volts. A mosfet has a similar parameter known as $R_{DS\text{-}on}$. This is the resistance between the drain and source when the drain current is at maximum. It therefore behaves in the same way and develops a voltage between the drain and ground when the current is a maximum. So we also have to take this into account when design the output matching network. Typical values for $R_{DS\text{-}on}$ vary depending on

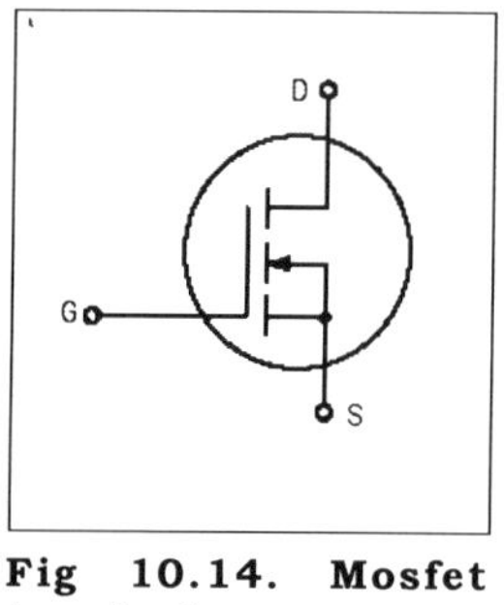

Fig 10.14. Mosfet terminals

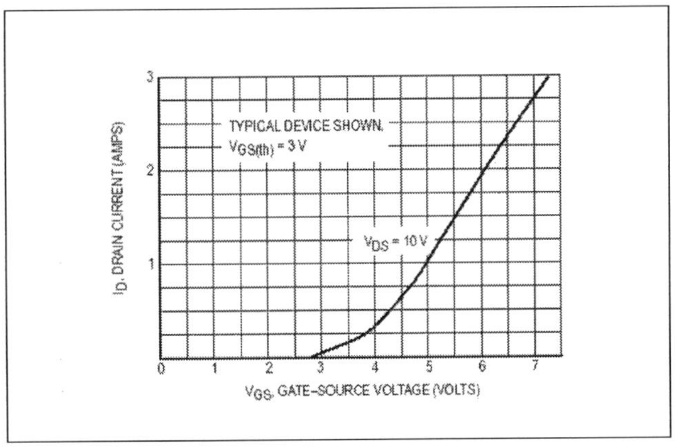

Fig 10.15. Drain current versus gate voltage

the number of die-cells paralleled up in the package. A 50W mosfet will be approx 0.5Ω, which sounds quite low. If the drain current is 3A this a voltage of 1.5V, so it is quite similar to a bipolar transistor.

The major difference between the two types is that the mosfet requires a positive gate voltage to turn it on whereas a bipolar needs a base current. Because the gate appears as an open circuit to DC the biasing schemes can be a lot simpler. The mosfet has a gate threshold voltage of about 3 to 5V, if the gate voltage is below this it will not conduct. Setting the required quiescent current then only requires the correct gate voltage to obtain the required current. The curve shown in **Fig 10.15** is a typical RF mosfet gate-source voltage against drain current. To obtain a quiescent current of 500mA requires about 4.5V. This parameter is not well controlled in manufacture and it is necessary to ensure the bias voltage generator has sufficient range to cater for production spreads. In a push-pull or parallel amplifier using two or more devices it is normally necessary to arrange a separate bias control for each device to cater for this variation. This complicates the matching networks because we have to arrange extra DC blocking capacitors for each gate.

MATCHING

To use an RF mosfet we need to follow the same steps we did for the bipolar. The first thing is to get the manufacturer's data on the device we have chosen. We have chosen a 30W-output device made by Motorola, the MRF137 designed for a 28V supply voltage as a linear amplifier for use up to 400MHz. This device is packaged in the 380 SOE-Flange style (Strip-line Opposed Emitter - because it was originally used for bipolar transistors) with a flat mounting flange. (The 380 = 0.380 inch or about 10mm diameter across the ceramic top cap.) This is shown in **Fig 10.16**.

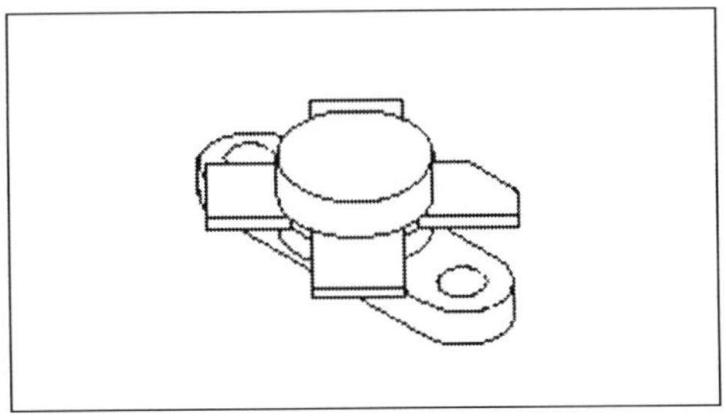

Fig 10.16. MRF137 in 380SOE-Flange package

The maximum safe dissipation is 100W when the flange is maintained at +25°C on a suitable heatsink. Very rarely can we attain this but it is how the manufacturer specifies the device dissipation using a water-cooled heatsink block to characterise it. From the data sheet we can obtain the de-rating factor for other heatsink temperatures. This is 0.571W/°C. So if the heatsink is at +60°C the safe power dissipation falls by:

$$(60 - 25) \times 0.571 = 20\text{W}$$

So it can only safely dissipate 80W at this temperature. Assuming an efficiency of 45% when operated as a linear amplifier and we run at 30W output power then the DC input power is (30 / 0.45) = 66.6W and hence 36.6W will be dissipated as heat. So the 80W limit is well covered by this assumption.

The manufacturer gives some details for operation at 150MHz as a linear amplifier. The typical gain in a narrow band circuit is ≈13dB (20-times) and the efficiency is ≈60%, so it will make a nice 144MHz linear amplifier, requiring only 1.5W drive for full output. The manufacturer's recommended quiescent current with a 28V supply is 25 to 100mA for optimum linearity. It is also tested at full output power into a 30:1 VSWR without any damage, so it is quite a rugged device.

To begin, let us get the input and output impedance values. These are shown below:

Freq MHz	Zin Ω	Zo Ω
100	2.11 - j11.07	8.02 - j2.89
150	1.77 - j7.64	5.75 - j3.02
200	1.85 - j3.75	3.52 - j2.67
400	1.74 + j3.62	2.89 - j1.52

MRF137 data

Like the bipolar transistor these are in series equivalent form and so we need to convert them to parallel values. But before we do, notice that the input and output are both capacitive reactive right up to 200MHz and the input goes through zero somewhere between 200 and 400MHz; by interpolating the input data it seems to about 300MHz. So from the 150MHz data, which is our closest

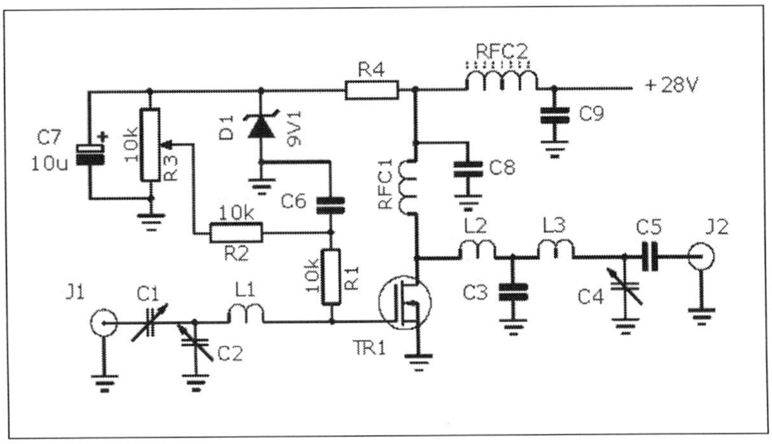

Fig 10.17. 150MHz test circuit for the MRF137

data point, we already have the values for 145MHz to within a few percent.

Tackling the input first we convert the value of 1.77 -j7.64 into the parallel values. These are R_p = 34.747Ω and X_p = 8.05Ω at 150MHz. R_p is close to 50Ω so the matching should be quite simple and low loss. The equivalent component values for the reactance X_p at 150MHz are 131.8pF and 9nH. Hence, as the gate is already capacitive we require the inductive component.

An inductance of 9nH is a very small value, about 9mm of 16SWG wire, and so we need to either make the inductor a strip-line on a printed board, or increase the inductor value and then compensate by additional capacity in the network. Often the manufacturer gives a test circuit with component values. The data sheet shows the 150MHz test circuit (**Fig 10.17**) so we can get some ideas how to tackle the problem.

The test circuit component values show that L1 is an inductor of 2-turns with an ID of 7mm and wound with 18AWG enamel wire close wound. Capacitors C1 and C2 are compression mica trimmers with values of 35pF and 115pF maximum. Resistor R1 is the gate bias feed and is a value of 10kΩ and so does not affect the matching.

Looking at the output matching we go through the same routine by converting the series equivalent values to parallel. The series values are 5.75 -j3.02 at 150MHz, converting to parallel gives the values as R_p = 7.336Ω and X_p = 13.968Ω. The inductance required to cancel the drain capacitance is 15nH, also a small value.

In the test circuit the output match is a double L-match network with two series inductors and shunt tuning capacitors. Hence, the impedance is transformed in two steps, firstly from the 7.336Ω up to some intermediate value and then to 50Ω. The total transformation ratio is 50 / 7.336 = 6.8:1. The drain RFC needs to present a reactance of minimum value 73Ω, in the test circuit it is wound with 20-turns of 18AWG enamel copper wire with a 5mm ID.

However, a transformation ratio of ≈7:1 is feasible with a single L-match stage. The final circuit for the output match is shown in **Fig 10.18** with the values. These are the calculated values and do not take into account circuit strays: in practice a bit of 'PCO' will normally be required to get the amplifier working correctly. However, they are probably within 10% of the actual, so tuning up should be simple. (PCO = Post Computation Optimisation or, in simple language, a bit of tweaking.)

The tuning capacitor is made up from a fixed and a variable connected in parallel to handle the high RF current flowing to ground. C7 is a low inductance capacitor such as a 100B monolithic made by ATC. C6 is a 5 - 40pF trimmer. C8 is also a 100B ATC and merely functions as a DC block; it has no affect on the impedance matching as long as the value is greater than ≈330pF. The inductor is wound from 18SWG enamel copper wire to give an inductance of 21nH. This is approx 2-turns with an ID of 5mm. RFC1 and RFC2 are 6-hole ferrite chokes.

BIAS GENERATION

The MRF137 requires a stable bias supply to feed the gate. In the test circuit a 9V zener diode fed from the raw 28V supply provides the voltage to a potentiometer R3. This preferably should be a multi-turn type to allow fine adjustment. Resistor R2 and C6 are 10kΩ 0.25W and 10nF/50V respectively to decouple the gate bias voltage. To set the quiescent current, adjust the gate voltage control R3 with no drive applied until the drain current is 25 to 50mA. This is about the optimum for gain and IMD.

If the IMD needs to better then the quiescent current can be raised to around 500mA maximum at the expense of more dissipation. **Fig 10.19** shows the typical power out versus power in curve at various frequencies.

Here we can see that the gain figure of 13dB quoted earlier was a bit pessimistic. At 150MHz with 1W of input the amplifier delivers ≈36W, a gain of 15.5dB, so a 1W transmitter will drive it nicely. I have used the MRF137 in several designs for SSB transmitters for VHF and the 3rd order IMD at the 25W level is about -33dBc or -39dB PEP, which is quite respectable. As the graph in Fig 10.19 shows, they can deliver up to 50W output if pushed and the IMD is then about -24dBc.

HARMONIC FILTERING

All practical RF power amplifiers will require some form of low-pass filter to reduce the level of harmonics. These are covered in a separate chapter with practical circuit designs and tables to obtain the correct component values. The inductors and capacitors need to be carefully chosen to reduce the circuit losses as far as possible. The circulating RF currents flowing in the filter will be high and low loss components are essential for reliable operation.

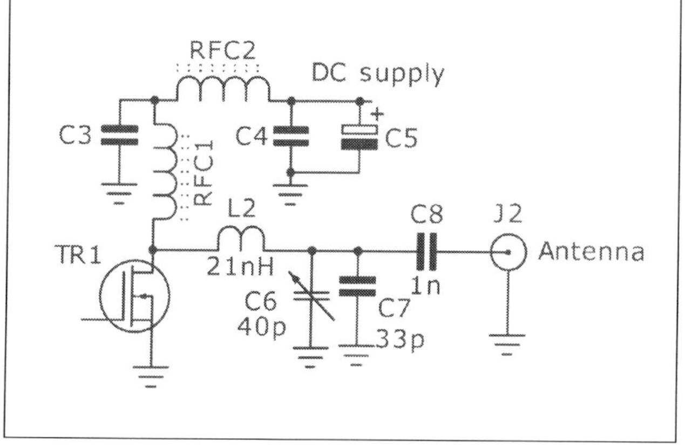

Fig 10.18. Drain matching for MRF137 at 145MHz

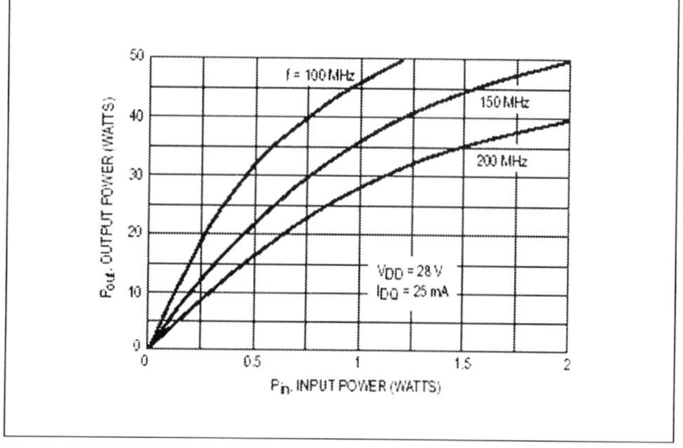

Fig 10.19. Power in - power out graph

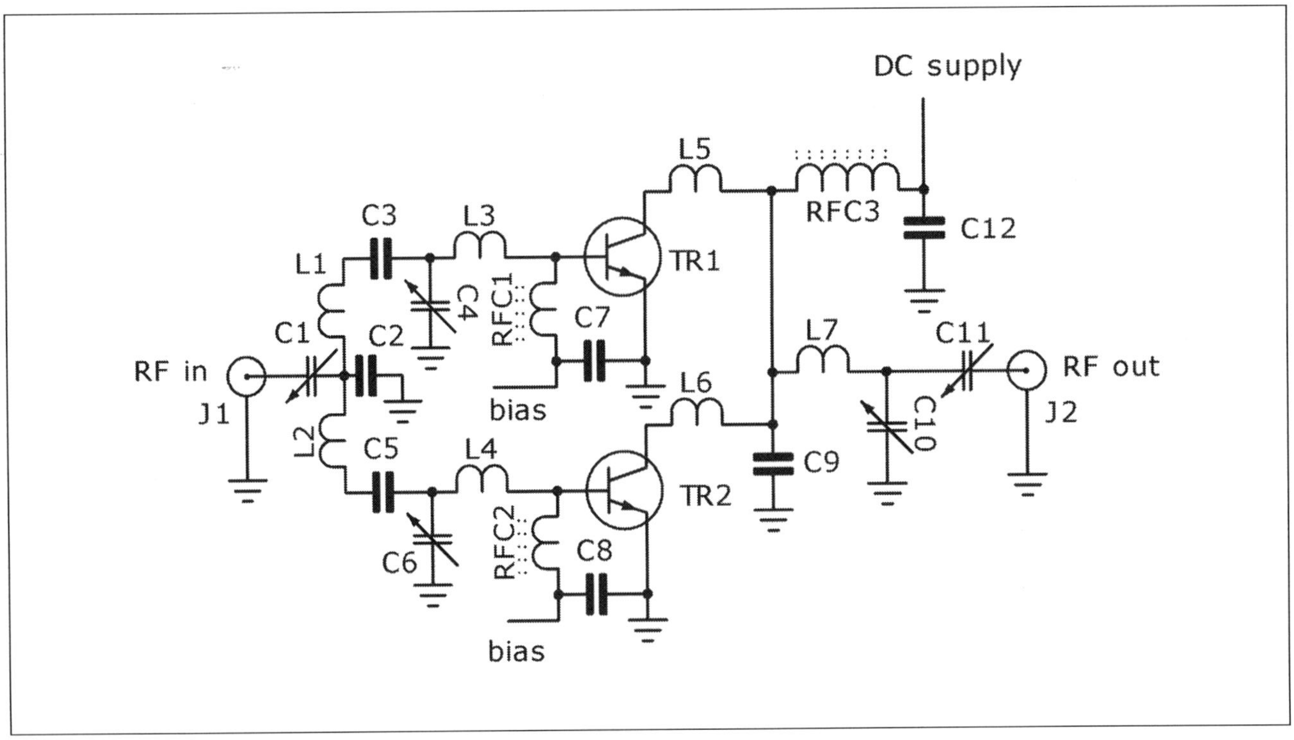

Fig 10.20. Parallel connection using L-match networks

MULTI-TRANSISTOR AMPLIFIERS

When we require more power an economic method is to use two or more transistors connected as either push-pull or parallel. For the simplest approach the parallel transistor amplifier is often the best option. However, arranging a suitable mechanical layout can be difficult. The physical size of the transistor package precludes very close spacing unless great skill is used. It is also a benefit to spread the heat generation over a larger area of the heatsink. Hence, a close spacing is not really necessary and a convenient and logical placement is a great assistance to making a working amplifier.

If parallel connection is required then connecting the bases / gates or collectors / drains together in a low inductance manner is often not simple. It often results in parasitic resonant 'loops' that can cause a problem with instability. The technique favored by the writer is to use separate L-match networks for each device brought to a common impedance point and then match to 50Ω. The chapter on valve amplifiers shows the parallel connection of two valves for 70cm with L-match networks and this also works well for transistors. Often two lower power transistors are less expensive than one high power device. The basis of the technique is shown in **Fig 10.20**.

CURING INSTABILITY

Very often when a bipolar transistor or mosfet is operated as an RF amplifier we experience instability at some frequency. The common semiconductor devices are very similar in characteristics to triode valves operated in grounded cathode configuration. The inter-electrode capacity between the anode and grid causes some of the output power to be fed back to the input. Bipolar and mosfet transistors are no different. To ensure stability it is often necessary to use a type of neutralising technique and to load the input heavily with a resistor to damp any tendency towards oscillation. With mosfets this resistor

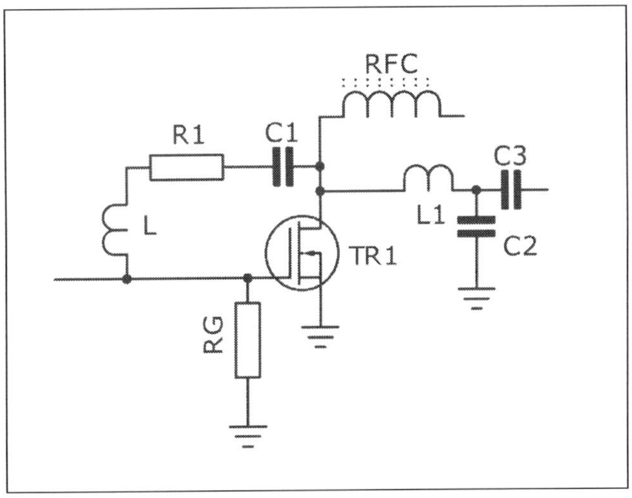

Fig 10.21. Negative feedback network

can be usefully be used as the means of feeding in the gate bias voltage. A common technique is to connect a series inductor, resistor and capacitor between the drain and gate (collector to base) which applies an equal and opposite phase signal to the internal capacitance path. This is especially necessary when the amplifier is designed to cover a wide bandwidth.

EMPIRICAL DESIGN METHODS

Sometimes we cannot find data on a device we have to hand. Rather than finding another transistor we can empirically design a circuit with a few assumptions. We have a fair idea of the collector or drain load required from the calculations of R_L. What we often don't know is whether the device is capacitive or inductive at our design frequency. In a case such as this we begin by building an output-matching network. If we drive the input via a suitable DC blocking capacitor we should see some output power, although only about 20% of the full power when correctly matched. It is important to get the output net-

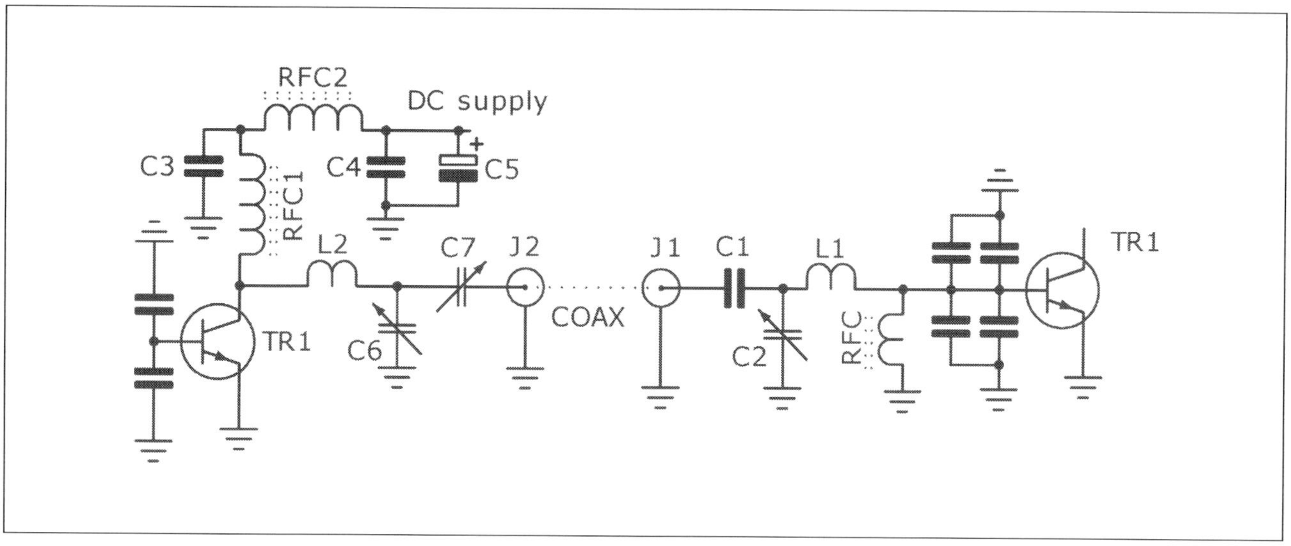

Fig 10.22. Inter-stage matching using 50Ω cable

work somewhere near the correct value before we drive the transistor up to full input. If the output is grossly mismatched we run the risk of over dissipating the device and it is likely to expire. By building the output network first we avoid this problem. Having got the output network close to optimum we can then build an input network and repeat the exercise. Now with the output somewhere near optimum we can safely drive the transistor up to its full DC input power and adjust both the output and input for best results.

SERIES CONNECTION OF TRANSISTORS

Normally when we are designing a transmitter we need to start with a fairly low power and raise this to the required output level. Often this needs more than one transistor and hence two or more stages are connected in series to obtain the required gain. When this occurs we need to match between the output of one stage and the input of the next stage. This means we are only using 50Ω at the input and output of the final amplifier and the inter-stage matching is between two complex impedances.

Although this sounds difficult it is a better method than bringing the impedance to 50Ω and back to some other impedance. Often we can simplify the circuit and reduce the losses with this technique. It may prove more convenient during design to treat each stage as a 50Ω block for testing purposes, but the transformation losses will be higher than a correctly designed inter-stage network. We can often pick up several dBs of extra gain by correct inter-stage matching rather than a 50Ω impedance between stages. 1dB in round numbers is 20% more power.

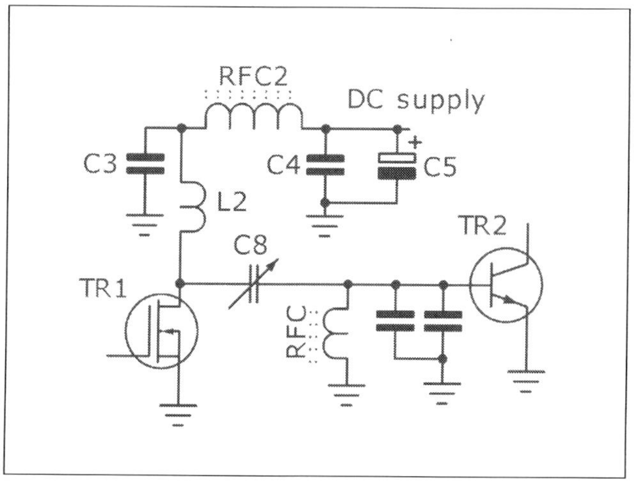

Fig 10.23. Simplified inter-stage matching directly between transistors

Depending on the transistors being used we can often reduce the inter-stage matching network to just two components, one being the collector feed choke of the first transistor and the second being the series capacitor from the collector to the next transistor base. By dimensioning these correctly we can normally obtain a low loss match over a wide bandwidth.

HIGH-POWER PIN DIODE ANTENNA SWITCH

When very fast switching of a transmitter and receiver is required on the VHF or UHF bands, a PIN diode antenna changeover switch (T/R switch) is the answer. For packet radio and other data transmission modes, such as high speed meteor scatter, the slow operation of conventional relays reduces the throughput speed considerably. A typical antenna changeover relay requires about 50ms to switch reliably, a PIN diode switch can switch in about 5μs, an improvement of 10,000 times. Relays also have a limited life, 100,000 cycles being a typical specification.

However, PIN diode switching of high RF power does have a few problems. If the receiver is connected to the transmitter via a PIN diode that is biased off, the capacity of an off state PIN diode can be as high as 5pF. At VHF and UHF this is a fair bit of power being dumped into the receiver front end and may be enough to cause damage. To resolve this problem we can use the characteristic of a quarter wave transmission line. This circuit is from an application note by Unitrode Corporation, manufacturers of PIN diodes. The company name changed to Microsemi Corporation some years ago. It offers a bandwidth of approx 20%, which would suit any of the amateur VHF or UHF bands.

If the circuit is examined the PIN diode switch consists of two diodes. Capacitors C1, C2, C3 and C4 are DC blocking or coupling capacitors and have a very low reactance to the RF. D1 is in series with the transmitter port and D2 is effectively across the receiver input. When D1 is turned on, by connecting a positive supply to R1,

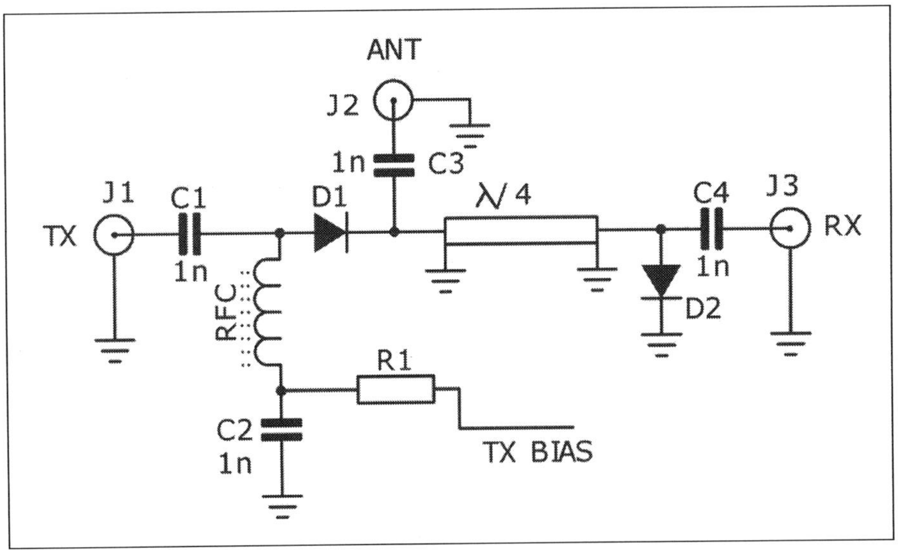

Fig 10.24. PIN diode T/R switch with quarter-wave line

D1 conducts and presents a low insertion loss to the RF signal from the transmitter to the antenna. With a good PIN diode we can expect as little as 0.1dB of loss. The current flows to ground in D2 and this is also turned hard on. This places a short across the receiver input and clamps the level of RF to a low level. Connecting D1 and D2 is a quarter wave transmission line. When D2 is providing a short across the receiver input the impedance at the other end of the line is an open circuit, so very little power tries to travel through the quarter wave line. The higher the DC bias-current the lower the insertion loss of the diodes. For a 1kW switch the current required as at least 500mA. Diodes can be the popular UM9401 for power levels up to 400W into a perfectly matched antenna (VSWR = 1:1) and 100W into an infinite VSWR antenna. This is with a forward diode current of 100mA. For higher RF power a higher power diode is required. The UM7000 diode will easily handle 5kW into a 1:1 VSWR and about 1500W into an infinite VSWR load. This diode requires about 1A bias for lowest insertion loss. Running the calculations on the UM7000 series shows that it can handle peak pulse powers of over 60kW in short pulse radar applications with a 1% transmit duty cycle.

On receive the PIN diodes are both biased off. The load presented to the signal from the antenna by the transmitter is very small, the PIN diode in the *off-state* looks like a high value resistor of greater than 10kΩ shunted by the junction capacitance of about 2 to 5pF. The signal from the antenna passes through the low-loss quarter-wave line to the receiver input with virtually zero insertion loss (see **Fig 10.24**). With careful construction and good PIN diodes this changeover switch can handle powers in excess of 1kW with receiver isolation as high as 70dB.

The switching time is almost entirely dependant on how quickly the PIN diode bias current can be switched on and off, this normally being the limiting factor and not the switching time of the PIN diodes, as some types can switch in about 1ms or less. The drive circuit must be able to charge and discharge any decoupling or coupling capacitors rapidly, this requires a driver that can source and sink several Amperes of current for the time period to charge and discharge the capacitors. The best choice is the 'totem pole' configuration (see **Fig 10.25**). The value of the dropper resistor needs to be calculated with the supply voltage to achieve the correct bias current. Often a driver similar to that used for switch-mode mosfets gate drive can be modified to give the required switching speed. If we analyse Fig 10.24 there is a total of 4nF of capacity effectively in parallel with the switching voltage and this needs to be charged and discharged each time the switch toggles from receive to transmit or back again.

When the TX / RX signal is high TR1 is turned on. TR2 is hence turned on supplying bias current to the PIN diodes via the current limiting resistor R4. When the TX / RX signal goes low TR1 and TR2 turn off and TR3 is turned hard on, so discharging the decoupling capacitors via R4. This ensures a rapid switching of the PIN diodes. TR2 and TR3 need to be adequately rated for the peak current and heatsinked.

For the low VHF bands and HF the quarter-wave coaxial line becomes quite long. If RG-58/U is used for the transmission line for a 50MHz T/R switch this requires a length of approx 1m of cable. We can use an alternative circuit that simulates a quarter-wave transmission line with an inductor and two capacitors. This is shown in **Fig 10.26**.

This circuit works in exactly the same way as Fig 10.24, except the quarter-wave line is replaced with a π network consisting of L1, C2 and C4. The formula to calculate the values of L1, C2 and C4 is:

$$L1 = \frac{Z_o}{2 \pi f} \qquad \text{and } C2, C4 = \frac{1}{2 \pi f Z_o}$$

where Z_o is the system impedance and f is the operating frequency in Hertz.

For 50MHz the values are L = 159nH, C2, C4 = 63.6pF in a 50Ω system.

At HF, where the inductor value is quite large, L1 can be made very compact if wound on a ferrite toroidal

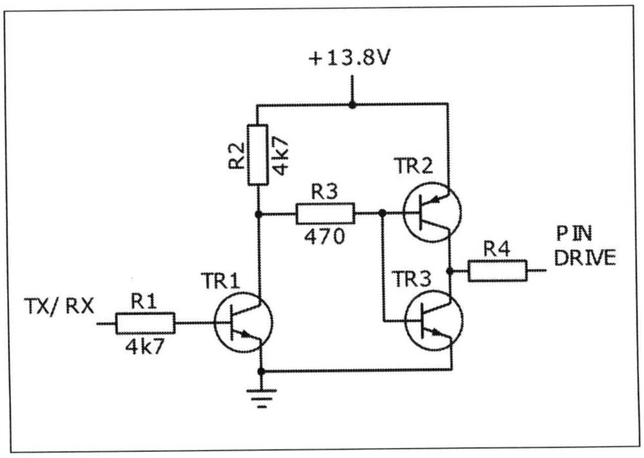

Fig 10.25. Totem pole driver

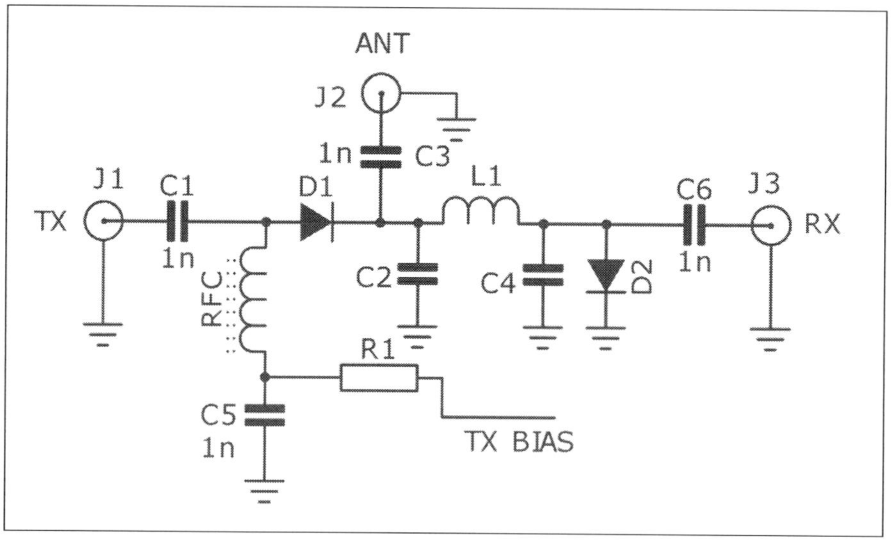

Fig 10.26. PIN diode changeover switch with lumped line

core and so the switch can be made quite small. The layout and grounding is critical to get the best isolation and lowest insertion losses on transmit and receive. Screening will probably be needed between the various sections to prevent RF signals from bypassing the quarter-wave lumped section.

The inductor RFC used to feed the bias current into the network must be rated to handle the DC bias current and also capable of withstanding the RF voltage developed by the transmitter. The self-resonant frequency of the RFC must be well above the operating frequency; the shunting capacitance of the coupling capacitors lowers the SRF considerably. The reactance of RFC should at least 10 times that of the system impedance at the operating frequency.

To give some indication of what can be achieved with the right components and a good circuit layout, in the 1980s I built a switch like the one detailed for a 1kW VHF frequency hopping jammer. The transmit / receive switching times were less than 12.5μs, being solely the limitations of the diode driver. This design used 2A forward bias on the diodes and the measured insertion loss on transmit was 0.06dB at the 1kW level. Receive insertion loss was slightly higher at 0.09dB at 80MHz and transmitter to receiver isolation was approx 75dB mid-band.

To increase the transmitter to receiver isolation even more, a low power PIN diode can be connected across the receiver input and driven from the same control signal as the main diodes. This will increase the isolation by about 20 to 30dB. Lower power versions of this circuit using a double pole LC network (π section followed by an L-section) with two shunt diodes on the receiver port exceeded 120dB isolation when correctly built.

NEW RF POWER MOSFETS FOR AMPLIFIERS

Recently a range of new high-power mosfets have appeared on the market from Advanced Power Technology (APT) which work on supply voltages up to 300V. These devices are packaged in the cost-effective TO-247 plastic case and some in special RF packages with BeO thermal links. Several articles and application notes have been written about this new range and the latest uses a pair of 450V devices in a 1.5kW HF amplifier with good IMD performance that is now being manufactured by Tokyo High Power in Japan. In May 2006 APT was acquired by Microsemi. The pioneer of the amplifier design is Richard Frey, K4XU, senior design engineer at APT.

Using a high DC supply voltage has a number of advantages, but also brings about certain limitations. With a high DC supply voltage the impedance matching to the drain is simplified because the impedance is quite high, in most cases about 12.5Ω, and a simple 4:1 transmission line transformer can be used to obtain a wide bandwidth match to 50Ω. The disadvantage of the high DC supply voltage is the possibility of excessive drain voltage causing breakdown in the drain-source junction. This needs to be countered by good matching to eliminate voltage transients that can cause the device to fail. Also at the very high power output levels the dissipation in the device needs careful attention to heat sinking. At the predicted efficiency level of ≈50%, typical of these designs, for a 1.5kW output amplifier there is about 1.5kW being dissipated in the mosfets.

Most of the published application notes use a simple DC supply with an unregulated bridge rectifier and a large smoothing capacitor across the bridge output. In a 50MHz design giving 300W PEP the peak drain current is about 6A at 80V DC. This required a smoothing capacitor of 50,000μF to reduce the ripple voltage to manageable proportions. A better method would be a linear regulated supply but this is more difficult to arrange at high output voltages.

The other factor apparent in the designs is that the mosfets are not self-protecting with a rise in junction temperature. Normally the drain current falls at elevated die temperature, this is not so apparent in the high voltage types and some protective circuitry is mandatory to prevent run-away under high dissipation conditions. If the gate bias voltage is held constant the drain current rises rapidly above a certain die temperature and a run away condition similar to bipolar junction transistors occurs. Therefore it is necessary to reduce the gate voltage with increasing die temperature or the device will fail. Fortunately this is not too difficult to arrange with a thermistor bonded to the heat sink that pulls the gate voltage lower with increase in temperature. This however needs to be proportioned correctly or the danger exists of excessive IMDs being produced if the compensation is too aggressive.

A novel feature of the new range in the TO-247 package is the availability of two devices that have the gate and drain pins transposed in the package. This allows a much better symmetry in the layout and eliminates imbalance in the matching networks. This is especially important for the VHF designs where small track length differences have a profound effect due to the imbalance this produces. Most of the published designs

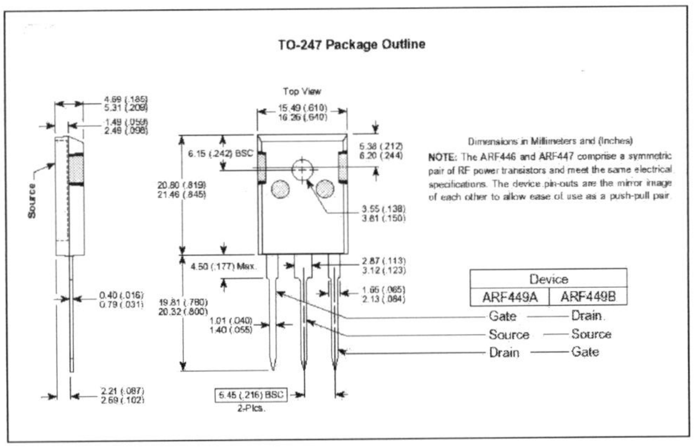

Fig 10.27. RF mosfet package details (courtesy of APT/ Microsemi)

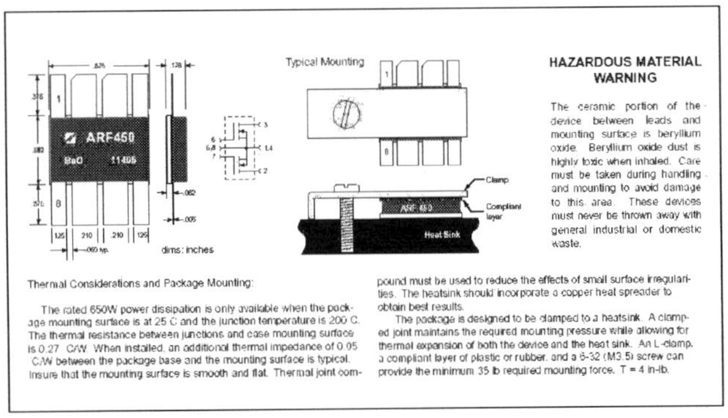

Fig 10.28. ARF450 package details (courtesy of APT / Microsemi)

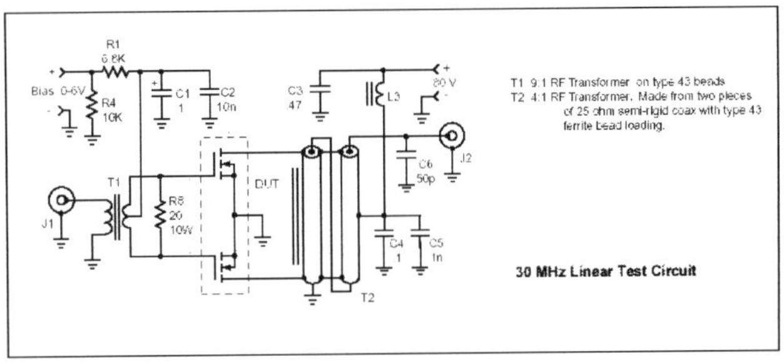

Fig 10.29. ARF450 application circuit for 500W PEP HF amplifier (courtesy of APT / Microsemi)

are push-pull and an imbalance causes high losses in the drain matching networks with consequent excessive dissipation in the devices, ferrite cores used in the transmission line transformers and a large loss of gain and power output.

Although the new APT devices seem to be ideal there are several problem areas that need to be overcome. One relates to the gate impedance. In an ideal mosfet the gate would appear as a pure capacitance with no resistive part. The gate capacity is high for the larger devices, typically about 1nF. Matching to this type of device is not as simple as other devices, especially when a broad bandwidth is required, as in a 2 - 30MHz HF amplifier. However, with step-down transmission line transformers and gate damping resistors a good VSWR match is normally possible with not too much loss. As with all broadband amplifiers the gain at the lower frequencies is normally too much and a 'gain-sloping-network' (GSN) is often required to flatten the gain across the wanted bandwidth. Another technique often used is a low value power attenuator in series with the input to absorb some of the

Freq. (MHz)	Z_{in} (Ω)	Z_{OL} (Ω)
2.0	23.00 - j 7.0	93.0 - j 10
13.5	4.30 - j 9.1	63.0 - j 43
27.0	1.00 - j 4.2	32.0 - j 43
40.0	0.42 - j 1.7	17.5 - j 34
65.0	0.35 + j 1.1	7.7 - j 22
80.0	0.56 + j 2.5	5.1 - j 16
100.0	0.90 + j 3.8	3.4 - j 12

Z_{in} - gate shunted by 25Ω

Z_{OL} - conjugate of optimum load impedance for 150W at 150V

Fig 10.30. ARF450 input and output impedance - per half (courtesy of APT / Microsemi)

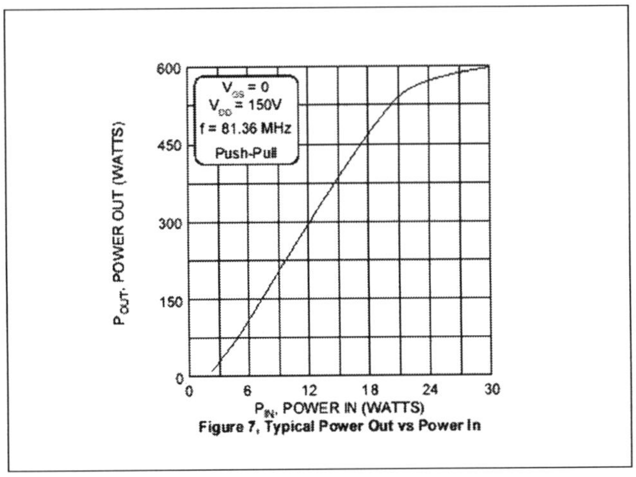

Fig 10.31. Typical power out vs power input for ARF450 at 81.36MHz - Class-C (courtesy of APT / Microsemi)

mismatch. This is no different to other types of transistors used to date. The typical input and output values for the ARF450 is shown in **Fig 10.30**.

500W output requires approx 20W of drive, hence the gain is about 14dB.

'HALVING'

One of the odd effects with solid-state power amplifiers when operated with zero bias (Class-C) is an effect known as 'halving'. In this a spurious signal is generated that is half the input frequency and is caused by the parasitic varactor capacitances in the device. Halving is most prevalent when the input drive level is a little below the level to cause full saturation and it manifests itself by producing a spurious carrier at sub-harmonics that can be hard to detect by conventional means. To detect halving requires a spectrum analyser.

The obvious cure for halving is to ensure the drive level is at least enough to drive the device into full saturation. Normally the only negative of applying too much power is a small increase in the true harmonic power generation. When an amplifier is halving, the spectrum contains a multitude of spurious carriers between the normal harmonics and these cause the spectrum to become very 'dirty' and difficult to filter out the unwanted harmonic components. Often a very small forward bias will reduce the tendency to halving. Halving can also be aggravated by an input network which has a spurious response near the half input frequency and so care must be taken not to introduce this inadvertently when constructing a Class-C amplifier.

Chapter 11

Valve RF power amplifiers

THE purpose of this chapter is not to present complete designs, there are already many books and articles published that cater for this, but rather to discuss some of the subtle points often omitted. By understanding the limitations and good points of various techniques we can gain a better insight into why a particular valve type or circuit configuration is used.

Valves are still a very good means of generating high levels of RF power. Although solid state power amplifiers are available for operation with low DC supply voltages, when the power level required exceeds about 100 to 200W the valve is often a better proposition. The problem today is that many amateurs have no knowledge of how valves work, because this is no longer covered in the Radio Communications Examination syllabus. Also, valves are looked upon as specialised items because hardly anyone uses them commercially and the number of different types available has decreased markedly over the last 20 years. In the period before and after WWII valves were everyday items and all amateur equipment was constructed with the popular types. With the steady decline in valve manufacture and use, the skills needed have largely been lost. However, when high power is needed, for example in broadcasting, commercial, amateur equipment and radar transmitters, valves are a cost-effective method.

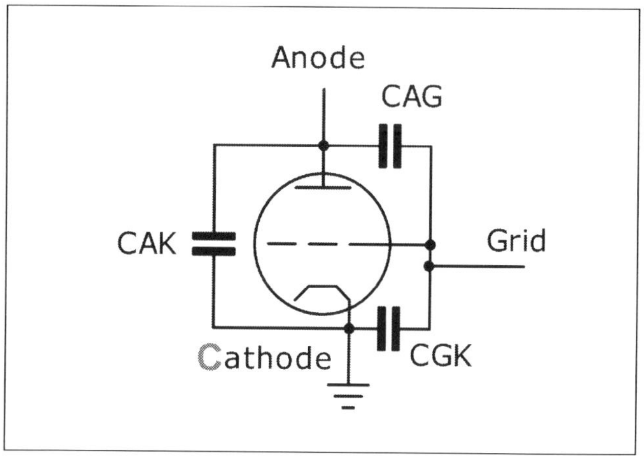

Fig 11.1. Inter-electrode capacitance of a typical triode valve

TYPES OF VALVES

Broadly speaking we can divide valves used for RF power amplifier duty into several different categories or types. The most basic type is the triode, followed by the tetrode and pentode. The types can also be sub-divided into the cathode construction, indirectly heated and directly heated. We could also divide valves into classes according to frequency: HF, VHF, UHF and microwave. Also we could divide them again into types of cooling: radiation cooling, forced air cooling or liquid cooling. Also the physical construction varies, some are all-glass envelope, some a mixture of glass-metal or ceramic-metal and some all metal. The anode type can also be a way as classifying them: metal, graphite, internal anode, external anode etc. Some types are available as two identical valves in one envelope, typically dual-triodes and dual-tetrodes. However, today there is not such a large number to choose from and many types can be used on different frequencies, so it is better to examine them in electrode number form.

TRIODES

As the name suggests these are valves with three electrodes, and these are the anode, cathode and control grid. They can be types designed for grounded-cathode operation or grounded-grid operation. Which type is the favoured one often alters the physical construction and visual appearance. Also the cathode construction and the heating filament also can be two different types, directly heated and indirectly heated.

The upper frequency limit of useful operation is largely a function of the inter-electrode stray parasitics and hence the physical construction. These can be the lead lengths and capacity between the various electrodes. Some types are limited to quite low frequencies, as little as 30MHz, whereas others can be used up into the microwave regions. Generally it is best to choose a type more suited to the operating frequency, although with experience even microwave triodes can be made to perform satisfactorily at HF.

Gain figures vary widely; large HF triodes typically at best will only attain about 13dB (20 times) maximum due to stability problems. Often the larger types are designed for grounded-grid and then gains may not exceed about 8dB (six times). Grounded-cathode operation is often not viable because of the high value of internal feedback capacity between the anode and grid (CAG) and hence grounded-grid is the only option.

Fig 11.2. 2C39 microwave triodes. The original 2C29A glass seal version on the left and the later 2C39BA ceramic seal on the right.

TETRODES

These are four electrode valves and so have one extra electrode compared with the triode. This new electrode is known as the 'screen grid' or grid2. The inclusion of the screen grid is similar to the grounded-grid triode and the screen grid *screens* or isolates the anode from the control-grid (grid1). Because the tetrode inherently is more stable than a triode, it is normally operated in grounded-cathode. This allows higher gains to be achieved for a similar size valve. Because the tetrode is capable of higher gain, it requires less drive than a comparable triode. Typical gains vary from about 16dB (40 times) to as much as 30dB (1000 times) depending on frequency and power level, however the upper frequency limit is lower than a triode. Hence, their use on frequencies much above 1GHz is rare. A practical upper limit is about 600MHz for the types available today.

Tetrodes require an extra power supply apart from the anode supply. This is normally a lower voltage and usually requires a stable voltage.

A special version of the tetrode is the 'beam tetrode': these use beam-forming plates to funnel the electron stream towards the anode and away from the screen grid. Many of the more modern tetrodes are beam tetrodes. Beam tetrodes provide better operation than normal tetrodes.

PENTODES

The final type is the pentode, which has five electrodes. Like the tetrode it has a screen grid and also a grid known as the 'suppressor grid' which screens the screen grid from the anode. The suppressor grid is normally internally connected to the cathode. Hence, it is a bit like a 'double grounded-grid' triode in operation. The inclusion of the extra grid allows slightly more gain than a tetrode requiring less drive power. However, pentodes are today quite rare and new types are usually only made for very high power applications such as broadcasting. Because of the cost and high power rating amateurs do not often

Fig 11.3. 6146B HF pentode

Fig 11.4. Siemens RS-1006 HF SSB transmitting valve. This is equivalent to 2 x 6146 in power output. The anode is of graphite construction

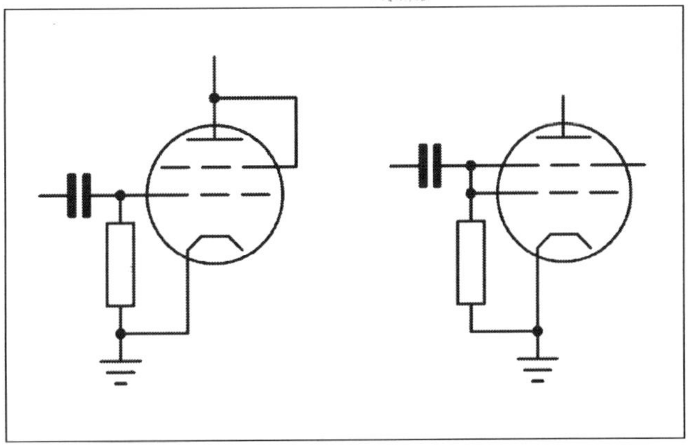

Fig 11.5. Two methods of connecting a tetrode as a triode

use them. Many of the popular traditional valves such as the 5763, 6146, 6JS6, 807 and 813 etc are sometimes incorrectly listed as tetrodes whereas they are in fact pentodes.

Pentodes require the same power supplies as a tetrode, anode and screen grid, again with good voltage stability.

TRIODE CONNECTED TETRODE

In some applications the tetrode is operated as a triode with the screen grid connected to the anode. Because the screen grid has a lower safe operating voltage rating than the anode this means the anode voltage applied is lower than normal. This lowers the gain a bit but eliminates the extra supply needed for true tetrode operation. Several of the older amateur designs use the triode connected tetrode to get around the problem of excessive gain which the tetrode has. This allows a 100W type transmitter to be used without having to reduce the drive level. Another scheme is to connect the control grid and screen grid together and drive them both. This often requires more voltage swing on the grid to achieve full power. Hence, it is a popular way to reduce the gain of a valve so that it can be driven by a 25 to 40W transmitter safely.

VALVES AS AMPLIFIERS

Valves are voltage-driven devices. The input signal needs to develop an AC voltage between the grid and cathode. The anode current varies with the changing grid voltage. This is termed the conductance (g_m) and is measured in mA/V, where the mA is the anode current variation and V is the voltage appearing across the grid-cathode. A typical power amplifier valve might have a conductance of 25mA/V. For each volt of grid variation the anode current will change by 25mA. Suppose we have an input signal of 10V DC. Hence, the anode current will vary by 10 x 25mA = 250mA. If the anode current is flowing in a resistor of 1kΩ the power developed is I^2 x R, in this example $(0.25)^2$ x 1000 = 62.5W average. This will develop a voltage across the 1kΩ anode load of 250V.

If the effective grid resistance the input signal experiences is 1kΩ the input power is V^2/R which is (100/1000) = 0.1W. So for 0.1W input power we obtain 62.5W output power, which is a gain of 62.5/0.1 = 625 in power ratio terms or 28dB in power gain terms.

The above example only looks at the average voltage and current. A sinusoidal waveform varies plus and minus about zero volts. The peak voltage is 1.414 x the average for each half of the sine wave, so the total peak-peak voltage is 2.828. For a 10V-rms signal the swing is 28.28V. The grid of a valve needs to be driven more positive with respect to the cathode to cause the anode current to rise. For 10V-rms input signal the anode current will vary from the quiescent current up to a value of 14.14V x the conductance value of the valve. If we use the figure of 25mA/V this is an anode current variation of (14.14 x 25) = 353.5mA peak. Assume the valve quiescent current is adjusted to 50mA with no drive signal by adjusting the negative bias voltage, then the anode current will swing up to (50 + 353.5) = 403.5mA.

METER ERRORS

The anode current meter shows the *average current* flowing and this will indicate the average current of approx 50% of the peak anode current. If the input signal is a steady carrier then the meter should read exactly 50%. If the input signal is varying in amplitude the meter will be unable to follow it accurately and the meter indicates the average effective anode current. If the input signal is a square wave with a 50% duty cycle the meter will indicate 50% of the average plus the quiescent current. In our example this will be (50 + 125) = 175mA although the true peak anode current is 403.5mA. Because meters inherently are heavily damped they cannot follow a rapidly varying current and hence give false results. The only time an anode current meter indicates correctly is when the valve is drawing the quiescent or idle current, that is the input is not driven.

GRID IMPEDANCE

When a valve is operated in grounded-cathode the impedance seen looking into the grid is a function of the resistive portion and the grid-cathode capacity. For some valves the resistive portion can be high and for others it is much lower. The grid-cathode capacity modifies the impedance so values between about 500Ω to as much as 25kΩ are usual. If substantial grid current flows the resistive portion can fall to very low levels for large valves. Often due to stability reasons it is necessary to shunt a resistor across the grid-cathode to reduce the effect of feedback from the anode to grid, even though the screen grid in a tetrode supplies some additional suppression. If the grid and anode can physically see each other then RF generated at the anode can be coupled into the grid circuit causing instability. Hence, the physical construction of the amplifier needs to ensure the two parts cannot couple unwanted energy from the anode to the grid. In some amplifiers this is insufficient to prevent oscillation on the higher bands due to the inter-electrode capacity of the valve and then a technique known as 'neutralising' is used. This is an external network that add an equal amplitude but an opposite phase signal between the anode and grid to cancel the internal feedback.

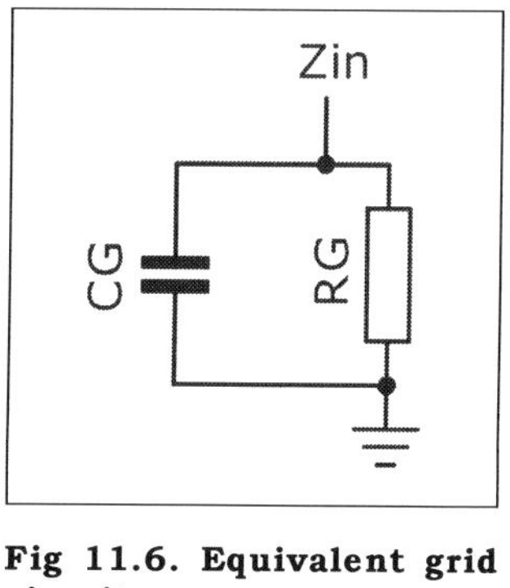

Fig 11.6. Equivalent grid circuit

CATHODE IMPEDANCE

When a valve is operated in grounded-grid the cathode becomes the input terminal. Generally, the resistive portion when looking into the cathode of a valve operated in grounded grid is lower than grounded cathode operation. This makes matching to the cathode simpler than to the grid and often with less loss in the matching network. Because the ratio between the 50Ω source and the cathode impedance is quite low the transformation can be achieved simply and with minimal loss. In some HF amplifiers no matching is used, the 50Ω being fed directly to the cathode terminal. This however results in a loss of efficiency compared to a matching network scheme. The capacity presented to the driving stage is the same as a grid driven valve operated in grounded cathode. This is shown in **Fig 11.7**.

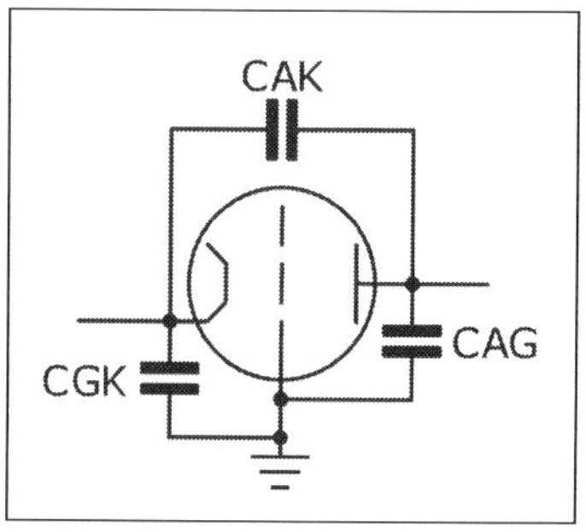

Fig 11.7. Grounded-grid input capacitance

GRID CURRENT

When a valve is driven hard the grid-cathode region acts like a rectifier diode and grid current flows. If the grid current is high, the rectifier diode effectively shunts the inherent grid resistance and lowers the apparent resistance RG. Hence, when zero or very little current flows the grid resistive part is high, falling to a lower value when grid current increases. Designing a suitable network becomes difficult because the impedance changes radically between low drive and full drive.

The conduction voltage of the grid diode is quite high, not like a normal silicon diode, and it may be as little as 1V to as much as 5V. The application of grid bias effectively moves the diode forward voltage to a higher value. So if the grid has a bias voltage of -50V with respect to the cathode to obtain the correct anode idle current the grid will begin generating grid current if the drive signal cancels the bias and exceeds 0V. Hence, as long as the driving voltage is less than 50V p-p no significant grid current will flow.

When operated in Class C the grid bias voltage is high and sufficient to cut-off the flow of anode current during the negative half cycle of the input waveform. To make anode current flow the grid needs to be driven positive with respect to the cathode to overcome the bias voltage. The anode current flows in pulses of clipped half sine wave shape and hence is very non-linear. For Class C amplifiers we can use the grid current flowing to 'self bias' the valve. If we drive the grid with sufficient power the negative grid voltage developed across a high value shunt grid resistor to ground will provide the

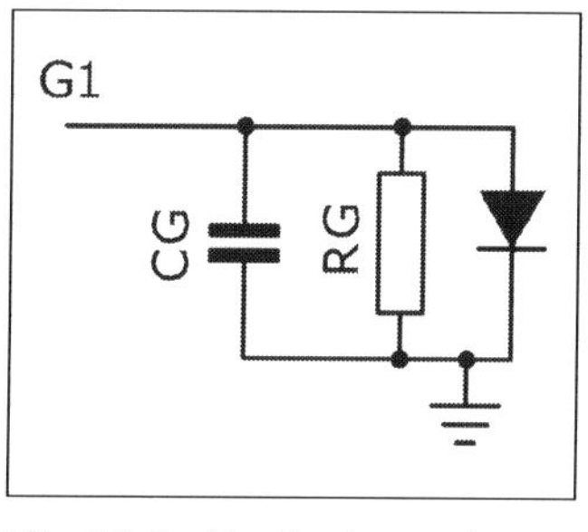

Fig 11.8. Equivalent circuit of a valve grid with grid rectifier included

necessary bias voltage, the negative grid charge is stored in the grid-cathode capacitance. The danger with this scheme is if the drive fails the valve will draw maximum anode current and the valve can be destroyed. Hence, it usually necessary to have a small fixed bias from a power supply to ensure the anode current cannot exceed a certain maximum.

If the valve is a tetrode a popular method is to use the lack of grid bias voltage to control another small valve to reduce the screen grid voltage; this lowers the anode current to a safe level. This is known as a 'clamp circuit' and was popular with CW operators because the drive was switched on and off by the keying. When the key is up the clamp valve holds the screen grid voltage to a low value and when the key is down and a carrier is being sent to the amplifier grid the negative voltage developed biases off the clamp valve and the screen voltage rises to the normal level.

When operated as a linear amplifier the anode current is adjusted to some small value with no drive and the grid voltage developed by the drive swings positive and negative about the grid bias voltage. Suppose the valve requires -35V to obtain the correct anode idle current for optimum linearity. The drive signal if it is 20V p-p will cause the grid voltage to swing between -55V and -15V. The voltage of -55V will cause the anode current to fall to a very low value and the voltage of -15V will cause the anode current to rise to a very high level. This also causes the anode current to flow in pulses of half sine wave shape but the difference between Class C and this mode is that more of the sine wave will be used. In Class C only the upper portion of the sine wave is used, with linear biasing the whole half wave is used.

The fact that we only have a half sine wave signal is not of too much concern because the anode tank circuit is a resonant circuit and fills in the missing half of the sine wave. This causes the AC voltage appearing at the anode to look more like a true sine wave and not pulses of half sine wave. This is one of the features of a resonant circuit. If we send just one very short pulse of energy into the network it will cause the network to 'ring' at its resonant frequency and the result will be a decaying sinusoidal signal. If we do this often enough the result is a continuous sine wave signal. This is also the basis of oscillators.

> The same effect happens if we take a ruler and place one end sticking out over a desktop and hold the end on the desk down. If we now push down on the ruler so that it deflects and then suddenly release it, it will vibrate at a frequency determined by the length of ruler protruding over the desk. We have all done this sometime. The longer the ruler the lower the frequency. The ruler is a resonant network.

This means of course that we are only effectively using half the input voltage swing. To get around this problem we can use two valves in a *push-pull* circuit where each valve operates on a half wave. The two anodes are connected to a common network and arranged so that the two out of phase signals combine into a common signal with equal phase. The advantageous of the push-pull circuit is that inherently the non-linear characteristics of the two valves cancel out and a more linear output signal occurs. In push-pull amplifiers the even order harmonics tend to cancel out and so we obtain better 2nd and 4th harmonic suppression than a single ended stage. In theory this occurs but in practice it is not so easy, if the valves have slightly different gains then the push-pull output voltage is skewed in favour of the valve with the higher gain. It is also necessary to ensure each valve is driven with the same amplitude of grid voltage and exactly 180° phase difference which is not as easy as it sounds, especially at higher frequencies such as 144MHz and above.

HOW MUCH GRID CURRENT?

Many amateurs using linear amplifiers using high gain tetrodes such as the QQV06-40 and 4CX250B are labouring under the misapprehension that any grid current is a bad thing. This is not strictly correct. Although high levels of grid current in a grounded cathode tetrode can lead to poor IMD performance, the real reason is usually because the DC grid bias voltage is moved due to excessive drive.

If the resistance feeding the bias voltage to the grid terminal is a high value then the grid current flowing adds to the bias voltage and causes a DC-shift when drive is applied. (This resistance is not only the one used as a RFC but also any other resistors in series between the RFC and the bias generator circuit). If the grid current flowing is, say, 1mA and the grid resistor is 10kΩ the DC shift will be 10V. The QQV06-40 with an anode voltage of about 750V and a screen voltage of 300V requires a grid bias voltage of about -35V for optimum idle current biasing. If the drive causes grid current to flow in a high value resistor used as a RFC feeding the grid bias to the grid terminal this moves the operating point from Class AB1 towards Class C. Hence, the value of resistor used to feed the bias voltage to the grid needs to be low. It is preferable to use a RFC rather than a resistor for this application, although some valves will require some damping resistance to maintain stability or to broadband the input network. If this is the case then often a wire-wound resistor is a better option. (A 5W wire-wound resistor of about 47Ω works well as a low Q RFC for VHF power amplifiers.) An alternative is to shunt the RFC with a resistor; this does not affect the DC conditions. With a well-designed grid stabiliser circuit (shunt stabiliser) a QQV06-40 can tolerate as much as 10mA of grid current, which enhances the output-power, efficiency and linearity.

Several manufacturers of valves have written engineering bulletins about the grid bias shift problem and how to counter it. One manufacturer of the popular 4CX250B valve gives details of the optimum grid current when operated as a Class AB1 linear. If the grid current is allowed to rise to a small value, typically 250 to 500μA the IMD products actually fall about 6 to 8dB. Increasing the drive so that the grid current rises more brings the IMD products back to the original value and then with increasing drive they start to rise rapidly. As long as the grid current remains below about 1 to 2mA we should have no cause for alarm.

> By connecting a cathode ray oscilloscope (CRO) to the DC grid bias rail and observing what happens when we apply RF drive will often show a DC shift of several volts. This is causing a degradation of the IMD and a lowering of the gain due to the higher effective grid voltage. Taken to the extreme it makes the amplifier saturate at a far lower level than the valve is capable of.

Perusing the various data books on the 4CX250 is quite enlightening. Mullard, in an old data book published in 1969, give information of their version, the QV2-250C. When operated in Class AB1 with 2kV-anode voltage and a screen voltage of 350V the grid bias voltage required is -55V. When fully driven the peak RF grid voltage is 50V, hence it stays below the grid current point. The RF output power was measured to be 300W single tone (CW) with a quiescent anode current of 100mA at 175MHz and an average indicated anode current of 250mA. This attained an anode efficiency of 60% with an anode dissipation of 200W.

Details are also given for television transmitter service. The anode voltage is 2kV and the screen voltage 350V. The grid voltage is -70V for Class B operation. With an average indicated anode current of 360mA the power output is 440W at 216MHz. The grid current under peak drive is 5mA. This amplifier has excellent IMD performance as required by the television signal.

For maximum efficiency when the mode is CW or FM, linearity is of no concern. To enhance the efficiency, and hence how much power we get out for a particular input power and a reduction in power dissipated in the anode, we can change the grid biasing more towards full Class C and compensate for the slightly lower gain by driving the valve harder. Typical efficiency figures for a modern valve operated with a good anode network can reach values as much as 70% in Class C. The maximum theoretically possible is 78% but over 80% have been achieved in some very efficient amplifiers with a high drive power being used.

The reason for the apparently higher efficiency is because part of the input power is coupled through the valve inter-electrode capacity to appear at the anode. Grounded grid amplifiers operated in zero-bias typically transfer up to ≈10% of the cathode drive power through to the anode, because the cathode and anode signals are in phase in a grounded grid amplifier this power is added to the anode power. In a grounded cathode amplifier the two signals are out of phase and hence the feed-through power is subtracted causing a slightly lower power output.

A simple way of switching an amplifier between Class AB1 and a quasi-Class C mode is to include a high value resistor in series with the grid bias supply. For Class AB1 this resistor is shorted out and left in circuit for FM or CW. It is also useful for tuning up, firstly with the resistor in circuit to limit the anode current for initial adjustments and then switched out for the final adjustments. This same technique is also used to insert a resistor in series with the screen grid feed to limit the screen grid voltage as the drive is increased. A combination of the two is a good way of switching between SSB and FM / CW to limit the input power under FM operation to reduce the anode dissipation under continuous carrier operation.

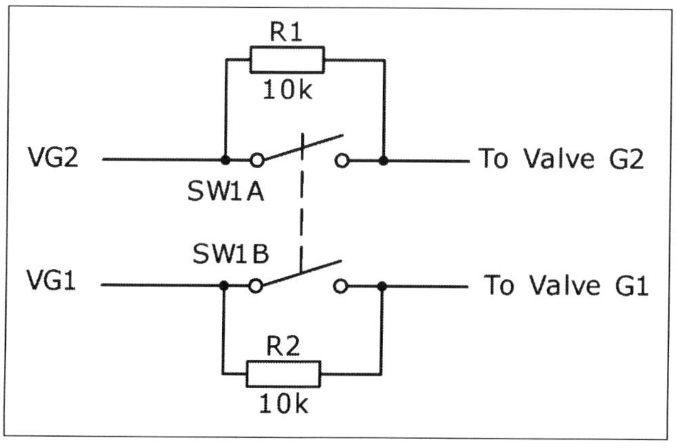

Fig 11.9. Reducing anode current with screen and control grid series resistors

DEFINITION OF OUTPUT POWER

It might be a good time to mention what the 'output power' quoted in a data sheet actually means. We often assume this is the power measured at the output of the amplifier. But in some cases we would be wrong.

The way the manufacturer determines the output power is to construct a test amplifier using the highest quality components. Even these will have some losses associated with them. Therefore the amplifier power output is measured and then the valve is removed and the output-matching network analysed to measure the insertion loss between the output connector and the valve anode. Say this is 1dB, this in round numbers is 20%. So if we measured 100W output using the amplifier the power exiting the anode is 120W and this is the figure quoted in the data sheet. This also makes the efficiency seem better than it actually is in a practical circuit.

Some manufacturers state this and others conveniently omit this fact. This makes the valve appear to be better than it actually is a practical circuit. Some manufacturers give two output power values. One is called 'useful output power' or 'load power' and is the actual power measured under test. The second is called 'available output power' and is the calculated power at the anode with the matching network loss taken into account. Be sure you know which one the manufacturer is quoting! At 70cm typical network losses can be as high as 2dB, which is 40% of the power. The better the output network the more power we will see into the antenna.

DEFINITION OF CW

Manufacturers of valves and semiconductors use a different meaning when CW is stated. As amateurs we use the term to mean Morse code operation with the carrier switched on and off. This is, however, incorrect. CW stands for *Continuous Wave* or *Carrier Wave*, that is a continuous carrier of infinite duration. The manufactures use the correct interpretation, generally speaking amateurs use the incorrect one.

HOW MUCH DRIVE POWER?

This depends on several factors. The type of valve chosen, the operating frequency and the way the circuit is configured either grounded cathode or grounded grid. As it is impossible to cover every type of valve and frequency we can only give a few examples. A typical HF transceiver using 2 x 6146s will use a driver valve such as a 5763, 6BW6 or a 12BY7A. These are small valves and can deliver about 5W maximum. On 80m a pair of 6146s need about 500mW of drive, most of which is dissipated in the grid damping resistors. On 20m the drive level is about 1W and at 10m it rises to about 2W.

The reason for the greater drive level as the frequency rises is due to two things. The first is that the

Mullard give some figures for the 4CX250 in a test amplifier at 175MHz in Class C, this requires 1.8W for full output. They also quote the two power output values, 235W and 200W so the anode network is dissipating 35W at 175MHz to deliver 200W of useful output power, this is ≈ 15% lost power or about 0.75dB. The anode efficiency is given as 78%.

gain drops with rising frequency and hence we need to apply more drive power. The second is due to circuit losses; at lower frequencies the losses are quite small, but as the frequency increases the losses rise.

ANODE LOAD

The optimum anode load is a function of the anode supply voltage, anode current and output power. The anode load is the resistance the matching network needs to present to the anode at resonance for a particular power output, the *conjugate match*. All practical power amplifiers will be designed for only one output power, the maximum required. This causes problems but not ones that we cannot overcome.

The anode at resonance appears to behave as a pure resistance, as does the matching network. Hence, for maximum power transfer the two resistances need to be the same value, the conjugate match criteria. The anode load is approximately the anode voltage divided by the anode current, using Ohm's law. R_L = Va/Ia. (See a later section for a more accurate method of obtaining the anode load resistance.) If the anode voltage is 1000V, with an anode current of 1A, the anode load resistance is 1000Ω. This is an input power of 1kW. At lower anode current the value of anode load increases, at 500mA with 1000V the load is 2kΩ etc. Raising the anode voltage to, say, 3kV require less anode current to flow for the same 1kW input power. It is now 333mA and the anode load is now ≈ 9kΩ. Hence, if a published design is copied and the anode supply used is radically different there is very good chance that the component values used for the anode matching will no longer be able to extract the power with low loss.

Now consider the design of an HF amplifier popular some years ago. This used four low-cost television 'sweep tubes' (PL-504) connected in parallel with an anode supply of ≈ 650V, obtained by rectifying the 230V AC mains directly in a voltage doubler rectifier without requiring a high voltage transformer. To attain an output power of 400W requires an estimated 750W DC input, an efficiency of approx 55%. (When operated as a linear amplifier for SSB the efficiency is lower than for CW operation where linearity is not important.) Hence, at 400W output the power dissipated as heat is 750 - 400 = 350W, almost as much as we are sending to the antenna.

But the real problem with this design is the low anode voltage used. To deliver 750W to the anode requires an anode current of 1.15A DC, or about 300mA per valve. The optimum anode load is therefore only 560Ω. Compare this to another amplifier using 1500V to attain the same power output and with the same efficiency. 750W DC input at 1500V is 0.5A and so the anode load is now 1500/0.5 = 3kΩ, a difference of almost six times. So the pi network values would be vastly different for these two amplifiers.

MISMATCHED ANODE LOAD

Now consider the anode load when the amplifier is not driven to maximum output. For the highest efficiency the matching network resistance and the anode load resistance need to be the same value. If we use an anode voltage of 1500V and drive the amplifier to half its rated anode current we can calculate the optimum load. Say the anode current can be driven to 1A without exceeding the linearity criteria. This is an input power of 1500W and the load needs to be 1500/1 = 1500Ω.

With an anode current of 500mA the load is now 3000Ω. If we adjust the matching network for the 500mA input and then drive the anode current up to 1A the mismatch is very high and excessive anode dissipation will occur. Let us assume that at the 500mA input level we achieve an efficiency of 50%, that is for every watt of DC input we attain half a watt of RF output, which is fairly typical for a normal HF linear amplifier. Now if we leave the matching network as it was for the 500mA input case the efficiency will fall to about 25% because the anode is mismatched at the new current level. Now for every watt of DC input power we only a manage quarter of a watt. This means the vast majority is being dissipated in the valve and it is likely to fail under this high stress level.

Hence, we have no option but to adjust the amplifier matching at maximum input power and suffer the consequences of poorer efficiency at the lower input levels. If the mode were CW then this is of no concern because when the key is up the only current flowing is the idle current, which is typically only about 5% of the maximum current. But if the mode is SSB the idle current will about 10 to 15% of the peak anode current for lowest IMD products. When speaking softly the efficiency is quite poor and the valve will be dissipating more power than is necessary. In practice it isn't too serious a problem because the valve is capable of dissipating far more power at the higher level than is being dissipated at partial output power levels. However, if someone were able to come up with a scheme where the anode always had the correct load applied across the whole anode current range it would save a lot of power.

'TUNING-UP'

This leads to a common mistake made by some amateurs when setting up a linear amplifier. They tune up at low power and then use it at high power without readjusting for this condition. They believe they are doing things correctly but they do not understand what is going on. They use low power so as not to stress the valves during the tuning-up process. Some are slower at tuning-up than others. This is fine if the tuning is simply semi-optimised at the lower power output and is then quickly readjusted for the full output. But this method is counter-productive: why do it twice when you can do it only once and get it perfect?

A far better method is to use an electronic keyer set to send a string of fast dots. This is a 50% duty cycle and the valve will only dissipate half the power during tuning up. If the power meter is a peak-reading type, the power output indicated will be correct. If the power meter is an average reading type the indicated power will be approx 50% of the peak power. The reason it isn't *exactly* 50% is because of errors in the duty cycle of the keyer and the keying circuits in a typical transmitter have a slow rise and fall time to eliminate key clicks. As long as the matching network is adjusted for maximum output power the setting will be correct. In 'the old days' the accepted technique to tune up a pi network for a transmitter was this:

Set the anode and loading capacitors to fully meshed and apply a little drive power, enough to indicate about 20% anode current. Then swing the anode tuning capacitor through its range until you find the biggest dip in anode current. Now reduce the loading capacitor to about half-meshed and re-dip the anode current. Increase the RF drive to about 50% anode current. Readjust the anode tuning capacitor for the biggest anode current dip again. Keep on reducing the loading capacitor value whilst increasing the RF drive to obtain about 75% anode current, readjust the anode tuning again for lowest dip. Then increase RF drive to obtain 100% anode current or until the anode current is at the required value for the power input desired. Finally, readjust the anode tuning again for best dip and adjust the loading capacitor a bit more to bring the anode current back to the required amount, reduce the RF drive a bit and so on *ad nausea*.

I don't know about you, but this is a painfully slow process and if you take too long the sweep tubes make strange crackling noises, glow red in the face and finally go pop! What we are trying to achieve is to set the transmitter to a wanted output power. Why go through all this routine? When the anode tuning and loading controls are set for maximum output power the amplifier is at its most efficient operating point, surely that is more important? The transmitter RF drive control, not the dip in the anode current, should control the output power. All that the dip tells us is that the network is at the resonant point. For best linearity the amplifier should be heavily loaded, i.e. tuned for maximum output power, and *not* lowest anode current dip.

Now take the case of a station where not only does the transmitter needs to be tuned-up, but also an ATU because the antenna is a long wire, G5RV multi-band dipole or some other type that has a poor VSWR on some bands. These two items inevitably interact and you can really chase your tail at times. Getting into the habit of writing down the knob setting for different bands, once you have worked them out, saves a lot of time, time you should be using to chase that rare DX instead messing about with adjustments.

ADJUSTING A TRANSMITTER AND ATU IN SERIES

To adjust a transmitter and ATU correctly one should firstly tune up the transmitter into a dummy load to suit the system impedance, normally 50Ω. Having set up the transmitter to the required output power the ATU is then connected between the transmitter and the antenna feeder. The transmitter controls are not touched. The ATU is now adjusted so that the VSWR meter inserted between the transmitter and ATU shows a perfect match. Now although the antenna VSWR hasn't changed the ATU is providing a low VSWR match to the transmitter.

> I know one amateur who was so scared of blowing up his new valve linear, bought at great expense, that another experienced amateur had to go around to his shack and set it up on 20m. Some time later I heard him on 40m, but his signal wasn't the same booming one as on 20m. He eventually confessed that when he needed to operate on 40m or any band other than the 20m SSB portion where the linear was set up, he turned off the linear! No wonder 'black box' operators have such a bad name. It would take him about 30 minutes maximum playing with the settings to learn how to do it.

GROUNDED GRID OR GROUNDED CATHODE

This very largely depends on two factors. The first is the valve chosen and the operating frequency. There is one other factor and that is the drive power available from the transmitter driving the valve.

As we have already mentioned for best stability, that is the valve not becoming an oscillator, we often have to settle for grounded grid with its lower potential power gain. Sometimes this suits the transmitter driving it better. If the amplifier is for HF and the transmitter is a 100W type it can be difficult to reduce the drive power to suit a grounded cathode amplifier. Let us take a popular VHF tetrode, the 4CX250 series and examine its characteristics.

4CX250 SERIES

This is a compact metal-ceramic construction with an external anode designed for forced air-cooling that easily fits in the palm of your hand and is rated at 250W anode dissipation. It is designed for grounded cathode operation and it requires a special airflow socket for operation at VHF and UHF with a built in screen grid decoupling capacitor. There is also an HF airflow base made without the screen grid decoupling and it is a popular valve for HF duty. It is usable up to about 600MHz at reduced power output and the common upper limit at full rated power is about 450MHz. Hence, it is a very popular choice for 144 and 432MHz linear amplifiers. Its efficiency is quite high, at 432MHz we can expect at least 50% in a good circuit even when used for SSB. At HF it has masses of gain, 26dB typically at 30MHz, so it needs very little drive power when operated in grounded cathode configuration at HF, 1W input can develop 400W of RF output.

It requires a very stable and stiff power supply for the screen grid of about 300 to 400V maximum. There are several versions made, the original valve was the glass-metal 4X150 which had an anode rating of 150W. The

Fig 11.10. Glass-metal 4X150A (left) and ceramic-metal 4CX250B (right)

later versions are the 4X250, 4CX250A, 4CX250B, 4CX250F/G and the specially modified for military service 4CX250R which is the premium version. There is also a bigger version known as the 4CX350 which has 350W of anode dissipation, but this is only rated at maximum output for use up to about 175MHz. It can be made to work on 144MHz but the gain is a little less than a 4CX250. It finds most applications as a FM broadcast driver valve in Class C for bigger valves, as does the 4CX250 family, as well as HF operation where it very popular.

The part numbering system adopted by Eimac who developed the 4CX250 series and others is quite logical once you understand what each number and letter denotes. The 4 means it has four electrodes, i.e. it's a tetrode. The C means it is of ceramic construction. The X means it is an external anode type. 250 denote the continuous anode dissipation in watts. The final letters denotes the filament voltage, A or B = 6V, F/G = 26.5V. The odd one out is the R; this denotes not the filament voltage but that it is a rugged construction capable of handling high vibration and G shock loads. There are also special derivatives such as the 4CX250BC with a 6.0V heater rating which is a specially selected premium grade and the coaxial version, the 4CX250K designed for use up to 1500MHz in radar transmitters.

Another oddity is the 4CPX450K designed for grounded grid operation, the only version for this mode, and it is used for airborne jammer service at up to 500MHz rated at 10kW peak power in Class C with a gain of 10dB, hence it requires 1kW of drive! For this mode it requires an anode voltage of 5kV. (The P in the part number denotes pulse operation.) In CW mode it can develop 500W output at 450MHz. It requires only 3W of drive at 450MHz.

Other valves follow the same system. The 4-400 is a glass envelope construction tetrode of 400W anode dissipation. The 4CX400 is a tetrode with ceramic construction, external anode and 400W anode dissipation. The 3-500Z is a glass envelope triode with 500W anode dissipation etc.

The 4CX350 family consists of two versions, the 350A with a 6.0V heater and the FJ version for a 26.5V heater supply for aircraft service.

A pair of average 4CX250Bs with an anode voltage of 2kV at 144MHz can deliver about 450W of reasonably linear power under SSB service. If the anode voltage is wound up a bit to about 2.5kV it is possible to see as much as 600W out on 2m but the linearity suffers quite a bit. For CW this is not a concern. If you get adventurous and crank up the anode supply beyond the manufacturer's recommended maximum of 2.5kV to around 2.8kV they could deliver about 800W for short periods with CW. When scratching for the last few watts when operating EME they can be pushed quite hard without failing, as long as the forced air-cooling is adequate. The '250' in the part number is the anode dissipation and with a really efficient anode circuit they can be persuaded to dissipate quite a bit more if the blower is large enough. For CW operation the idle current can be adjusted towards the zero current point and this enhances the efficiency by about 10%, but is hopeless for SSB as it will generate high IMD products. When biased towards the zero current condition the drive power required goes up quite a bit, in other words the gain falls. Even so, with about 5W of drive at 2m they generate lots of power. In an air-borne jammer made in the 1950s for the US Air Force four 4X150A valves were used to generate over 10kW pulsed power at 200 to 400MHz.

Using a 4CX250 series valve at HF means we need almost zero drive to achieve full output if the grid is matched. Typically as little as 1W can drive one to full output at 30MHz. This is a problem when the transmitter is a 100W type, what do we do with the other 99W? The simplest way is to use what we term a 'passive grid' mode. This does not use any matching at all, but simply a high power 50Ω load resistor across the grid to generate the required RF voltage. This is very wasteful because we are dissipating all the transmitter power into effectively a dummy load, but it is often the only way around the problem.

Another method would be to use a high power attenuator between the transmitter and the valve to reduce the power to a few watts and then loosely couple via a low value series capacitor into a higher value grid-loading resistor, perhaps as high as 500Ω. The benefit of these two schemes is that the transmitter is terminated in a perfect match for all frequencies, so a wide-band amplifier covering the amateur bands is not too difficult to arrange. The only problem is then the anode network, which follows standard methods. Loading the grid with a low value resistor makes the amplifier very stable and practically impossible for it to oscillate.

BRITISH VALVE PART NUMBERING SYSTEM

Valves developed in Britain and Europe use a different part numbering system to American valves. They use part numbers like QV06-20 and QQV06-40 for example. The two systems are really quite similar. The QV06-20 is the British version of the American 6146. (American valves in the early days simply issued part numbers in a numerical sequence with no thought to convey parameter information in the part number.)

Q denotes a RF tetrode, if it is a dual tetrode then QQ is used. T means it is a triode. P means it is a pentode. The V in the part number denotes it is an indirectly heated thoriated-tungsten filament. Z denotes a directly heated oxide-coated filament. X denotes a directly heated tungsten filament.

The numbers after are the maximum anode voltage in kV, hence 06 means 0.6kV = 600V. The final number after the hyphen symbol is the anode dissipation in watts. A QZ35-3000 is a directly heated cathode tetrode with a maximum anode voltage of 3500V with an anode dissipation of 3kW. A final letter is sometimes used to denote other characteristics. The letter A means it is the air-cooled version. The letter W denotes the water-cooled version.

For example, the QQV03-20A is a twin tetrode with an indirectly heated thoriated-tungsten filament rated at 300V anode voltage and 20W anode dissipation designed for air-cooling. Because the valve consists of two tetrodes in a common envelope the 20W anode dissipation is split equally between the two halves, 10W per anode. The larger version of the popular QQV06-40A is the QQV07-50A, a 700V anode voltage rating and 50W total anode dissipation.

These anode voltage and dissipation are the limits

for maximum valve life, the CCS Ratings (Continuous Commercial Service), which is for continuous transmitting for an indefinite period, i.e. 24hr-day. A second rating system is the ICAS (Intermittent Commercial and Amateur Service) which allows more power, voltage and anode dissipation at the expense of a slightly shorter life. As the *intermittent* signifies the valves are not operated continuously but periods of transmit and resting periods on receive. '*Intermittent*' is defined as not exceeding 50% duty cycle, e.g. 30s transmit and 30s receive but often not clearly defined generally. 1hr transmit followed by 1hr receive is also 50% duty cycle. If the manufacturer does not specify the meaning it is safe to assume the times should be minutes rather than hours! All of the valve manufacturers' ratings assume adequate cooling: if the cooling is inadequate the valve life will be shortened or it may fail during service.

The original British numbering system has since been superseded by a pan-European numbering system. For example, the QQZ06-40 twin tetrode is now known as the YL-1030. The QQV06-40 changed to the QQE06-40, the formalised European standard design parameters and is also available as the American 5894.

The one oddity in this system I have come across is the 813 equivalent. This is the American pentode developed just before WWII by RCA. Following the British part numbering system it should be known as a PY2-125. The P meaning pentode, Y denotes a directly heated cathode, 2 is the maximum anode voltage of 2kV and 125 is the anode dissipation in watts. However, Philips the successor to Mullard who developed the numbering system messed it up and gave it the part number of QB2/250. The Q meaning a tetrode, B is normally reserved for receiving valves of greater than 1W anode dissipation, the 2 denotes the maximum anode voltage (the only part that is correct) and the 250 normally denotes the anode dissipation. Hence, many amateurs are under the belief it can dissipate more power than it actually can! The 813 has two different anode dissipation ratings. The CCS rating is 100W and the ICAS rating is 125W.

L-MATCH APPLIED TO VALVES

A common problem when using large triode valves for RF power amplifiers is establishing a good match between the driver and the amplifier. Most RF power triodes are intended for grounded-grid operation, for stability reasons, and hence the input terminal is the cathode. Depending on which book you read will give different values of the

Fig 11.11. Dual-tetrode valves. The small valve on the left is a QQV03-10, next is the QQV03-20A, QQV06-40A, QQZ06-40 and YL-1030

cathode impedance to expect. Unfortunately some of these are incorrect because they assume the triode is operated in Class AB2 or B where substantial grid current flows, for example the Eimac 3-500Z and the 572B types. The higher the grid current the lower the impedance looking into the cathode from a 50Ω source.

When the triode is operated with zero bias, which is how the 3-500Z and many other valves are used, the effective cathode resistance at full drive is low because high grid current flows, as much as 40% of the anode current value.

Eimac in the data for the 3-500Z give the cathode resistance for a single valve driven to 750W output under single tone conditions with an anode voltage of 3kV, this is 115Ω. This requires 30W of drive at 2MHz in the test circuit and hence requires an impedance step-up network from 50Ω.

With high conductance triodes the bias is not zero, in fact it is quite a high value in cathode voltage to achieve the correct idling current, this reduces the grid current to a very low value. Hence, the cathode impedance is much higher than a 3-500Z, which is typically 100Ω when driven with 30W of RF signal. The other thing is the gain a high conductance valve has. Svetlana gives details of the 572B operated in grounded grid for HF amplifiers. With an anode supply of 2.4kV, a quiescent current of 45mA and anode single tone current of 250mA it delivers 300W of useful output power. This requires a drive power of 50W; hence the gain is six in power ratio terms or 7.8dB in logarithmic terms. This is for a single valve. With two valves in parallel we can expect 600W output and a drive power of 100W. This gain figure sounds quite low but the reason is historical.

When external linear amplifiers were first used the licence conditions in most countries still used the 'DC input' method of limiting the maximum power allowed. Before the days of SSB the predominant modes were AM and CW. The amateur licence conditions for the UK stated that the '*maximum DC input to the final stage may not exceed 150W*', this with the assumed output stage efficiency of 66.66% gave an output power of exactly 100W. Hence, transmitters tended to be built so that they were typically capable of delivering about 100W carrier output.

When SSB operation first started the regulators did not know how to handle this new mode and so fell back on the older AM specification. As you should know, if we amplitude modulate a carrier to a depth of 100% the output power rises from the non-modulated condition to a peak value of four times the carrier level. Hence, for a nominal 100W carrier the peak envelope power (PEP) is 400W.

In America, where the 'linear amplifier' first became popular, they had a different power rating and this effectively meant that it was possible to use output peak powers of up to 1kW. By reading between the lines of the regulations, which allowed up to 1500W DC input for AM and SSB, it was possible to run up to 1000W output power. Hence, the default output level for linear amplifiers became 400W in the UK and other countries and 1kW in America (this was later increased to 1500W).

So a suitable valve needed to be found that could accept about 100W drive power and deliver about 1000W in a simple amplifier. Eimac designed the 3-500Z, and others, to suit this market. Often in the interests of longevity two valves are used in parallel so the stress is reduced. So the *de facto* standard linear amplifier for many years was the dual 3-500Z triode operated in Class B zero bias and hence why most textbooks state the cathode impedance as being about 30Ω.

During the design of a triode amplifier for 50MHz the standard textbook formulas were found to be far from the impedance found in practice. This prompted the investigation of the typical cathode impedance to be found for different values of grid current. The particular valves used were the Russian Gi7bt, which operate under virtually zero grid current condition at 50MHz and make superb linear amplifiers with a gain of about 17dB (50x) in a practical circuit, so a drive level of 10W gives at least 400W output. In fact a single Gi7bt at 50MHz or 144MHz can deliver about 1kW CW output in Class C if correctly matched and cooled. Under radar short pulse duty (7μs pulses), for which it was originally designed, it can deliver 12kW of peak power at 3GHz so it is quite a potent valve. The air-cooled anode is rated at 350W dissipation but when modified to water-cooling this increases by at least two times.

The cathode impedance was established by building a dummy amplifier with a valve powered up and the cathode driven with various power levels and bias conditions. A matching network was empirically constructed to match the 50Ω input to the valve. The VSWR was adjusted to 1:1 and the cathode RF voltage and grid current were measured. The matching components were measured out of circuit to calculate the real part of the cathode impedance.

Having established the expected cathode impedance for various drive-levels it was then necessary to design a suitable input matching network. The best option was a modified version of the L match because the cathode impedance was still relatively high. The resistive part of the cathode input impedance at 50MHz was determined to be approx 300Ω at maximum drive level of 10W. The cathode capacity is ≈12pF; this at 50MHz is a reactance of 265Ω that requires an inductance of 0.85μH to cancel. The cathode of the valve needs a RFC to carry the cathode current to the biasing circuit so this means the RFC can serve two purposes, one being to cancel out the capacitive input and also as a RFC to carry the cathode current to the biasing generator circuitry. As it is necessary to choose an inductor with a reactance of ten-times the connection point impedance this meant the 0.85μH

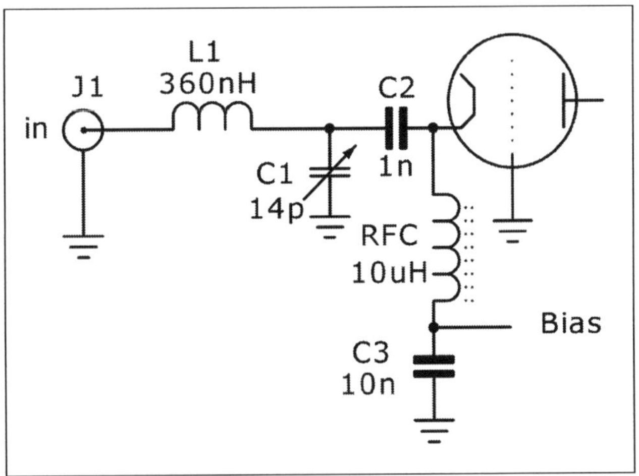

Fig 11.13. Schematic for Gi7bt cathode matching for 50MHz

value calculated is too low. So a higher value of inductor is required.

It is preferable to choose a value that can be purchased than having to wind a custom inductor. The value of choke required is (3000/265) x 0.85μH, which is 9.62μH, a 10μH choke will be suitable. A standard radio interference suppression choke of 10μH wound on a ferrite rod intended for commutator type electric motors is available that is rated at 5A which is suitable for the cathode peak current of approx 1A maximum. This choke has a SRF of about 100MHz so it will be suitable.

Having now been forced to use an inductor much higher than the first calculated value is actually a bonus because we have 'over compensated' the cathode capacitance. This makes the cathode now a net inductive input, so we will not need as high an inductor for the series part of the network. By calculating the component values to match to the inductive cathode we find we require a shunt capacity of ≈14pF and a series inductor of 360nH. If we had used the 0.85μH RFC we would not only need far more drive level but also the value of the series inductor would need to be ≈700nH. The shunt capacitor needs to be variable to allow us to tune the network to resonance as we move about the band; an air variable of 25pF maximum will suffice. The final circuit is shown in **Fig 11.13**: this gives a maximum VSWR of 1.05:1. Capacitor C2 is a DC block to prevent the bias voltage shorting to ground.

The graph in **Fig 11.14** shows the typical input VSWR against frequency. If the input is tuned at 50MHz the worst case VSWR at 52 MHz is 1.19:1, so having to extend the shaft of the tuning capacitor to the front panel

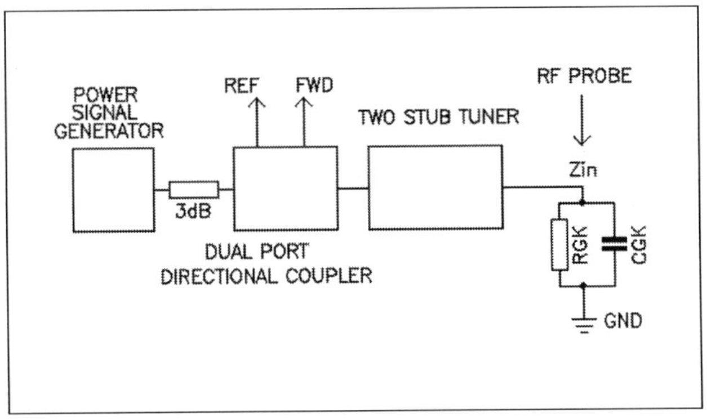

Fig 11.12. Test equipment set up to measure cathode impedance of UHF triode

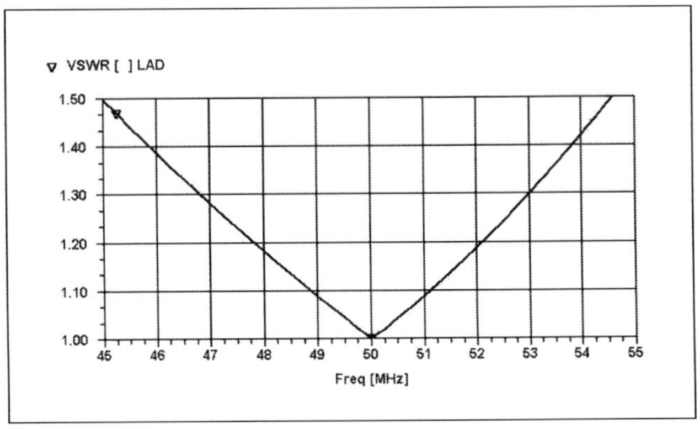

Fig 11.14. Input VSWR plot for the 50MHz amplifier

is not really necessary.

If a dual-valve amplifier is required then the cathode resistance falls to ≈150Ω and the capacity rises to 24pF. This requires a modification of the network values. Because at 50MHz the cathodes can be strapped in parallel we can use one RFC for the cathode biasing. Under this condition the value of C1 becomes ≈7pF and L is ≈233nH. All the other components remain the same as the single valve design. (See a later section for the anode circuit calculations.)

800 SERIES AMERICAN VALVES

During WWII several new valves were developed and manufactured in large quantities. Of these some became popular for amateurs because so many were available after the war on the surplus market at low cost. Some are still being manufactured today.

The lowest-power valve popular with amateurs was the 807. This is roughly equivalent to the modern 6146 but was almost three times the length. It was capable of about 75W output on HF with an anode supply of 750V so fitted nicely with the 150W DC input licence conditions in the UK. Two in parallel could generate well over 150W in a good circuit and being pentodes they needed modest drive levels. A valve such as the 5763 pentode (another American valve rated at 10W output) was well able to drive a pair of 807s to full output.

The next valve up in power level was the 811 triode. Physically this is about the same size as the 807. Today this is still very popular and several manufacturers still make amateur amplifiers using two, three or even four in parallel. Its power output is about twice an 807 so two in parallel provide close to the 400W limit under CW operation. Again the anode voltage requirements are quite modest and 1500V is enough to give substantial power with modest drive levels.

The big daddy is the 813, which is a pentode and quite a large valve physically. Like the 811 it used a solid graphite anode. This was developed by RCA and first made in 1938, just in time for WWII. The majority of 813 valves manufactured during the war were made by Westinghouse in Canada and bore the British Air Ministry service number of VT-144 (Valve Transmitting).

Several amateur designs used one or two 813s in HF amplifiers. But because of the peculiar licence conditions in the UK and other countries none of these designs actually achieved anything like the true potential of the 813. Some designs ran the valves as triode connected whereas for maximum power the valve was designed for operation as a pentode. They also ran the 813 with a low anode voltage of about 1500V. One famous design using two 813s in parallel claimed an output power of 500W when driven with 150W, a gain of 3.33 or 5.2dB at best. The anode dissipation is 125W for the ICAS rating and 100W for the CCS rating.

The manufacturer's recommended operation of the 813 is an anode voltage of 2 to 2.5kV with a screen voltage of 400V. Under CW operation in Class C at 30MHz it is rated at 375W continuous carrier output and under intermittent duty cycle SSB operation ≈450W PEP; this is for a single valve! Here in South Africa they were used until very recently in transmitters for outer marker beacons and air navigational beacons at airports running in a very much under-run mode of 100W carrier, 24h a day. Under this operation the valves would typically last for 5 to 10 years before requiring replacement.

Like most valves for linear amplifiers, the 813 and others are designed for high anode voltage and low anode current operation. Lowering the anode voltage and raising the anode current to achieve the same output power often means the cathode emission is insufficient to deliver enough electrons. If the drive is pushed too high the excessive cathode current can cause stripping of the cathode material with a consequent reduction in gain. This is a spiralling condition because to maintain the output power the drive needs to be increased, which exacerbates the problem.

> In this country (South Africa) the terms 807 and 813 also refer to size of beer bottles. Typically beer is sold in pint and quart bottles. The slang name for a valve is a 'bottle' hence if at a ham radio party someone asks what you would like to drink often the reply is an 807 or 813, an 807 being a small beer and 813 a large beer. One radio club in Johannesburg has a Friday evening meeting known as the '807 Club' at a local bar.

AUTOMATIC LEVEL CONTROL (ALC)

ALC is similar to AGC in receivers. A sample of the RF output in the linear amplifier drives a detector diode to develop a DC voltage (by convention a negative voltage) proportional to the output power. This is fed back to the transceiver and used to control the gain of the RF driver stages. In this way we can automatically limit the peak output power to whatever level we require. Even if we bellow into the microphone the ALC holds the peak output power to the set level. Virtually every commercial linear amplifier made and all HF transceivers and some VHF / UHF types have this facility but hardly any amateurs use it. Those who are either unaware what the connectors marked ALC do, or are simply too lazy or cheapskate to hook up a cable between the two items, are the 'splatterers' we hear daily on our bands. Using ALC to control the peak power output is a very good scheme. The equipment manufacturers went to the trouble to include it, so why pay for something and then not use it?

Most ALC systems allow the operator to adjust the peak output power down to low levels. On my 2m linear using 2 x 4CX250Bs I can set the power output anywhere from 800W PEP down to as low as 25W PEP by simply turning a knob on the linear front panel. Most countries' amateur licence conditions say you shouldn't use any more power than required to make the contact. Well I don't, I turn the power up and down to suit the conditions. If I am working someone 50km away I will often switch out the linear amplifier and run 'barefoot'; I simply do not need 400W to make the contact, 10W is plenty. For stations at a greater distance, which might have slow QSB, I run the linear amplifier ('put the boots on') and set the power to an appropriate level. For the really distant stations I wind the power up to a suitable level.

PUSH-PULL AMPLIFIERS

We have already mentioned push-pull amplifiers: the two valves are operated in a circuit where each valve only operates for half the input drive sinusoidal signal. Obtaining the necessary 180° phase difference between the grids can be achieved in several ways. Many early designs simply used a centre-tapped inductor and a coupling loop from the 50Ω driver. To bring the network to resonance a capacitor is required across the inductor. To ensure balance this is often made up from two variable capacitors with the centre point being ground. This is known as a split-stator capacitor. Often it is necessary use two smaller value capacitors, one from each grid to ground, to achieve perfect balance of the grid voltages.

Each grid requires a negative bias voltage; a simple way of achieving this is to feed the bias voltage into the centre-tap of the grid inductor so that both valves receive the same bias voltage. For dual valves this is acceptable because the emission of the two halves will be very closely matched during manufacture. However, when using two individual valves they often have slightly different current gains and so one valve might require a different bias voltage to obtain the correct anode idle current. In this case we need to DC isolate the two grids by placing blocking capacitors between each grid and the centre-tapped inductor and then feed each grid with a RFC or high value resistor to separate bias supplies.

If the valves have high potential gain, the problem always exists that any imbalance can cause the amplifier to turn into an oscillator. Balancing the grid AC drive and idle currents goes a long way to prevent this but internal capacitance between the anode and grid may also cause enough feed through signal to make the amplifier unstable. Often this occurs when it is driven by an input signal and we would not ordinarily be aware unless we kept a close watch on the anode current meter. To reduce the possibility of oscillation we normally have to do two further things. The first is to heavily load the grids with resistors to damp the tendency of the feed through energy. This lowers the grid voltage and hence requires more drive power. This isn't normally too big a problem as the average push-pull amplifier using tetrodes usually has an excess of gain. If we take a 2m amplifier using a pair of 4CX250s with the typical grid network it will require

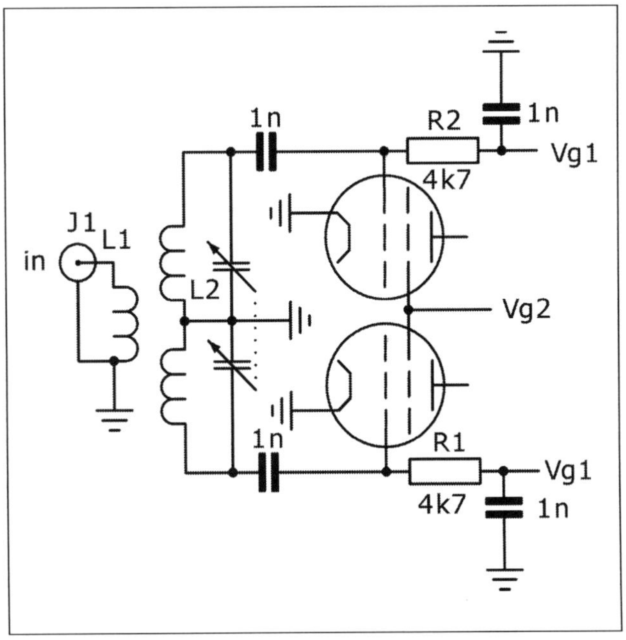

Fig 11.15. Typical push-pull grid circuit for grid driven amplifier

about 4W of drive to fully saturate it. Because the average 2m transceiver is about 10 to 15W we have ample drive in reserve; hence damping the grids with quite low value resistors helps to absorb the extra drive power as well as making the amplifier more stable. These resistors can serve the dual purpose of feeding in the grid bias to each grid so eliminating the RFC, which can become self-resonant and cause stability problems at some higher or lower frequency. The other benefit is that the grid resistors help to terminate the driver, hence it sees a low VSWR match to the grids.

The second thing that is normally necessary in this type of 2m amplifier is some type of external neutralising. In a typical 2m push-pull amplifier using 4CX250s, such as the 'plumbers-delight' design by Joe Ludlow, GW3ZTH, published in *Short Wave Magazine* in the 1960s, and the popular modern version, the W1SL design, these use small 'flags' of copper or brass shim stock positioned close to each anode to form a very low value capacitor. The neutralising flags are cross-coupled to the opposite valve grid terminal. By adjusting the position of the flags closer or further away from the anode the neutralising can be adjusted. However, this is not a simple adjustment to make.

Today many VHF transceivers are of the 25W category and some even as high as 80W or more. If we need to drive a typical 2m amplifier using a pair of 4CX250s we have a problem because the drive is now much more than the valves require. In order to get around this problem I came up with a different approach which eliminates the conventional grid circuit and makes the input broadband, requiring no tuning.

If we look at an antenna fed from an unbalanced coax we have a similar problem. A common method of obtaining the 180° phase difference and equal amplitude is to use a piece of coax which is electrically 180° in length, a λ/2 length. This has the property that it is a 4:1 step-up transformer or *balun* because it converts an unbalanced feedline to a balanced antenna, so the impedance appearing across the two ends is now 200Ω if we use a 50Ω drive source. By using non-inductive resistors from each grid to the bias supply we not only correctly terminate the λ/2 balun but also provide heavy damping to the grids to ensure stability.

I first published this design about 15 years ago in the *Radio ZS* magazine and soon after a Japanese manufacturer started using it in a 2m amplifier based on the famous W1SL design using 2 x 4CX250Bs. The drive power required to drive the amplifier to full output is about 25W and the linearity is excellent. The basis of the grid circuit is shown in **Fig 11.16**.

The grid damping resistors are 120Ω and not 100Ω because of the inherent grid capacity of the 4CX250 require a higher value resistor to correctly match to the 4:1 balun. These need to be rated at 5W or more. Paralleling 3 x 390Ω carbon composition resistors of 2W rating is the optimum solution.

A further variation is the method I now use, which dispenses with the long λ/2 coax, although this is simple enough to coil up. The balun is replaced with a 1:1 Ruthroff transmission-line transformer balun and individual L-match networks to each grid terminal. This allows a better adjustment of the balance.

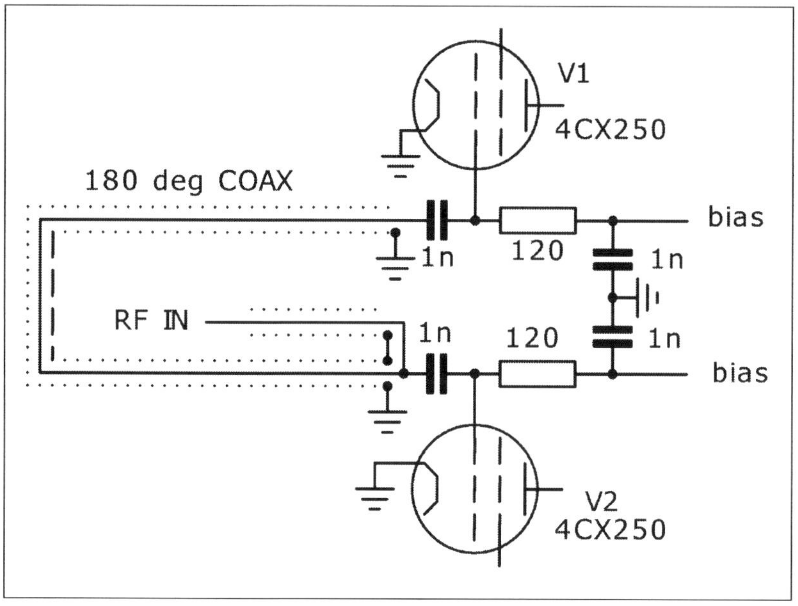

Fig 11.16. Push-pull grid circuit for 2m amplifier using 4CX250s

The transformer is wound on a two hole ferrite core with miniature 50Ω coax. The impedance appearing at the grids is approx 500Ω so giving a 10:1 voltage step-up from the 50Ω source. The 108nH inductors are wound with 20SWG tinned-copper wire on an 8mm former and then slipped off; they consist of about 4 turns. The 47pF series capacitors are Philips ceramic plate types with 500V rating. With the values shown the maximum drive required is ≈8W for full output, so suiting a 10W-category transceiver. Note that the grid resistors dissipate a fair amount of power and need to be able to handle this without heating up too much. Because they are situated in the grid compartment, which receives a high volume of air from the cooling blower, they run cool.

The bonus of both of these grid circuits is firstly they are both broadband and so moving about the band is quicker as we now only have the anode tuning to worry about. The second thing is now we no longer need those troublesome neutralising networks. The amplifier is completely stable without them because of the heavy grid loading caused by the grid resistors.

PARALLEL CONNECTED VALVES

For some applications it is better to use the alternative connection method, the parallel connection. Often when working at some high frequency mechanical difficulties preclude the use of push-pull, also for the higher frequency bands such as 432MHz and above the problem of constructing a well-balanced push-pull circuit is very difficult and can lead to spurious resonance and oscillation. The problem with strapping two valve grids or cathodes together is arranging to do this in a low inductance manner and this often means we have to look at alternative methods.

A classic example is the 432MHz amplifier designed by Richard (Dick) Knadle, K2RIW, in the 1970s that uses two 4CX250Bs. The grid network of this used a grid tuning line shaped like a letter T. The two ends of the top part of the T connect to the two valve grids and the bottom part connects to an air variable to bring the line to half wave resonance. This is a low impedance 'slab line' suspended over a ground plane formed by the amplifier chassis. Power is coupled in with a small value variable coupling capacitor in series with the 50Ω input coax.

The distance separating the two valves is fixed by the physical size of the airflow bases and this causes problems with spurious resonance modes and makes the setting up tricky. In fact it can be very 'twitchy' even when correctly set up. This is because of the high gain of the 4CX250Bs at 432MHz, the original K2RIW amplifier design only requires about 5W for full output. At the time the K2RIW was designed the average 70cm transverter only developed about 4W. I favour a different approach that eliminates these problems.

Each grid needs to be driven in-phase and with equal amplitude signals. Strapping together two or more valves at HF or low VHF does not normally present a problem because the inductance of the strap is not too serious. But at higher VHF, UHF and microwave frequencies the inductance becomes significant and ensuring equal drive to each valve something of a challenge because of potential spurious resonance effects.

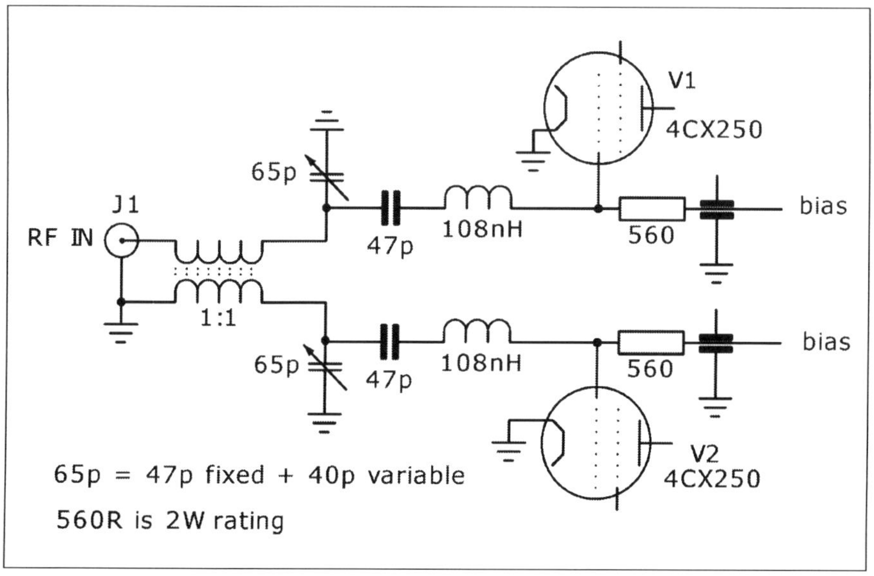

Fig 11.17. Modified 2m grid circuit using a 1:1 transformer and L match networks

PARALLELING VALVES WITH L-MATCH NETWORKS

The basis of the network is to treat each valve grid or cathode as a separate entity with its own L-match network and then to connect the low impedance ends of the L-match networks to a common point. This also allows some degree of balancing by adjusting each network individually.

This was used in my version

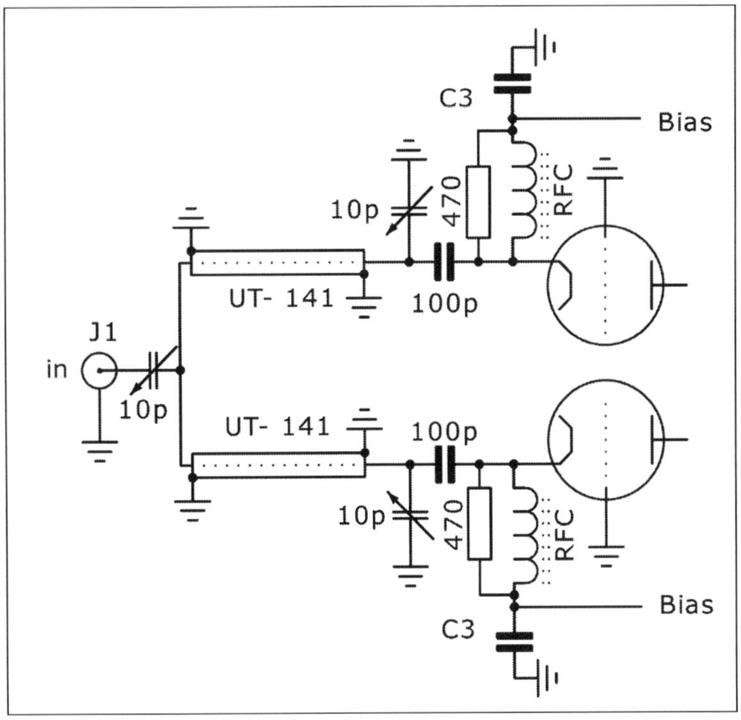

Fig 11.18. Paralleling two valves with L-match networks

of the K2RIW and gave superior performance. Not only is it more compact, easier to construct and adjust, it is lower loss. With 4W of drive the two 4CX250Bs could be driven to 1kW DC input at which the amplifier delivered about 550W output. Increasing the drive to 8W saw nearly 1kW output using Class C. The impedance matching is performed by sections of 50Ω transmission line. In this design semi-rigid coax UT-141 is used as a λ/4 transformer with the two ends brought to a common connection point and then matched to 50Ω.

I also used the same technique for a dual-valve amplifier for 23cm using Russian Gi7bt triodes in grounded grid. This scheme also used short sections of UT-141 50Ω semi-rigid coax to form the inductors. Details can be found in the RSGB book *Microwaves Projects 2*, edited by Andy Barter, G8ATD.

GRID CIRCUITS FOR UHF AMPLIFIERS

When using valves such as the 4CX250B for 70cm the design of a grid-matching network is often difficult. Losses at UHF can be high if unsuitable components are used. The main problem is deriving the actual grid reactance and resistance. Looking at the manufacturer's data is not of much help. The grid capacitance of a 4CX250B is quoted as being 15.6pF, but this is measured at 1MHz using a cold valve (no heater current) in a shielded fixture. The airflow socket and the grid pin add inductance which at 432MHz more than cancels the grid capacity. Using a Vector Network Analyser with a valve in an airflow socket allows the input impedance to be established. An airflow socket was fitted with a BNC connector with the centre pin connected via a low loss 1000pF capacitor with very short leads to the grid connection. A high value resistor was also fitted and decoupled so that a negative grid bias could be fed to the grid. The filament pins were connected via RFCs to a 6V DC supply to power up the heater.

The reactive and resistive series components were measured to be X_L = 485.94Ω and R_S = 131.69Ω. The grid at 432MHz looks likes an inductance of 179nH in series with 131.69Ω resistive portion. Converting the series components to parallel format gives the values of 192nH in parallel with a resistance of 1924Ω. Hence, to resonate the grid we need a capacitance of approx 0.7pF. This is not what we would assume if we simply accepted the manufacturer's data for grid capacity of 15.6pF. The same measurement was made at 144MHz and here the grid is capacitive. The grid bias has almost no effect on the measured values. Of interest is how the values change with the heater on or off, with no filament current the values are R_S = 84.375Ω and X_L = 425.64Ω converted to parallel format these are R_P = 2231Ω and X_P = 442Ω.

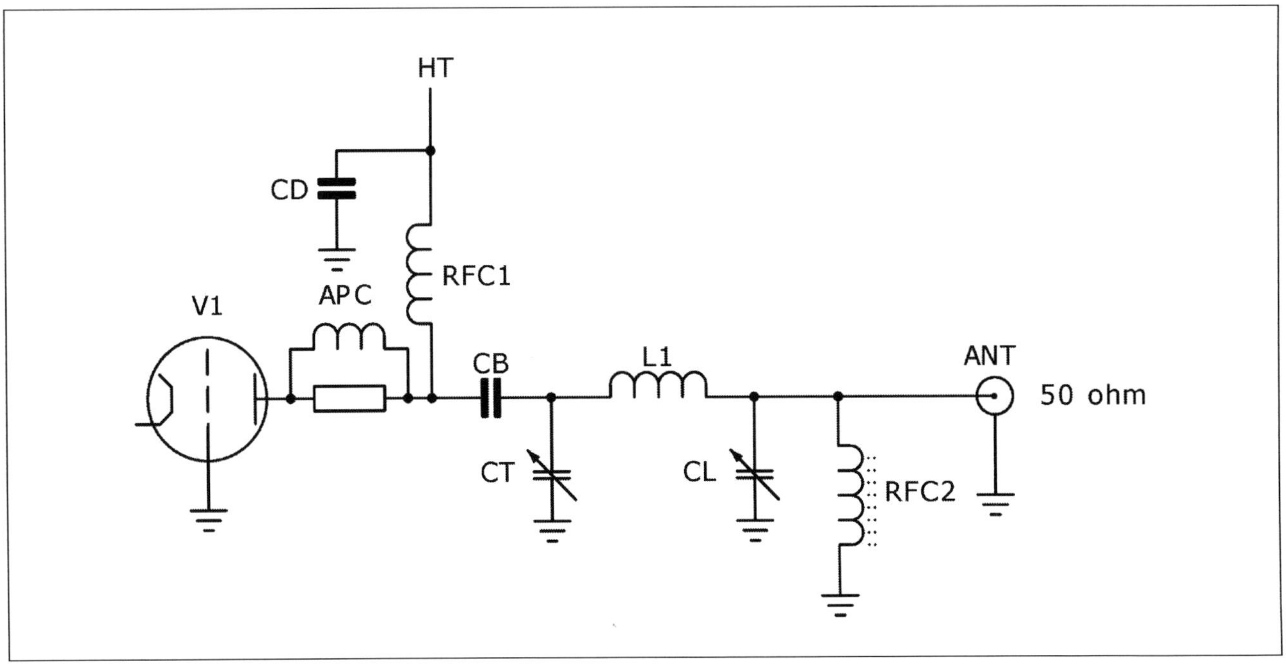

Fig 11.19. Pi tank network

THE PI TANK NETWORK

The vast majority of HF amplifiers use the pi tank network to match the anode load to the antenna. This consists of two air variable capacitors and a series inductor. Because the anode needs to be supplied with the high voltage this is often performed by an RFC connected directly to the anode and the cold end is decoupled to ground. This scheme has a number of disadvantages.

If we examine **Fig 11.19** we can see that the anode supply is fed via an RFC which connects to the high voltage supply. The reactance of this needs to be high, typically at least 10 times the impedance we are connecting to. Eimac, in their publication *The Care and Feeding of Power Grid Tubes*, recommend a value of 100 times the reactance of the pi tank inductor. If this is a single-band amplifier this makes the design of the RFC simpler. However, if this is a multi-band HF amplifier with a tapped inductor (L1), the RFC needs to be effective on all the bands. Taking an example, an amplifier for 80m using 2 x 6146 tetrodes will have an anode load of approx 1800Ω. At 80m this requires an inductance of 820μH, which is impossible to achieve unless a ferrite-cored choke is employed. If the reactance is much lower we need to consider the RF current flowing to ground. If the reactance of the RFC is 500Ω and the anode supply voltage is 1kV, the AC voltage across the RFC to ground is approx 1kV rms. This causes a RF current of 2A to flow to ground via the decoupling capacitor CD, and hence this capacitor needs to be rated to carry this high current without over heating. This 2A of RF current is lost power, power we could usefully use to send to the antenna, hence why the reactance of the anode RFC needs to be as high as possible to reduce circuit losses. If one considers an amplifier with an anode voltage of 3kV we get some idea of the potential losses and why typical anode RFCs tend to run very hot.

Most HF amplifiers use a single layer solenoid wound inductor on an insulating former. If the amplifier needs to cover several bands then the RFC must be effective on the lowest frequency band as well as all the others without any spurious resonance effects. Preventing a spurious resonance due to self-capacitance between the turns is very difficult. Consequently most HF amplifier anode chokes are lower inductance than optimum and a fair portion of the power is dissipated in the choke. The optimum condition is to operate the anode RFC just below parallel resonance with the anode shunt tuning capacitor that brings the pi network to resonance. In this condition the RFC appears as a high value resistance across the anode to ground. This can be made to work in a single band amplifier but not in a multi band type. Generally it is not a simple task to make the RFC resonant and it is preferable to make the RFC higher in value and suffer the consequences of RF current losses. In *Care and Feeding of Power Grid Tubes*, Eimac gives the diagram shown in **Fig 11.20** which illustrates the optimum condition for operating the anode choke at just below or at parallel resonance.

The skin loss resistance of fine gauge wire used to wind RFCs can be high and when the copper losses are included, due to the anode DC flowing, it is not unusual to find that a fair percentage of the power is lost as heat. Ideally we would wish to have the Q of the RFC approaching infinity, in the real world this is never likely to occur. Of interest is the single-band RFC shown in **Fig 11.21**, this at 10 MHz has a Q of about 200 when measured in a low power circuit using a Marconi Q meter. With the anode DC of 400mA the Q is expected to fall to about 150 due to the additional copper losses when the wire heats up. It is not unusual to find the enamel wire discoloured due to high temperature.

This choke when measured has an inductance of approx 25μH. The amplifier circuit used two 6146 pentodes with an anode supply of 850V and a current of 400mA. At resonance the anode present a resistance of 2125Ω. At 20MHz the RFC presents a reactance of 3140Ω which is only ≈50% more than the anode load value. At 3.5MHz it presents a reactance of 550Ω which is one quarter of the anode load resistance. Hence, it is not effective at 80m and it can be expected to introduce extra network losses. This perhaps explains why we need to input 340W DC to obtain 100W RF output: the RFC is absorbing a large part of the power. To be effective at 80m the RFC needs to have a reactance of at least 20kΩ, which is a value of 910μH.

For optimum Q when a single layer solenoid coil is wound the length to diameter ratio needs to be between 80 to 120% and the wire spaced at one wire diameter from the next turn. Many anode RFCs ignore this criterion to allow the RFC to fit in the limited space available in the anode compartment of most commercially-manu-

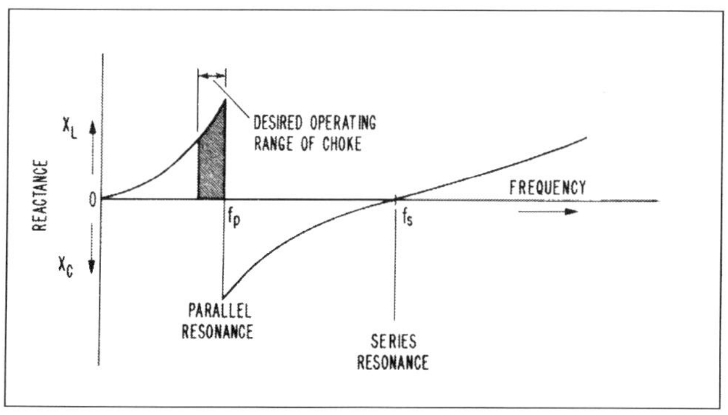

Fig 11.20. Optimum resonance for anode choke (*courtesy of Eimac*)

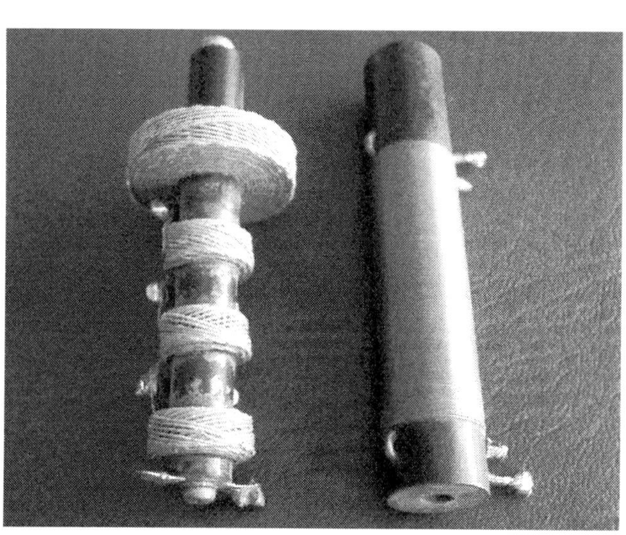

Fig 11.21 Typical anode chokes for HF amplifiers. The right hand choke is for a single-band design; the pile-wound choke is for a multi-band design. Both these chokes are for transmitters in the 100 to 150W output category.

factured amateur equipment.

Some of the techniques used to prevent spurious resonance are to pile wind the RFC in sections for different bands; this introduces extra losses although eliminating as far as possible spurious resonance effects it makes the inductor difficult to construct. With the advent of the new WARC bands of 10, 18 and 24.5MHz this now places spurious resonance's in between the traditional bands and caused a dramatic rethink of the anode RFC question. Previously these spurious resonance points weren't a problem, now they are.

The second problem with the standard pi tank circuit is the need for a DC blocking capacitor between the anode and the tank network. This is shown as CB in Fig 11.19. This needs to have a high voltage rating and also able to carry a high RF current. The current rating can be calculated from the output power and anode load resistance. Many high voltage blocking capacitors used by amateur constructors are not designed for RF applications and this increases the losses due to a higher value of series loss resistance. The minimum voltage rating required is at least twice the anode voltage. Off resonance the anode voltage swings up to twice the DC supply voltage. Because the anode blocking capacitor CB may fail short circuit a second RFC is normally placed across the output of the Pi network to hold this point at ground potential, a short to DC. This is to prevent the hazardous anode voltage from occurring at the antenna. Hence, if CB fails short circuit the RFC should cause the anode supply fuse to blow removing the high voltage. Because CB is effectively between the anode and ground when analysed for the DC condition its rating needs to be at least 3-times the anode voltage for a margin of safety, and a factor of 4 or more would be a better choice.

The value of CB needs to be large, if the standard rule of making the reactance less than one-tenth of the impedance point in the circuit to be blocked were heeded then the reactance of CB would need to be less than 180Ω in our 80m amplifier example. We will half this to 90Ω for a little more leeway. This is a capacitance of ≈520pF so a 1nF capacitor will be suitable. As the frequency rises the value of CB can be made smaller without unduly affecting the circuit operation.

ANODE CHOKE PARAMETERS

In a multi-band amplifier using a normal pi tank network the anode choke can absorb a large portion of the power. As we have already mentioned the reactance the choke presents to the anode needs to be high in comparison to the anode load resistance. But when operated in a tank circuit the inductor is partially resonated with a parallel capacitance. Part of this is the valve anode capacitance, circuit strays and the value of the anode tuning capacitor. The equivalent circuit formed by the anode choke and the capacitance is shown in Fig 11.22.

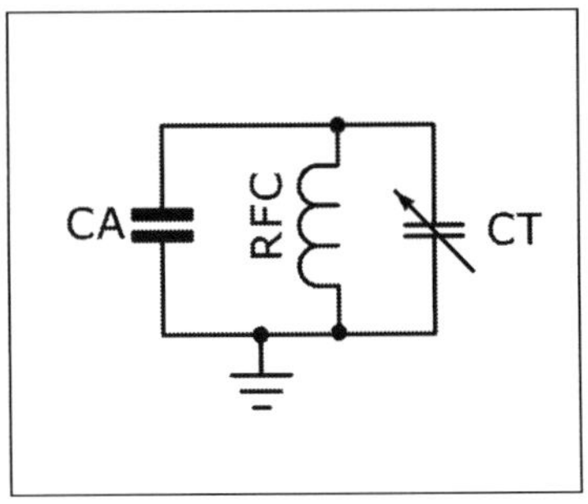

Fig 11.22. Anode choke and anode capacitance equivalent circuit

To gain some insight how these components interact we will consider the 25µH choke from earlier. We will assume a standard HF amplifier with two 6146 valves operating with an anode voltage of 850V and an anode current of 400mA; this gives an anode load of about 2kΩ. The value of the RFC was measured using a Marconi Q meter at a frequency below its self-resonant frequency (SRF). When inserted in a circuit with parallel capacity it will become parallel resonant at some frequency. The dominant term is the anode tuning capacitor in the Pi tank network. Taking some typical values for a HF amplifier, at 80m the anode tuning capacitor will be approx 100pF. If we run the calculations to find the parallel dynamic resistance (R_D) of the RFC when shunted by this capacity we find the value is 60kΩ when the total capacity is 84pF. This is much more than 10 times the anode load and so the loss will be low. However, suppose instead of 84pF we require 120pF to resonate the pi network, this gives a resistive load due to the non-resonant RFC to be 1250Ω. This is less than the anode load and so will dissipate power. If we continue with this analysis at other frequencies we can see the problem.

Assume we now change to 40m and the pi network now needs 75pF to be resonant, (because the tank inductor has been reduced in value by the inductor tapping being changed), with the 25µH choke this presents a resistance of 418Ω. The choke is actually resonant at about 3.7MHz where it appears as a resistance of 56kΩ. Arranging for the anode tuning capacitor to correctly track the RFC is wishful thinking in a multi band amplifier because the pi tank network needs to have certain specific values for maximum power transfer. A method where we could get close to optimum would be to also tap the RFC and arrange for the correct inductance to be selected as we change bands, a messy system needing an extra switch wafer on the main band switch. Some commercial amplifiers have a multi section RFC with parallel capacitors across some sections to form 'traps' to try and get around this problem.

One method of utilising the RFC at parallel resonance was used in a German marine HF SSB transmitter. This used a fixed anode tuning capacitor for each band chosen to resonate with the low value RFC. The pi tank network was tuned by varying the inductance of the series inductor and the loading capacitor. This, however, does not give much leeway for error and increases the network losses in the pi tank inductor because of the sliding contacts to tap the inductor to the correct value, and is not likely to be replicated by amateurs. It is doubt-

> In the 1980s I visited the Marconi works in Chelmsford to assist in performing the acceptance testing on a VHF television transmitter with the Chief Engineer of the SABC (South African Broadcasting Corporation). In an adjacent test bay was a 100kW AM broadcast transmitter. The covers were off this transmitter and I took a good look at it to see the technology being used. This transmitter used two large Eimac air-cooled tetrodes, the anode RFC consisted of a 100mm diameter ceramic tube about 1m long wound with 16SWG close-wound enamel copper wire. The pi tank coil used 20mm copper tubing through which water was pumped to keep it cool. This is real QRO! A spare heater transformer was adjacent to the bay, it was about 0.5m cube and served as a convenient seat whilst examining this monster. The heater current was about 500A at 10V, i.e. 5kW just for the heaters.

ful if the saving of cost on the RFC or space reduction was effective because the variable inductor mechanism was quite complicated. It was, however, interesting to see it being used commercially.

This type of variable inductor is known as a 'roller inductor' and a grooved pulley wheel presses on the wire and the inductor former rotates to move the wheel along the helically wound wire coil. This type of inductor is popular in ATUs and was invented during WWII. Often the inductor is motor-driven automatically to find the lowest VSWR.

INTERPRETING PI TANK TABLES

When designing an amplifier with a pi tank network we often use tables such as those in RSGB or ARRL publications to find the optimum values for the anode tuning, loading capacitors and the tank inductor. These often give values for several different anode load resistances and a loaded Q of about 10 to 12. What is not obvious from these tables is the effect that the anode capacitance or any circuit strays have. If the anode has a significant amount of capacitance or the circuit stray capacity is high this affects the value of the anode tuning capacitance. The value given for the anode tuning capacitor is the *total effective capacity* at that point in the circuit. Hence, we need to subtract the anode capacity and any estimated stray capacity from the value given. In a HF amplifier this is not normally such a problem because the anode capacity is quite small in comparison to the value of the anode tuning capacitor, but if the amplifier is for VHF or UHF then the effect is significant.

For example, the 6146 valve has an anode capacitance of about 9pF so if two valves are connected in parallel we can estimate the total anode capacity will be about 20pF when circuit strays are taken into account. This will exhibit an anode load of ≈2kΩ. If the amplifier is for 160m or some other low band the 20pF is insignificant because the tuning capacitance is likely to be high, typically about 450pF. But at 28MHz with an anode load of 2kΩ and a loaded Q of ≈12, the capacity across the anode end of the network needs to be a total of 34pF and the 20pF the anodes contribute is more than 50% of the total. Hence, the air variable chosen needs to have a low minimum capacitance to correctly resonate the network. As the anode load rises the situation becomes more serious. If the anode load is now 4kΩ the total capacity required at 28MHz is only 17pF which is less than the anodes contribute. Because we cannot have a minus capacitance tuning capacitor, we then need to modify the pi tank inductor to compensate. At VHF and UHF with available valves the anode capacitance although lower is not zero and often the only way to get a good match is to deviate away from low loaded Q tank circuits and values as high as 20 are not uncommon. This inevitably increases the total circuit losses because of the higher RF current flowing in the network but sometimes it is the only way to reach a workable solution. Provided the tank inductor and the tuning capacitors are good quality (high Q) this is normally acceptable.

The other thing the tables do not take into account is the value of reactance of the anode RFC that is effectively shunting the anode end of the network. These tables were calculated assuming the reactance of the anode RFC is infinite, which we have seen earlier is not normally the case. They do, however, form a good starting point and some adjustment may be necessary to obtain optimum performance.

VALUES FOR PI AND PI-L NETWORKS

Tables for the component values for the popular pi and pi-L networks can be found in several publications. These are normally given for several different anode load values. Usually these are for a loaded Q value of about 10 to 12, which are the most practical types for HF amplifiers. The values can, however, be calculated if the value of loaded Q required is different. The formulas to calculate the value of the anode tuning capacitor reactance are given in the Tuned Circuits chapter.

The value of R_A to a first approximation can be found from the anode supply voltage and current, $R_A = Va/Ia$. Normally a more accurate method is to assume the anode voltage swing is somewhat less than the anode DC voltage. The factor determining this is the anode saturation voltage and conduction angle. If the valve is a tetrode or pentode the anode voltage cannot safely swing much below the value of the screen grid voltage. If the valve is a triode the anode voltage can swing further towards ground potential, the valve type and applied anode DC voltage determine the exact value. For simplicity taking the anode voltage and current values will give a slightly higher anode resistance than that will occur in practice. If the network is able to match to this higher resistance it will generally have sufficient leeway in the variable components be able to match to slightly lower impedance. If the valve is operating with a lower anode DC voltage than normal then due allowances need to be made to cater for the lower anode load resistance. In general it is preferable to use the highest anode voltage the manufacturers recommends for best linearity, efficiency and matching.

A better approximation of the anode load resistance when tetrodes or pentodes are used is to factor in the screen grid voltage. If we take the popular 6146 for HF amplifiers the screen grid voltage is normally chosen to be about 150V. If the anode voltage is 850V the anode can only safely swing down to about 175V to prevent excessive screen grid current flowing. Hence, the effective anode voltage swing is 850 - 175 = 675V. If two valves are connected in parallel this will need an anode current of approx 400mA for the two valves. Since $R_A = V_a/I_a$ the effective load resistance is approx 1700Ω.

If valves using a greater anode voltage are used, such as the 4CX250 series, the screen grid voltage of ≈350V becomes less significant and an approximation for two valves connected in parallel with an anode supply of 2.2kV and a total peak anode current of 500mA would be about 3700Ω (see the Pi Tank chapter for a fuller appreciation of the anode load calculations).

LOADING CAPACITOR VOLTAGE RATING

In a pi network used for a transmitter, or ATU using a pi network, the RF voltage appearing across the loading capacitor is a function of the antenna impedance and the RF output power. For most applications the use of a broadcast radio air variable may be suitable with several gangs connected in parallel to obtain the required maximum capacity. The formula to calculate the peak RF voltage occurring across the loading capacitor is:

$$V = \sqrt{2P_O \times R_L}$$

where
V is the peak RF voltage in volts
P_O is the power output in watts.
R_L is the antenna load impedance.

For a transmitter supplying 400W to a 50Ω antenna feedline the value of V is approx 200V peak, or 141.4V rms. If the antenna feedline is 300Ω the RF voltage rises to ≈490V peak or ≈350V rms. However, this assumes a perfect match (VSWR = 1:1) and in the real world this almost never happens. So the voltage rating needs to be adequate to handle reasonably high VSWR loads. If the VSWR is infinite the RF voltage increases to twice the matched line condition, so for VSWR up to, say, 5:1 we will require about 50% more voltage rating to be safe.

TUNING CAPACITOR VOLTAGE RATING

The tuning capacitor off resonance may experience a peak RF voltage of up to twice the effective anode voltage. Since we need a good margin of safety it is prudent to increase this value to three times to ensure flashover does not occur. The voltage rating is a function of the air-gap between the fixed and moving plates and for high power applications a gap of 3mm of more may be required.

IMPROVED PI TANK NETWORK

A way to improve the standard pi network for HF amplifiers is to adopt a different approach suggested by my ex-Chief Engineer, Greg Smith, (now ZL3IX). This dispenses with the troublesome RFCs and DC blocking capacitor but provides the same safety features. This basic circuit is shown in **Fig 11.23**. Note that the usual anode anti-parasitic choke networks (APC) have been omitted for clarity. These are often needed for HF amplifiers to suppress VHF parasitic oscillation.

In this circuit the two air variable capacitors are raised above DC ground and mounted on ceramic or PTFE pillars and then decoupled with a large value capacitor, CD. This allows the elimination of the DC blocking capacitor and the two RFCs. The anode supply is fed via the primary winding of a 1:1 transmission line transformer wound on a ferrite toroid. Capacitor CD also decouples the cold end of the primary and the secondary winding carries the RF output power to the antenna. The anode supply is fed into the 50Ω point with the centre of the coax.

Fig 11.24. 1:1 transmission line 1:1 transformer

The primary inductance of the transformer needs to be large enough to act as a choke to the 50Ω connection point. Because the choke now connects at low impedance a reactance of 500Ω or more is adequate, this is far different to the normal circuit. The minimum primary inductance required is approx 22µH for operation down to 80m with a 50Ω antenna. A typical transformer is shown in **Fig 11.24**. This is wound with 50Ω PTFE coax of 4.5mm diameter on a Micro-Metals Inc T200-2 ferrite toroid core.

This transformer can easily handle 1kW at HF continuously. A disadvantage of this arrangement is that the tank inductor is now at the anode high voltage potential. This is of little concern because the tank inductor already has the RF voltage on it as normal and hence will be enclosed in an RF-tight enclosure to prevent radiation, so there is no real difference in the construction. A simple method is to support the tank inductor on insulating stand-offs such as ceramic pillars.

The capacitor control shafts also carry the full DC anode voltage and so the shafts need to be equipped with insulating extensions or shaft couplers. In many cases the shafts will need extending to reach the amplifier front panel so this is not too serious and simple to arrange with paxolin rod or some other insulating material and a shaft coupler.

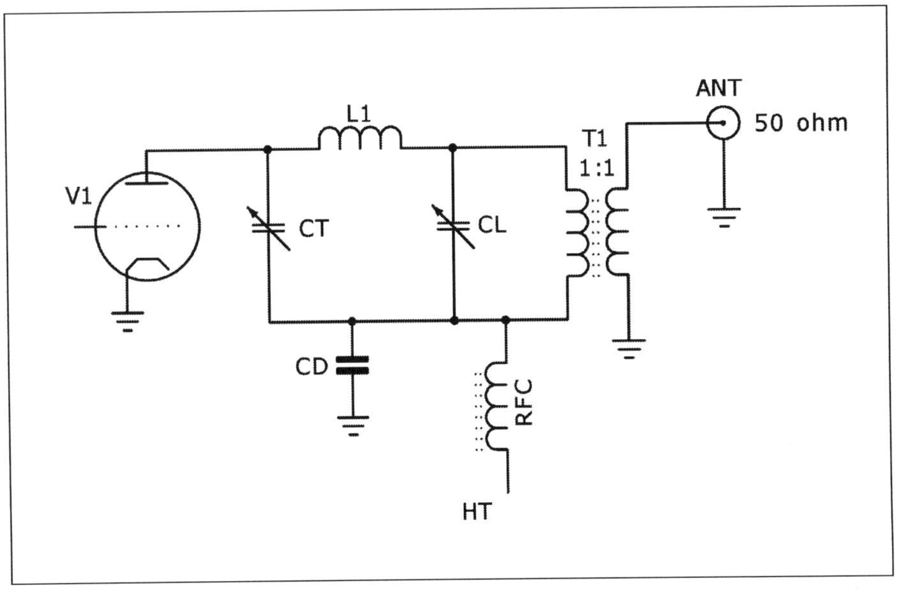

Fig 11.23. ZL3IX improved pi tank network

PI-L NETWORK

One of the problems with the normal pi network for HF amplifiers is the limited attenuation of harmonics. Often we need to incorporate an external low-pass (harmonic rejection) filter between the amplifier and the antenna to comply with emissions regulations. This can be improved by the inclusion of a second inductor to form a low-pass network. This is known as a Pi-L network. The network transforms from the anode load to some lower impedance, known as the *image-impedance*, which is often chosen to be 300 to 700Ω and the final L-match performs the final impedance conversion. The capacitor for the pi network and the L-match is shared and compromises of the normal loading capacitor CL. Using a Pi-L network gives considerable extra attenuation to the harmonics and this is often enough to comply with regu-

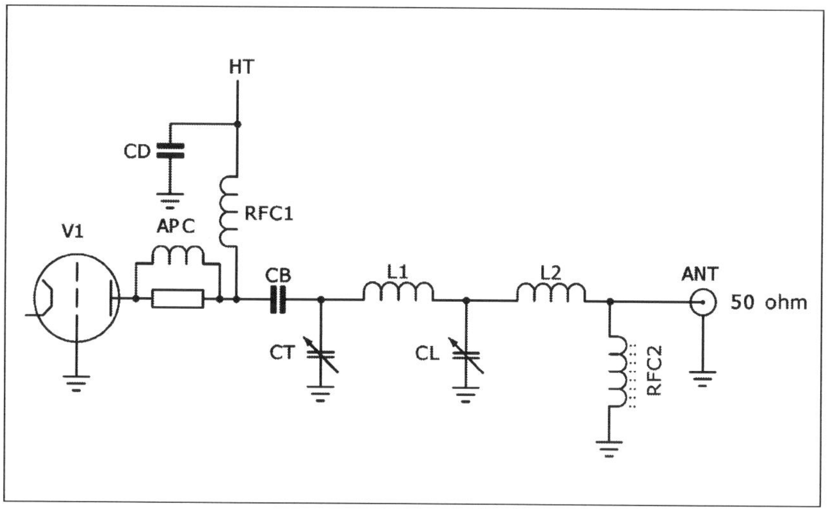

Fig 11.25. Pi-L network

lations on spurious emissions. Typically 2nd harmonic attenuation figures of ≈60dB are achieved with a Pi-L network.

COMMISSIONING A NEWLY-CONSTRUCTED AMPLIFIER

After they have built an amplifier many amateurs go about commissioning it in the wrong way. Many assume you simply turn it on, apply drive power and tune up into the antenna. In some cases this will work if the components are exactly the same as the published design, but if it is a design where substitutions needed to be made to suit available components you could be in for a rude shock. So there is only one safe way to commission an amplifier which has some unknowns.

To tune up any amplifier you will need certain items of test equipment: a through-line power meter or VSWR meter that can also measure power and a dummy load rated for the anticipated power. You will also need some low inductive resistors to simulate the anode load. These do not need to be high power types because we will be performing this measurement at low power; several 1W or 2W carbon composition types connected in parallel will normally suffice. The test set-up is shown in **Fig 11.26**.

The valve or valves are left in place and the power supply disabled so that no high voltage is present on the anode (and screen if the valve is a tetrode or pentode), but the heater is powered up with any cooling fans and relay supply. ***Ensure that the high voltage supply cannot, under any circumstances, be energised whilst you are testing***.

The calculated anode load resistance is made up from carbon composition resistors and connected between the anode connection and ground with short leads. In the diagram this assumed a 3kΩ anode load, made from four parallel 12kΩ resistors. If you have a diode probe or some other high impedance RF voltmeter this is also connected between the anode and ground. If you do not have a diode probe or a means to measure RF voltage then the detector diode and additional resistor shown in the diagram will be required. Connect this to a DC voltmeter to indicate the anode RF voltage. The 56Ω resistor need only be a low power type, 0.5W will suit. (For VHF amplifiers often the diode probe adds too much capacitance to the anode and the tapped resistor RF voltmeter is a better option.)

The components RFC1 and RFC2 are the normal pi network components as are CT, CO, CB and L. (RFC1 is shown as being connected to ground. From an RF point this is how it appears to the anode because the 'cold end' is decoupled with a capacitor.) The capacitor CA is used to simulate the valve anode capacity if the valve is removed; we do not need this component here. **Fig 11.27** shows how a 'dummy amplifier' would be tested without a valve being in place.

The components shown are not the ones that will be used in the final amplifier because the measurement is made at low power. This set-up uses the proposed anode tank inductor and anode RFC to assess performance. The other capacitors are lower voltage rating but the same capacitance as the final design.

To check the pi network is correctly transforming the anode load to the antenna output, we operate the network in reverse by applying some power from a low-power transmitter to form a step-up in impedance.

Connect a VSWR meter between the output connector and the transmitter so that the amplifier output connector is now the 'load / antenna'. If the amplifier has an antenna change-over relay arrange for this to be switched to the transmit condition. Set the driver to mid-band frequency.

We now apply some power and adjust the anode tuning and loading controls; the VSWR should be able to be brought to 1:1. When the VSWR looking into the amplifier is the lowest value the RF voltage at the anode should show a maximum value. Swing the anode tuning capacitor through resonance and check that the voltage peaks when the VSWR dips to minimum. Now change the frequency to the bottom of the band and repeat the measurements,

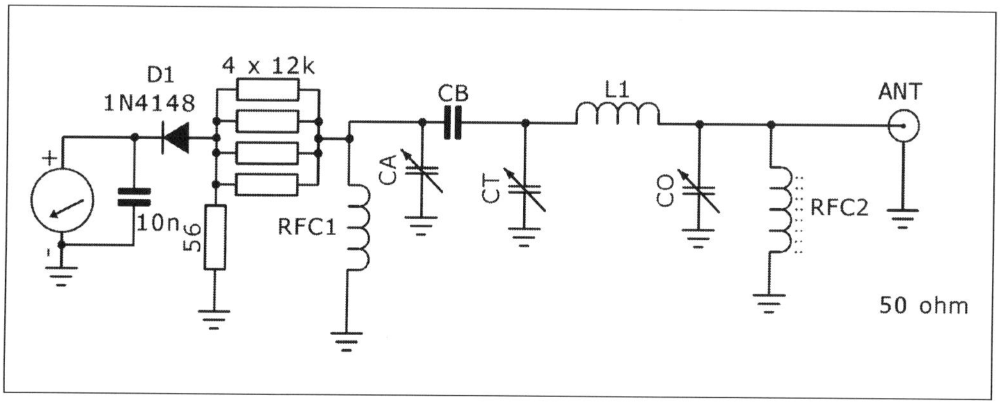

Fig 11.26. Test set-up for anode network

Fig 11.27. Typical 'dummy amplifier' anode network breadboard.

then change to the band top frequency and repeat the measurements again to ensure we have adequate tuning range in the network. If the measurements are correct then we now know that the pi network should be able to correctly match the valve to the antenna.

If the amplifier is a multi-band type, repeat the measurement for each band. Write down the capacitor dial settings for each band so you can preset these when changing bands.

One other thing to check is where the tuning capacitors are at resonance. Ideally the anode tuning capacitor should be approx half-meshed and the loading capacitor about three-quarters unmeshed to allow for reactive antenna loads. If the anode tuning capacitor is almost fully meshed or fully unmeshed then the inductor value is incorrect or the anode load resistor is incorrect. If the capacitor is towards fully meshed the inductor value is too low and requires more turns. If the capacitor is towards fully unmeshed the inductor value is too high and requires some turns to be removed. If the inductor is a tapped type, with a band switch to select the tapping points, then the tapping point needs moving to either increase or reduce the number of turns. This we can safely adjust with low power applied but we could never do it at high power because of the dangers of high RF and DC voltages. Working at only a watt or so makes it safe to make adjustments, although you should switch off the drive power before making any changes.

Having confirmed that the anode network can correctly transform the anode load to the antenna we can remove the test components and now tune up the amplifier normally. For this we ensure all safety covers are firmly in place before switching on and any high voltage

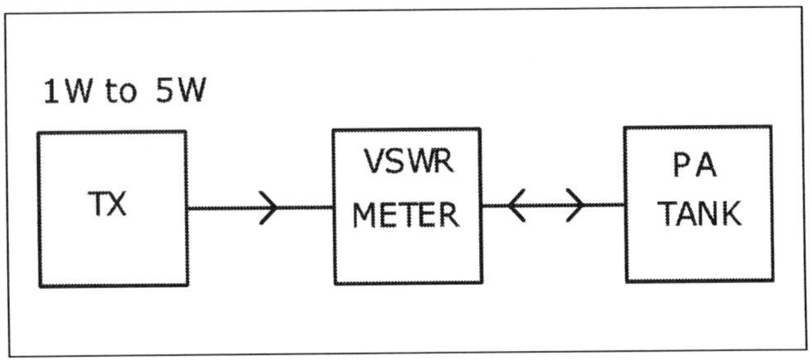

Fig 11.28. Connection of VSWR meter for testing an anode network

interlock we needed to 'cheat' previously are working as intended. Running a high-power amplifier at full power with the covers removed is a risky business! Not only will high RF voltage give you a nasty burn if you come close to a point carrying it; the anode DC supply is also a very hazardous item.

Under no circumstances attempt to make adjustments to the anode network when the amplifier is switched on or delivering RF power.

POWERING UP FOR THE FIRST TIME

Connect the power meter to the output connector with a short coax and the dummy load to the power meter output connector. Do not use an antenna for initial commissioning! Connect a suitable transmitter to the input connector. Switch on and preset the anode tuning and loading capacitors to the approx positions ascertained during the dummy run at low power. Now key up the amplifier and apply a little drive, not more than 20% of the nominal. You should see some output power already, so fine-tune the anode and loading to peak the power. Now increase the drive to 50% and repeat the fine-tuning to bring the power to maximum. Keep increasing the drive in small steps and retuning. When you have reached full drive the amplifier should be developing the full power. It is a good idea to switch off the drive every few seconds to allow the valve(s) to cool.

Many amateur amplifiers will not tolerate high power for extended periods and the valves may get too hot if you attempt to do this. If the power meter is a peak-reading type, an electronic keyer set to send a string of fast dots takes a lot of strain off the valves during tune-up at high power.

PLOTTING GAIN SLOPES

It is a good idea to connect another power meter in series with the input to measure the input power and VSWR presented to the driver transmitter. With two accurate power meters we can plot the power / gain slope at different output power levels and establish the gain at different drive levels. It is very rare to find a truly linear amplifier across the whole useful power range. If we can determine the -1dB compression point this sets the practical upper limit we can expect the amplifier to be linear and suitable for SSB operation. (1dB might sound like an insignificant drop in the power output but in power ratio terms it is about 20% so it is quite a lot.)

Often we find that a *phantom* -1dB compression point also occurs very low down on the power output scale. This is because the gain at very small anode current levels is often quite a bit higher than near full output, as much as 3dB in some cases. In a 100W output amplifier it is usually around the 1 to 2W output point. However, we can safely ignore this spurious lower point and find the true -1dB point, which will be somewhere near the top of the power output curve. From about 10% output power measure the input and output power in small increments of power, say every 10%, and then plot the points on a graph. The curve should be substantially a straight line from about 10% to about 80% output and then it will start to slowly bend over towards a horizontal line.

The plotted curve will often shows a extended S shape across the 25% to 75% power output region. This

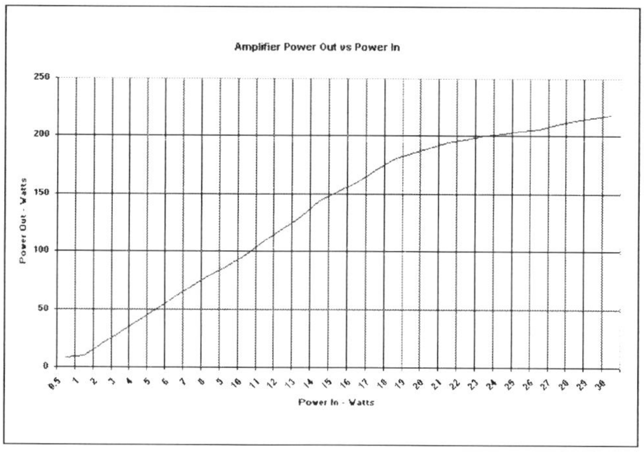

Fig 11.29. Typical plot of power output Vs power input

is partly due to measurement errors in reading the exact power levels. It is also due to the non-linearity in the device as the power level changes. Power transistors have very pronounced S shaped curves due to the parasitic capacitance acting like Varactor diodes at high RF currents and voltages, valves in general have much shallower S shaped curves. The general trend for all power amplifiers is that the gain slowly falls as the output power rises.

If we drive the amplifier well past the -1dB compression point the curve will flatten out to a straight horizontal line. The point where the curve becomes the flat line is the *saturation point*. Once we have passed this point no matter how much extra drive power we apply the output remains constant, this will generate horrendous IMD products and we need to stay far away from this point. Transistor power amplifiers, especially ones operated on 12V, have a very fast change between the -1dB point and the saturation point, often only a few dB extra drive is required to cause them to saturate. Valves have a more gentle transition into full saturation and hence tolerate minor over-driving far better with less IMD products being generated.

The gain reaches the -1dB point at about 206W output. Hence, we would normally specify this as a 200W amplifier. Note how the gain at very low output levels is about 2dB more than over the major portion of the graph where it averages 10dB, this is fairly typical of all amplifiers. The S shaped curve is more apparent in this plot than the power out vs power in plot.

One of the reasons a valve power amplifier's gain tends to fall away as the power output increases is due

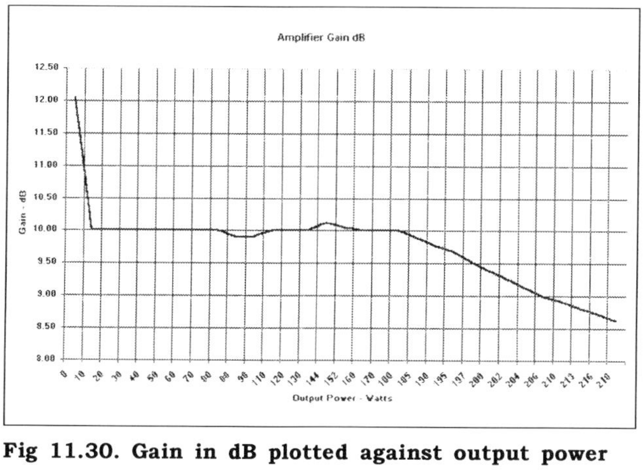

Fig 11.30. Gain in dB plotted against output power

to the cathode not being able to supply enough electrons and this is called lack of *emission*. A cathode can only 'boil-off' so many electrons per second due to its limited surface area. If we use a valve with a larger cathode area, i.e. higher cathode emission, it would be able to supply sufficient electrons to a much higher power level. Often the only way to provide sufficient cathode emission is to use a higher-power valve. A cheap and cheerful way is to increase the heater voltage slightly, but this is not recommended. It was an accepted dodge in the days of black and white televisions to restore the brilliance of a tired picture tube, but it was only ever a temporary cure!

Another reason is the high voltage power supply being unable to maintain the anode DC voltage with increasing anode current. The anode DC voltage will be at maximum at idle current and gradually decrease as the anode current increases. This is a function of the power supply regulation and is to be expected. If we try to draw more anode current than the power supply is capable of supplying the anode voltage will take a steep drop as we exceed the transformer rated power. Running a valve in a starved anode voltage condition causes excessive IMD products to be generated and causes excessive heating in the valve and high voltage transformer. Some commercial manufacturers of power amplifiers cut corners on the power supply and these have poor voltage regulation near full output power, this degrades the IMD performance more than necessary.

Another cause of poor IMD when tetrodes are used is that the screen grid voltage wanders all over the place with varying drive levels. For lowest IMD the screen grid and control grid supplies needs to be very stiff, that is the voltage should not vary as the drive level changes.

INTERMODULATION VERSUS POWER OUTPUT

If we desire really good figures for the IMD products one way to achieve this is to construct a much higher-power amplifier and then run it at perhaps only 50% of its potential. This is an expensive option, but one which gives excellent IMD performance. If the device chosen for the output stage can only just achieve the rated output power required it will much closer to the -1dB compression point than an amplifier of twice the potential output. If a suitable device can be found at a reasonable cost this is the option I would choose. By under-running a large device we can attain very low IMD products and far longer valve life. However, in some countries this is not a viable option.

In South Africa until a few years ago we had a set of amateur regulations that had been changed so many times over the years by different people that many anomalies existed. One clause in the regulations required the IMD products of amateur SSB transmitters to be -30dBc or better. That is a fair limit and I fully agree with such specifications. However, elsewhere in the General Radio Regulations was a clause stating that it was illegal to be in possession of equipment capable of exceeding the licensed power, in our case 400W (26dBW), even it you never connected it! Simply having it sitting in your garage was sufficient cause for a prosecution under the Radio Act. Needless to say, I thought this was a stupid regulation. When it became common knowledge amongst the local amateur population that I had equipment that could not only exceed the regulation power, but that I was using it to achieve low IMD products, a few amateurs started to point fingers and make snide remarks. I would rather have a large amplifier under-run giving low IMD than a 'splatter-box' pushed to the limit.

To demonstrate this, one day I was in contact on 2m SSB with a local about 10km away and line of site. We got on to the subject of linear amplifiers and how he would like to build one, but he was unsure which one to build. I was running the 10W transceiver barefoot. I asked the other amateur to tune carefully across my signal and look for IMD products. He reported none of any significance. I then told him to take an over while I switched on the linear and let it warm up. This uses 2 x 4CX250Bs and can give 1250W on CW in Class C for EME if needed. On SSB the ALC system allows me to turn down the power to 400W to comply with the regulations. I turned the antenna away from him while he was talking to place him in the first null of the Yagi. When I replied I was running the linear at 400W PEP. I asked him to tell me if he noticed any difference in the signal. He told me it was about the same strength as before. I then asked him to have another careful look across the signal while I was transmitting to look for any IMD products. He reported none. While he was transmitting I turned the Yagi back on to him. When my turn came to transmit he nearly fell over such was the signal level. I asked him to tune across my signal again and look for IMD products; he still couldn't find any. The signal with the linear in circuit is just as clean as the barefoot transceiver, which is what a 'linear amplifier' should be.

My 2m amplifier has excellent IMD performance, even at 800W PEP, because it uses state-of-the-art screen grid and control grid stabilisers and I have determined the optimum biasing conditions using a spectrum analyser to measure the IMD products at various power levels. I therefore know the optimum settings and power output for best IMD performance, something the average amateur is unlikely to know without expensive test equipment.

When I was approached in 1998 by the authorities to rewrite the South African Amateur Regulations (because they had no technically competent clerks!), I took care to eliminate the silly contradictions the original version contained. To the best of my knowledge, as someone who has been professionally involved in the design of linear amplifiers for about 30 years, there is no way to make a linear amplifier that can only just reach 400W PEP and still maintain a good IMD figure. The problem was that various Post Office clerks had amended the regulations over many years adding bits as they saw the need, but not one of them had sufficient technical background to appreciate the import of what they were doing, a very dangerous situation!

I came across similar anomalies during my professional career when vetting RFQ (Request For Quote) specifications. When drafting specifications for a contractor to quote on clerks often included stupid errors. In one case an upgrade was required for some older in service equipment, this had a receiver 3dB Noise Figure, the request for quote specification listed it as 6dB. When I queried this the clerk told me 3dB wasn't acceptable anymore and the customer wanted an improvement to 6dB!

In fact the amateur regulations in South Africa are a sub-set of the General Radio Regulations that cover all the other radio services, including the commercial users. The IMD specification for commercial HF users gave a figure of -26dBc, 4dB worse than the amateur specification. When I questioned why amateurs had to attain better IMD figures than the commercial users they couldn't give me a valid explanation. So I changed the IMD specification to be the same as commercial users to make the two sets of regulations consistent.

THIRD ORDER OUTPUT INTERCEPT POINT (OIP^3)

In linear amplifiers we also have another method of characterising the linearity. If the curve of the fundamental power (1st order) is plotted it will intercept at some point with the 3rd order IMD products. This is a theoretical point and will never occur in practice. But it gives us a means of calculating the IMD levels for different output powers. A typical diagram of the intercept point is shown in **Fig 11.31**.

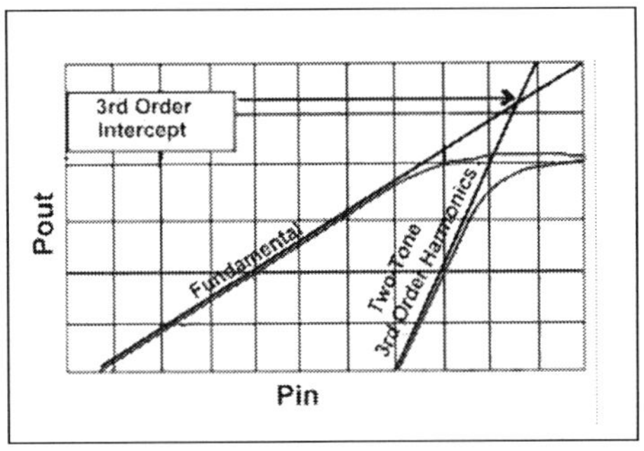

Fig 11.31. Plot of linear output power versus OIP^3

The fundamental output power slope will tend to *fall over* as the power exceeds the 1dB compression point. If the fundamental linear gain slope is extended in a straight line it will cross with the 3rd IMD products at some point. This is known as *the output 3rd order intercept point* (OIP^3).

Third-order IMD products rise at three times the rate of the linear fundamental (1st order) rate, which is why they are called '3rd order'. Hence, for 1dB of increase of the fundamental the 3rd order IMD products rise 3dB. At some theoretical power output they would eventually cross at the OIP^3 point. For most practical amplifiers the OIP^3 is about 11 to 15dB higher than the 1dB compression point. Once the amplifier approaches the 1dB compression point the 3rd order IMD products start to rise rapidly and this sets the peak output power that any amplifier is able to deliver with acceptable IMD products. The 5th order product rises 5dB for a 1dB increase in the output power so will overtake the 3rd order ones somewhere before the 1dB compression point is reached. The optimum condition for maximum power output under linear conditions is for the 3rd and 5th order IMD products to be equal amplitudes relative to the carrier and about -30dBc (dB below the peak carrier power). Normally when this is achieved the higher order IMD products fall by about 2 to 5dB.

If the biasing is incorrect, even slight overdriving can cause a sudden rise in IMD products to become equivalent to the carrier peak power. To attain the best linearity normally requires quite a large quiescent current, as much as 40% of the peak anode current in most cases. The optimum biasing conditions for a 4CX250B is approx 100mA quiescent current and the anode current then driven to around 300mA; hence the anode dissipation is quite high under idle conditions. In practice with a valve amplifier correctly biased it is normal to find that the IMD products only rise to -6dBc worst case for a severely overdriven amplifier. Solid state amplifiers are normally much worse and it is not unusual to find the IMD products reaching the same level as the carrier peak power, as is demonstrated in **Fig 11.31** where the two curves meet.

An amplifier rated at 100W output PEP will have an OIP^3 value of about 1.5kW for good linearity.

To demonstrate how the IMD levels change, consider a 100W PEP amplifier running at full rated output. 100W PEP is obtained by applying two equal amplitude drive signals separated by a small frequency difference each of which develops a carrier of 25W average power. Since PEP is four times the individual carrier power this is 100W PEP (see **Fig 11.32**).

The 3rd order IMD products will be quite high, let us assume they are 23dB below one of the carriers when viewed on a spectrum analyser, -23dBc. If we now reduce the output power by 1dB the levels should drop by 3dB to -26dBc. 1dB in round numbers is 20%, so we need to reduce the output power to 80W PEP or 20W per carrier. If we reduce the output power another 1dB this brings the power output to 64W PEP (16W per carrier) and the IMD products should drop to -29dBc. Reducing the power output by another 1dB brings it to 50W pep, where the IMD products should be -32dBc. (This type of measurement is normally performed by inserting a step attenuator between the two-tone driver and inserting 1dB steps.)

If we continue inserting 1dB attenuation steps then the IMD products we expect to obtain are:

100W PEP	(2 x 25W)	= -23dBc
80W PEP	(2 x 20W)	= -26dBc
64W PEP	(2 x 16W)	= -29dBc
50W PEP	(2 x 12.5W)	= -32dBc
25W PEP	(2 x 6.25W)	= -41dBc
12.5W PEP	(2 x 3.125W)	= -50dBc

If the linearity requirement is for -30dBc then we could only rate this amplifier at about 60W PEP.

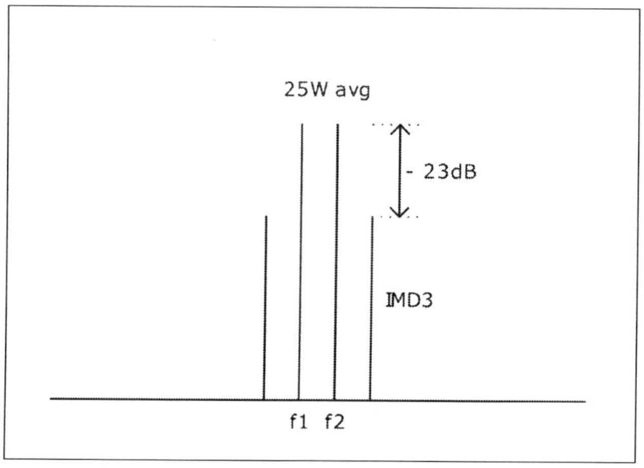

Fig 11.32. Measurement of IMD level between one carrier and one IMD product

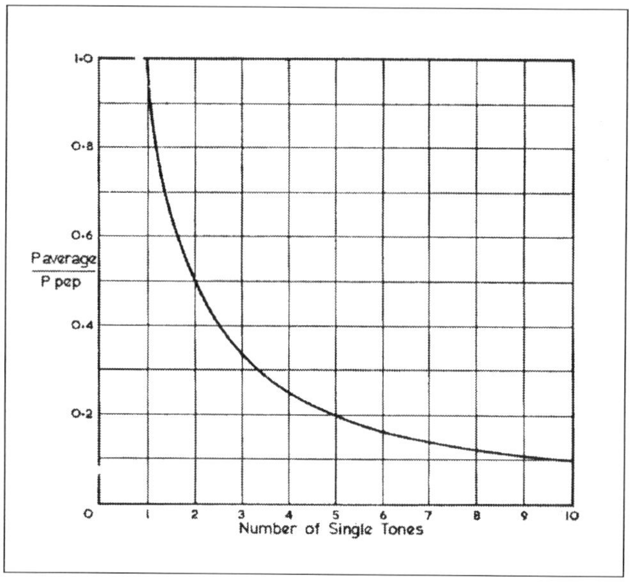

Fig 11.33. Average and PEP for different number of tones

One important thing to note from this is the relationship between the average power and the peak envelope power. Average power is determined by the PEP divided by the number of tones. This is what a normal power meter indicates; a peak-reading wattmeter shows the true PEP value.

$$\text{Average power} = \frac{\text{PEP}}{\text{N}}$$

where N is the number of tones.

To determine the amplifier PEP rating when multiple tones are used we use this formula:

$$\text{PEP} = \text{N}^2 \times \text{Ave Pwr}$$

Hence, if we have four tones each of 100W average power the amplifier PEP rating needs to be 1600W.

One thing to watch out for is how some manufacturers specify the IMD performance. Some use the '*dB PEP*' rating adopted by the American EIA organisation. This makes the amplifier appear to be better than it actually is. When we specify dB PEP we subtract 6dB from the spectrum analyser readings because the instantane-

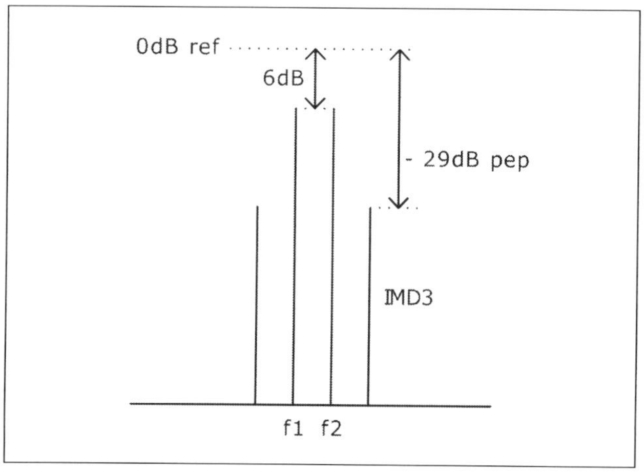

Fig 11.34. Measurement of IMD with reference to the PEP level

ous peak power is four times the individual carrier amplitude. Hence, if we measure the IMD level as -24dBc when expressed in dB PEP it is now -30dB, which sounds a lot better. (The method using one carrier and one IMD product and expressed in dBc is the MIL-STD measurement method.) This measurement technique is shown in **Fig 11.34** where the two carriers have been adjusted so that they are 6dB below the 0dB-reference line on the spectrum analyser screen.

IMD LEVELS

What most amateurs seem to misunderstand is what the actual levels of IMD equate to. If the IMD levels are -30dBc and the amplifier is operating at 1kW PEP the power contained in the IMD is 1W, because there are two IMD-generated spurious carriers, one either side of the main signal, this is a total power of 2W. If the signal being received is S9+40dB the IMD products are equivalent to a signal of S9+10dB.

INTERMODULATION AND DRIVER IMD

Many amateur transceivers today are solid state and hence employ transistors in the power amplifier. These generally are wideband fixed tuned amplifiers and do not like driving a reactive load. If the antenna is a long wire or other type that has a high VSWR this can cause the IMD to be really poor when driving such an antenna and the amplifier is disinclined to produce its full output power. This is the reason we need to use an ATU as an additional impedance matching device. The older valve transceivers do not normally need an ATU, unless the VSWR is really bad. This is because the pi network can usually cope with reactive loads and we can adjust the matching to suit the antenna impedance over quite a wide range, as much as 3:1 VSWR in most cases and some good designs can cope with up to 5:1 VSWR. Hence, the older valve transceivers have a distinct advantage over the modern types.

However, if a solid-state transceiver is connected to an amplifier that has a reactive or varying input impedance the transceiver may generate high levels of IMD. Many amateurs blame the linear amplifier. *When the amplifier is switched out the signal is OK, so it must be the amplifier causing it?* This is normally not the case and the amplifier, because it is linear, simply amplifies the input signal including the transceiver IMD products. I have measured the IMD of several modern HF transceivers when driving a reactive load such as a linear amplifier input. Some quite frankly are awful!

> The way we designed solid-state linear amplifiers for military applications to avoid antenna mismatch problems was to include a network between the amplifier and the antenna. This effectively dumped any reflected power from the antenna into a 50Ω load to prevent it being sent back to the amplifier. This circuit is known as a *constant-R* network but unfortunately it has some loss associated with it, as much as 2dB in some cases. This meant we needed to generate an extra 40% power to meet the specification when measured at the antenna port. This allowed the transmitter to operate into really bad VSWR antennas, typically wide-band short vertical whips used on small ground plane vehicles. We could even operate near full power rating with half the whip snapped off, albeit the antenna didn't radiate too well.

By placing a directional coupler between the transceiver and the amplifier input, to tap off a small sample of the drive signal, I can look at the signal going into the amplifier with a spectrum analyser. In virtually every case the IMD was generated in the transceiver and not the linear amplifier. Some solid-state transceivers are quite bad even when driving as little as a 2:1 VSWR load. The worst cases occur with grounded grid triode amplifiers using zero bias, such as the 3-500Z and the 572B triodes. The impedance change over the drive cycle can be quite high and the transceiver reacts by generating bursts of IMD as the power level ramps up and down. Into a dummy load the transceiver cannot be faulted. By contrast the older valve transceivers, using typically a pair of 6146s, have no problem driving such a poorly matched load. They do not generate any significant IMD even into a VSWR as high as 5:1.

Most 'linear amplifiers' are not truly linear; in fact I have yet to come across such an animal in over 30 years of designing which one could truthfully say is completely linear. But the amount of IMD they introduce to the signal should be minimal. There is a general rule of thumb RF designers use when cascading several linear amplifiers in a chain to get to a certain power level. This is the '-6dB per stage' rule. For the IMD at the output of the final amplifier not to be severely degraded the input to it needs to be approx 6dB better than the final figure required. If we need to maintain the IMD below -30dBc then the driver must not exceed -36dBc. If we have, say, three stages in series the first stage IMD needs to be approx -42dBc to ensure the final IMD is below -30dBc. This places quite a strain on the early amplifier stages in a transceiver and getting to figures like -50dBc and lower can be quite a challenge.

To give some indication of the difficulty of achieving a really low IMD in a series connected chain of amplifiers, many years ago I worked for part of the Marconi Company and I was involved in the design of a high-linearity broadband VHF amplifier system. The final amplifier IMD specification was -45dBc worst case at 100W PEP. (This amplifier was the driver for a large valve amplifier running approx 10kW PEP for a naval installation.) The output stage consisted of two large 28V mosfets each capable of 250W output operated in push-pull. The quiescent current in the output mosfets was about 10A (5A per device), that is they ran in almost full Class A, and the devices were bolted to a water-cooled heatsink to cater for the high dissipation. If pushed hard the amplifier gave over 600W under CW conditions. The driver stage used two 50W mosfets in push-pull running in Class A, also running a high idle current of about 3A (1.5A per device). The gain of the final stage was ≈10dB, so the driver needed to supply 10W PEP. The pre-driver used a 10W mosfet running 1W PEP in Class A to drive the next stage; this device used an idle current of about 1A. We had a nominal total gain of ≈30dB spread across three devices, so about 10dB per stage. The input to this amplifier chain was derived from a 5W mosfet as a pre-amplifier running 100mW PEP in Class A. Hence, the input IMD needed to be better than 45 + 18 = -63dBc to be able to measure amplifier IMD performance.

To test the amplifier IMD performance I needed to generate two carriers spaced about 10kHz apart with an

IMD of better than -65dBc and of 100mW PEP to be able to tell if the amplifier was generating any significant IMD. This is two equal level carriers each of 25mW. Not a simple task! It took several weeks to design and build the special IMD test set up to get to below the -65dBc point. If the drive signal IMD was -63dBc or worse I would not be able to tell if the amplifier was the culprit or not. Eventually I managed -68dBc and proved the amplifier met the specification with about 3dB to spare, -48dBc being actually measured.

By comparison, a typical colour television transmitter needs to achieve an IMD of better than -55dBc so as not to generate IMD products which affect the picture quality. A project I was involved with in the 1980s was to develop an IMD test system for TV transponders able to measure the transmitter IMD of UHF TV transmitters. We managed to measure to better than -70dBc directly off-air when the transmitter was running full power. These sorts of IMD make -30dBc seem terrible by comparison. These transmitters use complex pre-distortion techniques to attain the IMD performance required.

PRE-DISTORTION TECHNIQUES

For amplifiers that need to have exceptional IMD performance, such as television transmitters, a common method is to use 'pre-distortion'. Deliberately distorting the input drive signal by the correct amount can cancel out the inherent non-linearity of the amplifier. This is performed by *pre-distortion amplifiers* with adjustable gain and phase; a case of 'two wrongs making a right'. Unfortunately these are too complex for most amateurs to experiment with, but most GSM cellular telephone transmitters with solid-state amplifiers for base stations use this technique to linearise the multi-channel transmissions. Some GSM cell-phone amplifiers can amplify 11 channels simultaneously, each one having a different power output. Because the linearity required is quite high they use a method of backing-off the channel power with information gained from the companion receivers RSSI detectors. A portable handset close to the base station does not need as much power as one further away. The specification for GSM base stations is 320W PEP ±3dB, so the peak power can reach 640W.

Chapter 12

More on valve amplifiers

MOST amateurs purchase or build an external linear amplifier because they have a need for more power than the average transceiver can give. That is a fair argument and possessing a linear amplifier isn't a crime provided your licence class allows you to use one. What is a crime - or at least should be - is the indiscriminate use of a high-power amplifier when it isn't warranted.

Daily I hear stations here on 40m who are running a high-power amplifier to make a contact 500km away when 100W is more than enough. I have expended a lot of effort to make my antennas as efficient as possible to add to my station's effectiveness, so I don't often need to run high power for 40m or the other HF bands. The attitude of some of these amateurs is "I paid for this thing, so I am going to get my money's worth, and if you don't like it you can get lost". Not only is this extremely arrogant and anti-social but it is also not in keeping with the amateur creed. I find some of the amateurs guilty of this do not even know how to adjust the amplifier so it doesn't splatter all over the band. Or they are 'driving the pants off it' by winding up the RF drive well past the linear point, and not using ALC.

Sadly most of these amateurs are the older generation and do not take kindly to someone politely pointing out they are splattering over a large part of band. Very often they tell the station in a none-too-polite manner to QSY to a clear frequency. But when you get down to 7000kHz and you can still hear the splatter what is the station supposed to do?

I use high-power amplifiers because 10W or 100W or sometimes 400W or even 1kW simply isn't enough under poor band conditions. But I only use it when it is absolutely essential to make the contact. The amateurs who use it as a 'big-stick' all the time are in need of some counselling. For my EME contacts on 2m CW I admit I sometimes run more power than I should or I am licensed for, but at 3.00am with the antenna pointing at the moon I am very unlikely to cause TVI, and who watches television at 3.00am? Although I admit I am guilty of running illegal power very occasionally, I believe I am more correct with the 'spirit' of the licence conditions than amateurs on 40m who daily run all their licence allows, and then some! I don't splatter all over the band and many distant stations don't realise I am using an amplifier until I switch it out of circuit.

COMPARISON OF VALVE AND SOLID-STATE AMPLIFIERS

It is generally accepted that valve amplifiers, when correctly constructed and adjusted, generate lower levels of higher-order IMD products compared with solid-state amplifiers. The major difference is the amount of spectrum occupied by the IMD products and the amplitude of these. **Fig 12.1** shows the comparison between the two types.

The amplitudes of the 3rd and 5th order are quite similar at maximum output power. However, the solid state amplifiers IMD products of the 7th, 9th and higher orders do not fall away as rapidly and hence can spread across several adjacent channels. This causes interference to other users on adjacent frequencies. Unfortunately even

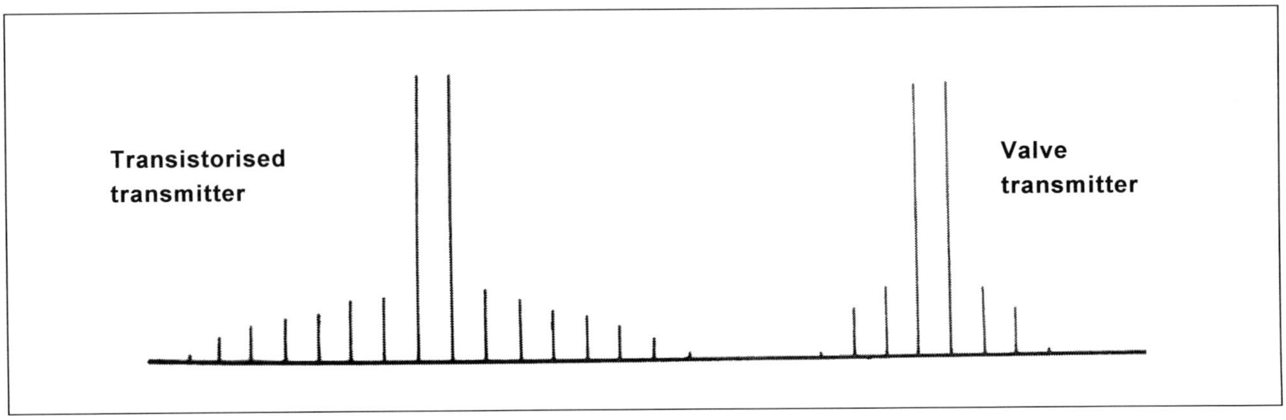

Fig 12.1. IMD performance of solid state and valve amplifiers

the best linear amplifier in the hands of someone who does not appreciate how to use it correctly is going to cause a problem.

SECOND-HAND VALVES

Or in today's terminology, 'pre-owned' valves. Many amateurs obtain valves that have seen service with commercial users. Often these are organisations such as broadcasters who prefer, if economic conditions permit, to replace a valve before it fails. In the days when money was less scarce, broadcasters would often replace a valve when its emission had fallen by about 10 to 20% and then dispose of them on the surplus market. So a good number of second-hand valves could often be found for sale at reasonable prices. These can, however, be a chancy prospect especially if they have been replaced because of some internal fault, which is often not obvious by casually inspecting them. For ceramic-metal version we cannot easily see inside the valve. In the broadcast industry these are known as 'pulls' because they have been 'pulled' from in-service equipment. Very often we can be lucky and get a 'nicely run-in' valve that has been carefully used ("Yes governor, only one careful lady owner since new. Never been over 50MPH.")

Commissioning a brand-new valve also has problems, when it is brand new the gain can be several dB more than when it has been 'broken in'. Running at low power for the first few hours knocks the rough edges off the valve and it is a bit like running in a new engine, it needs to be done gently before going for maximum power. Most commercial users are conservative and do not unduly stress the valves, rather under-running them to gain reliable operation. An unplanned breakdown is not good for broadcasters because many transmitting sites today are unmanned.

The way broadcasters get around this problem, especially with standby transmitters, is to run the cathode heater at what is known as 'black heat'. Switching a heater on from cold causes undue stresses and this often means the heater fails due to the high inrush current.

We have all seen a light bulb fail when we switch it on from cold, and the same thing can happen in large valves. By running the heater at a reduced voltage, normally about 30% of the full current, the cathode temperature is maintained at a level where it can safely be brought up to full emission in a few seconds when it needs to take over service because the other transmitter has failed.

Running at black heat keeps the vacuum hard even though no anode voltage is applied. Where the transmitter is only used for part of the day the standby mode puts the valve into black heat with the high voltages removed.

How can we tell if a valve we obtained at a bargain price is going to cause us problems? It may well be that the reason it was taken out of service was because it experienced a flashover during operation or experienced some other problem. This may be visible under a magnifying glass if you know where to look. Unfortunately, most amateurs do not know where to start looking. It could be that after spending a lot of money and effort constructing an amplifier using a 'pre-owned' valve when we come to 'fire it up' for the first time we have a rude shock and possibly a violent firework display in the valve. One thing you should do is to look carefully to see if any debris is floating around in the envelope. With directly heated cathode valves sometimes chunks of the deposited material will be shed off. If you find any debris give the valve a wide berth. For ceramic-metal valves a good test is to shake it in your hand rapidly and listen for anything rattling around. However, with some ceramic-metal valves there are glass or ceramic tubular insulating spacers fitted to the internal leads that are intended to float and these can sometimes be mistaken for a faulty valve.

The most important thing is not to slap it into an amplifier and switch on with full HT applied.

Many glass envelope valves are not fully sealed. Although when first manufactured they contain a high vacuum this often deteriorates over several years of service and then the valve has a problem. Even brand new valves, which may have been in storage for a few years without being used, suffer from a loss of vacuum. These valves are known as NOS - *new old stock*. The manufacturing process often leaves a few microscopic metal particles from the spot welding of the electrodes floating around in the envelope and from the *gettering* process. When the valve is first pumped down to create the hard vacuum it is impossible to remove the last few molecules of gas. To get around this problem the manufacturer uses a technique called a *'getter'*, so-called because it 'gets' the last few molecules of air.

Valve manufacturers normally issue engineering bulletins about the dangers of applying any voltages to a valve before the heater has reached full operating temperature. Most advise that at least three minutes must elapse before applying the high voltage. Failure to heed these warnings may cause a flashover in a cold valve.

If you examine any of the older glass envelope valves you will see a patch of silvery material on the inside of the glass that looks like a mirror. The getter causes this. This is normally a piece of magnesium tape spot welded across an open circle of metal shaped like a hoop. When the envelope has been pumped down as far as possible the valve envelope is sealed which leaves the glass pip where the vacuum pump was connected - the valve's umbilical cord. Now inducing a high current in the hoop with an induction coil ignites the getter or the tape is often connected across two of the electrode pins and a high current is passed through it to make it ignite. The burning magnesium grabs the last few molecules of air and in burning up deposits the metallic film on the inside of the envelope. The vacuum is now as good as it is going to be. Over several years the high vacuum sucks in minute amounts of gas via the glass envelope and also where the metal pins go through the glass. It is impossible to make a perfect glass-metal hermetic seal. Hence, the vacuum slowly deteriorates and some molecules of air are sucked in. If we slap the full HT voltage on the valve this gas can ionise and the valve behaves like a fluorescent tube and, with a blinding flash, the valve is no more.

Very high power transmitting valves used by broadcasters sometimes have a permanent vacuum pump attached to maintain a satisfactory vacuum in service. Because of the very high anode voltages used this is essential. Even a few molecules of gas can cause a flashover.

Some modern valves use a different getter system. Svetlana use titanium coating on the graphite anodes, which they call an *'active-getter'* because it contin-

Fig 12.2. Getter deposits on glass-envelope valves

ues throughout the life of the valve.

In **Fig 12.2**, the left-hand valve has the getter near the pins; the vacuum pinch off is also visible. The right hand valve has the getter at the top of the envelope.

COMMISSIONING UNKNOWN OR NEW VALVES

An important thing to do when commissioning an unknown or even a brand-new valve is to run it continuously for several days at full heater voltage and with no other voltages connected. The high cathode temperature will grab any stray gas molecules floating around in the envelope and convert them into extra material on the cathode. It may be necessary to run the cooling fans or blower to keep the valve envelope to a safe temperature. This also helps to 'pre-condition' the cathode and stabilises it for full power operation.

If the valve is a grounded cathode triode, tetrode or pentode the next step is apply full grid bias and run for another few days.

Never run a tetrode or pentode with just the screen grid voltage without the anode supply, it will expire very quickly.

After about three to five days it is normally safe to apply a reduced anode voltage (about 50% maximum) and tentatively bias up the valve to the normal idle current. This is normally about 5 to 10% of the maximum anode current. Run the valve for another 24 hours like this before bringing the anode voltage up to full value in steps. This last step hardens the vacuum back to somewhere near the original value and should not be omitted.

When you have achieved this last step and you have full anode voltage applied look very carefully at the anode current meter with the grid bias at normal running condition and the screen grid voltage at zero. There should be little or no anode current showing, if there is then the vacuum is still too low and it needs a few more hours to make it safe. If you see any significant anode current under this condition (greater than a few mA) it means the valve is not being controlled correctly and the current shown on the anode current meter is leakage current between the anode and cathode. This is an unsatisfactory state. As a final test, bring the grid bias voltage up to the cut-off condition and check the anode current again. There should less than a few hundred micro-amps of anode current in this condition.

GLOWING GLASS ENVELOPES

Many amateurs have noticed the ultra-violet / green glow when a glass envelope valve is running close to full output. Many believe this denotes the presence of gas in the envelope. In most cases this is incorrect. The glass envelopes of valves are made from a quartz-glass mixture to withstand the high temperatures; the trade name for this is Pyrex and it is similar to the type used for ovenware.

The anode of a high-power valve emits X-rays and this excites the quartz-glass material causing it to fluoresce. When the RF drive is reduced the glowing dies away. Older television sets when valves were common used high voltage rectifier valves to generate the EHT voltage for the picture tube. These needed to be contained in a metal shielding compartment to stop the X-ray radiation; if you looked at the sweep tubes they also emitted the same glow.

Most ceramic-metal valves also exhibit this fluorescence when pushed hard because the ceramic contains a small percentage of quartz. It is quite disconcerting to see a bright glow coming from inside a cavity until you realise what the cause is.

GLOWING ANODES

Some of the older glass envelope valves use a graphite anode. These are designed to glow dull red at full output - this is not a dangerous condition. The only way the graphite anode valve has to conduct the heat away is by radiation. A hard vacuum is a perfect thermal insulator. (The infrared radiation from the sun travels through a vacuum to reach us.) So the manufacturers design the valve anode to radiate the heat to the glass envelope where it can be conducted away by air. For graphite anode valves Mullard recommend a maximum anode temperature of +850°C, which is a dull red colour. If the anodes glow more than a dull red they are being over run. If they reach a dull orange colour only during speech peaks or keying this is normally safe, but if they approach a bright orange or dull yellow colour they are definitely being over stressed. The glass envelope is normally safe to about +200°C but plenty of air needs to be blown on to them to conduct away the heat for best valve life. You can never have too much air!

COOLING

All high-power amplifiers will dissipate a lot of heat. Because we cannot attain 100% efficiency or even get close to it we expect to have to deal with a lot of heat energy. If this isn't removed from the valve as quickly as possible the valve can reach unsafe temperatures and we can have a melt-down. There are two common types of cooling used for valves: air-cooling and liquid-cooling. Some very high power broadcasting valves use a form of vapour cooling with a closed loop refrigeration system to get the heat away, but these are normally beyond the bounds of amateurs.

There is one other type used by some UHF valves and that is conduction cooling where the valve has a dielectric block that is clamped to a heatsink. The dielectric block is a poor electrical conductor but a good thermal conductor, but these types are very specialised and rare today. The dielectric block is often made from beryllium oxide, a hazardous material.

Some external anode valves such as the 4CX family rely on cooling air being forced through a finned radiator. These normally use an *airflow socket* to mount the valve in. These generate a fair amount of restriction (back pressure) and the blower needs to have adequate volume to work against the restriction. Often these valves use a ceramic or PTFE 'chimney' to span the distance between the airflow socket and the valve anode. These are a potential source of additional restriction and air leakage. They are also a potential source of a flashover between the anode and the airflow socket if dust or other dirt collects, a better method is the scheme used in the K2RIW 70cm amplifier. In the normal system the grid compartment is pressurised by the blower and the air flows through the airflow base and then via the chimney to the anode. This is a high restriction system and needs a powerful blower, which is inherently noisy. In the K2RIW system the anode compartment is pressurised and the chimney is omitted, the air flows only through the anode fins and has far less restriction. The anodes have a PTFE chimney above the anode to conduct the air out of the anode compartment.

I modified my GW3ZTH 2m amplifier to this system and gained two advantages: the cooling is far more efficient, resulting in lower anode temperatures, and the blower noise is greatly diminished. The high volume of air forced into the anode compartment also conducts a lot of heat away from the anode network, which in the normal system do not get any significant cooling air. This allowed the valves to be pushed harder and over 1250W output is possible on CW in Class C without overheating the valves. The original Eimac airflow sockets are very restrictive. In this alternative method some of the air spills back through the airflow socket to cool the grid and cathode pins on the valves. The grid compartment needs an exit hole for this additional air and often a piece of mesh bonded to the metalwork covering the hole to prevent RF radiation is required. (The original cut out for the blower in the grid compartment can be used for this.)

WATER COOLING

By far the best system of cooling is liquid cooling, although a challenge to arrange for some amplifiers. This normally uses water flowing through a heat exchanger clamped to the anode. Some very power transmitting valves use exotic liquids such as glycol or Freon though these are hazardous chemicals. High voltage and water do not normally mix, but if the water is de-ionised the leakage current is quite low. Pure water is actually quite a poor conductor of electricity. The figure of 2.5MΩ/m is typical for distilled or de-ionised water. The main reason water is conducting is the dissolved salts and other minerals as well as metal oxides. Spa mineral waters are high in dissolved salts and so is not as good an insulator as other types. A good source of inexpensive almost pure water is from the refrigerator defrosting process. Adding some automotive antifreeze reduces the conductivity and helps to slow down corrosion. The writer uses water cooling for several amplifiers using several kilovolts on the anode and the cooling is greatly improved. Many high power broadcast transmitters use a form of liquid cooling to dissipate the large amount of heat generated. In one system seen by the writer an UHF *Klystrode* television transmitter amplifier with 28kV on the anode used water-cooling quite satisfactorily. This valve dissipated 50kW continuously.

A formula to calculate the approx power dissipation using water is:

$$P \approx 0.0698 \times F \times T$$

where P is the power in kW
F is the flow rate in litres/min
T is the temperature rise in Celsius/

For a flow rate of 2 litres/min with a temperature rise of 20°C, P ≈ 3kW.

The major benefits of using water cooling are the lower noise levels and a large inherent heatsink for unexpected dissipation. The heated water can be passed through a small liquid-air heat exchanger with some 120mm computer fans blowing air through the matrix. A good choice is an automotive heater matrix. These with water at +80°C are capable of dissipating in excess of 5kW continuously with sufficient airflow. As long as the flow rate is enough to hold the anode temperature well below +100°C we can extract a lot of heat energy from the anode. Normally a small aquarium or garden fountain pump is adequate. Most ceramic valves have a safe anode seal temperature of only +200°C, with water-cooling it is unlikely to exceed +100°C unless the flow is interrupted for some reason.

I developed a system for the Russian Gi7bt triode using a block of aluminium to which the valves were bolted (see Chapter 14 for full details). The block has several 6mm cross-holes through which the water flows. This with a flow rate of 2 litres per minute and a temperature rise of 20°C can safely dissipate 2kW continuously. This is shown in **Fig 12.3**.

This amplifier is used for a meteor radar system running 5kW-pulse power. The same cooling block is used for a 1kW SSB amplifier for 23cm. Single and triple valve amplifiers follow the same design methodology. The aluminium block is 12mm thick and the anode air coolers are removed and the peg nuts used to bolt the aluminium block to the anode flanges. A smear of heatsink compound fills any minute voids so ensuring a good path for the heat to flow to the heatsink block.

Fig 12.3. Water cooling anode block for 2 x Gi7bt triodes

Fig 12.4. N6CA water-cooled 2C39 triode amplifier for 23cm

The tubes that carry the water to and from the valve block need to be low current leakage material (high resistance). Normal aquarium clear plastic tubing is only rated to +60°C and will soften and may cause leakage of coolant. The best option is silicon-neoprene tubing used for peristaltic pumps. This will withstand up to +150°C; the peristaltic tubes are sterilised in a steam chamber to kill germs before use. Small hose clamps, P-clips or tie wraps are used to hold the tubing onto the ports. Avoid automotive black rubber or plastic tubing, this may contain carbon black which is a good conductor of electricity! Two manifolds are required for the inlet and outlet water. These convey the water to and from the circulating pump and the heat exchanger. The length of tubing between the anode and the manifolds needs to be longer than about 500mm to limit the anode current leakage to safe levels. The manifolds are made from metal and grounded to shunt the leakage current to ground. See the N6CA design for more details of the water cooling system.

Another popular microwave triode is the 2C39 family. These are normally air-cooled but several designs for water-cooling have been published. The best known is the N6CA design by Chip Angle for 23cm that removes the air-cooled radiator and replaces it with a water jacket. This raises the safe anode dissipation by as much as 400%.

ALTERNATIVE LIQUID COOLING

Some amateurs may regard the use of water as a bit scary although it is quite safe if the normal precautions are taken to reduce the dissolved salts and minerals. With tubing of about 500mm between the valve heat exchanger and the grounded water manifolds the expected leakage current when 2kV is used is below 2mA. Although the leakage currents are very low when the water is new contamination due to electrolysis is a problem, especially if dissimilar metals are used. A good plan is to measure the leakage current and when it exceeds 5mA the water should be changed for fresh distilled water.

One of the alternative fluids we can use is cooking oil, made from sunflower seeds. This has a greater calorific heat compared to water and is inherently a good electrical insulator. I used this for about five years in a high-power dummy load to conduct away 2kW. The major problem with cooking oil is its affinity for water; it absorbs moisture and becomes a sticky gel after a few years. If the system is closed (sealed from the air) this is not a problem. Another liquid I have experimented with is automotive automatic transmission fluid (ATF) as used in automatic gearboxes and power steering pumps. This is an excellent electrical insulator and being based on silicone oil is inherently immune to water absorption. The breakdown voltage of silicone oil is 15MV/m; compared this with air which is only 3MV/m.

The safe operating temperature of ATF is in excess of +150°C; some types are rated to +250°C in automatic transmission systems. The major disadvantage is the cost and the mess if a leak occurs. Cooking oil can be washed away with hot soapy water, ATF needs aggressive solvents.

Another option is to use paraffin / kerosene. Although this is more flammable it does have a high calorific content and can convey a lot of heat with an adequate flow rate. Paraffin is an excellent electrical insulator, typically it will withstand about 16kV/mm if it is free of water. Like cooking oil, though, it has an affinity for water and will absorb up to about 10% by volume if the relative humidity is high.

In all liquid cooling systems some thought needs to be given to the reservoir tank. The best option is make this out of a conducting material such as stainless steel or aluminium and bonded to the common negative chassis. Where the return line enters the reservoir tank the pipe should enter below the liquid level to prevent foaming. If the return line is allowed to spray the liquid above the water level then air will be trapped in the water and this sets up electrolysis, causing corrosion. If the return pipe enters the liquid no air in entrapped and the liquid will remain pure for a longer time.

LIQUID COOLING PUMPS

Many amateurs use aquarium pumps or garden fountain pumps for the cooling circulation. These are normally designed to be immersed in the water reservoir and often supplied with 230V AC directly from the mains supply. I have had two pumps fail with a blinding flash over the course of about 10 years and so sought a safer option.

A very good alternative is an electronic fuel injection pump. These are designed for high flow rates and high pressure, the typical 12V Bosch electric fuel pumps can supply about 190 litres per hour (≈3litres/minute) against a back-pressure of 3 bar (43psi) when pumping petrol / gasoline. Into an 'open-circuit' (zero back-pressure) they can deliver about three to five times the flow. The main disadvantages are the high initial cost and the current drain. The running current when delivering against a 3 bar pressure is about 8A at 12V DC but into an open circuit they draw a lot less. Because the flow rate is much greater than we require they can be run on a reduced voltage to cut down the flow. An alternative method is a 'bleed-off' to return the bulk of the fluid back to the reservoir. *The one thing they must not be used for is to pump water, as this causes rapid deterioration of the internal scroll pump.* Many electronic fuel injection pumps become noisy and drop in delivered pressure due to water in the fuel tank and the consequent rust that forms. This forms a 'grinding paste' and will rapidly abrade the close tolerance parts.

The clearances in the scroll pump are minute and any dirt will cause rapid wear to take place. The scroll pump and DC motor rely entirely on the fuel as a lubricant and must not be run dry. Fortunately they can be bought at low cost from breakers yards or even picked up for nothing from workshops because they have worn beyond a certain point. When the internal scroll pump becomes worn it emits a loud whine or screeching noise because the internal parts are rubbing themselves to death. However, although they are no longer suitable to pump low viscosity liquids such as petrol / gasoline they are fine for more viscous liquids such as ATF and cooking oil and will carry on working at lower pressure for years.

Many fuel injected vehicles use an additional fuel pump located in the fuel tank, designed for a high flow rate at low pressure. This is situated between the fuel tank and the main high pressure fuel pump and is known as a 'helper-pump' or 'lift-pump' and is intended to supply the input to the high pressure pump so it can never run dry. However, if the fuel tank level falls too low both pumps will run dry with consequent damage. These in-tank pumps are ideal for a liquid cooling system because they are low pressure and high flow rate. Many of the modern types have nylon impellers so can be used to pump water safely.

AN EXAMPLE DESIGN FOR 6M

I constructed a dual valve amplifier for 50MHz using Russian Gi7bt grounded grid triodes. This amplifier is rated at 1kW output for SSB or CW operation, a second version was modified for meteor radar applications and the anode network optimised to give 5kW peak output. The amplifier is made as a separate item from the power supply because a suitable supply already existed for another amplifier. This supply is rated at ≈2200V at 1A.

The design began by studying the manufacturer's data for the valve. This, being a Russian valve, was in Russian but fortunately a friend was able to translate it. Be aware that some versions of the Russian valve data sheets contain errors, in the version I used the value of the capacitance between the anode-grid (Cag) was transposed with the output capacitance (Ca).

The important data is the input and output capacitance and the tolerance for these. The anode capacity when operated in grounded grid is 4.6pF ±0.4pF, so for two valves connected in parallel the values are 9.2pF nominal, 8.4pF and 10pF minimum and maximum. Because there will be some stray capacitance this was estimated at ≈2pF. This occurs because the valve water-cooling block is fairly close to the top cover of the amplifier. The anode load was calculated for the anticipated peak anode current of 1A and a supply voltage of ≈2kV. This therefore gives an anode load of ≈2kΩ. The anode matching network chosen was the popular pi network that works satisfactorily at 50MHz if correctly constructed. The loaded Q was chosen to be approx 15. (See the Tuned Circuits section for details of how to calculate the circuit values.) The approx values for the pi network were determined to be:

CT = 20.6pF (including the anode capacity)
CL = 130pF
L = 570nH

A stud bolted to the anode-cooling block arranged the connection of the tank inductor to the anode. This has an inductance of ≈10nH that needs to be taken into account. The anode tuning requires a high voltage capacitor of low minimum capacitance. If one where available in a small version a vacuum variable would have been a good choice. These can handle very high RF voltages and currents. Unfortunately the smallest type found had too much capacity at the minimum setting and was also physically too large. Hence, it was decided to use an air spaced capacitor made from two circular discs of copper. The amount of capacity required is quite low for optimum operation. For best Q the anode capacity would form the main tuning capacity and the inductor value varied to resonate the network. It was calculated that a maximum value of 10pF would allow full band coverage. The minimum value was calculated as being ≈5pF and hence the mean value is 7.5pF. This allows a wide spacing, which will withstand the high voltage.

The formula to calculate the capacitance is:

$$C = \frac{0.0855KA}{D} \text{ Picofarads}$$

where:
K = dielectric constant of insulator (air is 1.0)
A = area of the plate in square cm
d = dielectric thickness in cm.

For a 50mm diameter disc with a spacing of 0.5cm, C = 9.8pF.

Dry air at standard atmospheric pressure (102 kPa) and temperature (+20°C) has a breakdown voltage of between 50V per mil (.001 inch) and 200V per mil when measured under laboratory conditions. The breakdown voltage per mil (or mm) for very close spacing is considerably higher than for larger spacings; in other words it doesn't scale linearly. This assumes the two electrodes are perfectly smooth (highly polished surfaces) and the moisture content of the air is essentially zero. In practice any sharp edge or surface roughness will lead to a tendency to corona discharge which lowers the breakdown voltage considerably. Similarly for air with a moderate to high moisture content the breakdown voltage decreases. At high altitudes (low air pressure) the breakdown voltage also decreases considerably and at high air pressure the breakdown voltage goes up dramatically. Hence, although an air gap of 5mm will theoretically withstand a voltage of ˜40kV, in practice it is often much less. It is prudent to allow a maximum of only about 2kV per mm when an air spaced capacitor is used.

As an example of the sort of breakdown voltage to expect under ideal conditions the data shown below is taken from a test laboratory report using two highly polished metal spheres of 100mm diameter with a variable air gap and dry air.

Air gap mm	Breakdown voltage kV	Volts per mm approx	Volts per mil approx
5	17	3400	86
10	31	3100	79
15	46	3066	78
20	59	2950	75
25	72	2880	73
30	85	2833	72

Because the anticipated maximum peak anode voltage is twice the DC supply this is about 4kV and so a 5mm gap will be more than adequate. The edges of the discs were chamfered and polished to prevent corona discharge. As a comparison of the breakdown voltage of common insulators the table below will be of interest.

Dielectric strength of various common materials - Material Dielectric Strength (MV/m)

Material	Dielectric Strength
Air (Dry)	3
Quartz	8
Strontium titanate	8
Neoprene rubber	12
Nylon	14
Pyrex glass	14
Silicone oil	15
Paper (Dry)	16
Bakelite	24
Polystyrene	24
Teflon	60

CIRCUIT VALUES

To determine the required circuit values an analysis software package was used. This treats the network as a 1-port network terminated by the anode load resistance, tuning capacity, anode capacity and the stray capacity. The anode tuning capacitor is in parallel between the anode and ground. The series Pi tank inductor connects between the connection stud and the loading capacitor. This capacitor CL has a minimum value of 25pF and a maximum of 150pF with a wide spacing to handle the anticipated RF voltage. The equivalent circuit is shown in **Fig 12.5** where the anode DC feed choke and output transformer have been omitted because these do not alter the AC analysis. See later for details of these components.

CA is the anode capacity of ≈12pF including circuit strays and RA is the calculated load resistance of 2kΩ. CT is the air variable for tuning. LS is the inductance of the stud from the anode block to the main tank coil. LT is the main tank coil and CL is the loading capacitor of 150pF maximum.

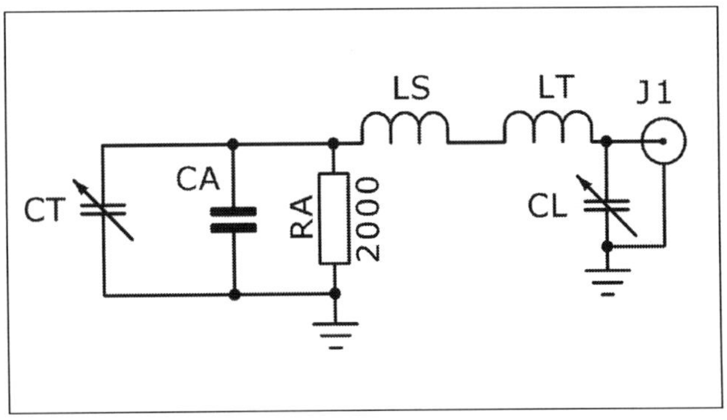

Fig 12.5. Equivalent RF circuit for anode matching network

Fig 12.6. Anode tuning capacitor and anode-cooling block

The capacitor CT is shown in **Fig 12.6** with the fixed half bolted to the anode-cooling block. The movable disc is fitted to the end of a threaded shaft that runs in a threaded bush (see **Fig 12.7**). By rotating the shaft the disc is varied in distance between the anode. The threaded bush bolts into a hole in the amplifier front panel to ground it. The diameter of the discs is 50mm. This requires a spacing of approx 6 to 10mm to obtain the correct anode tuning capacity.

Fig 12.7. Close-up view of the tuning capacitor movable portion

It is essential to have a positive mechanical stop to prevent the two plates touching; this would short out the DC supply and cause severe damage.

A PTFE button pressed into the centre of the fixed plate is one method; a second is to have a stop collar on the panel that prevents the shaft from being screwed past a certain point.

The spindle is threaded with a fine thread (32tpi) to allow fine adjustment. The control knob engaging with the mounting collet bush forms the mechanical stop.

With the network constructed on the analysis software the value of CT is fixed at the expected mean value of 7.5pF and CL and LT are allowed to vary between preset limits. The software 'looks into' the network from the 50Ω port and iterates the values to try and find a perfect match. This typically only takes a few seconds. The anode DC feed is connected via a decoupling capacitor effectively in series with CL, so it does not appear across the high impedance circuit as it would in a conventional pi network. This reduces the circuit losses by about 20% and equates to more output power.

COAXIAL STUB ANODE CHOKE

The RFC to feed in the anode voltage is a little unusual in this amplifier. The main reason for changing the topology was to reduce the value of the anode feed RFC in a normal pi network and to avoid spurious resonance. For the amplifier in question the anode resistance at resonance is approx 2kΩ and so the RFC would need to present a reactance of at least 10 times this to have little loss. This is a fairly large inductance, approx 64μH, and it is very difficult to wind a high Q compact inductor with this value whilst avoiding SRF and other spurious resonances.

It was decided to move the RFC to the 50Ω end of the network where the reactance required is only 500Ω. An RFC of 1.6μH is the lowest value that should be considered. Making the RFC approx 5μH would present a reactance of about 1.5kΩ, which is 30 times the impedance present. However, this RFC also needs to handle substantial RF voltage, at 1kW output the rms voltage is 225V AC or a peak voltage of 320V.

The properties of a shorted quarter-wave stub are used here not only to form the high impedance but also to use the 2nd harmonic trap effect. The RFC is formed from a piece of URM-43 coax cable cut to be λ/4 at 50MHz. (PTFE coax cable such as RG-142/U would be a better proposition for very high anode voltage applications.) Taking into account the velocity factor of the coax of 0.66 this is a length of ≈1m. This is wound on to a piece of insulating tube (electrical conduit) of 25mm diameter. The DC block is formed by a high current RF capacitor (CB) with a rating of 5kV AC and a value of 10nF. This connects to a transmission line transformer wound on a ferrite toroid with a ratio of 1:1. This forms the safety short circuit should the DC blocking capacitor fail. It also eliminates the troublesome output RFC that also needs to withstand the same peak output voltage.

The final circuit is shown in **Fig 12.8**, which includes the DC feed and the output 1:1 transformer. The value of anode tuning capacity CT is ≈6pF, L1 is 630nH and CL is 73pF.

The high voltage decoupling capacitor CD consists of two metal plates between which is sandwiched a piece of PTFE (Teflon) sheet of 0.25mm (10-mil) thickness. Nine screws pass through this plate to clamp it to the bottom plate. 8 are M4 nylon screws around the edges and the centre screw has a PTFE insulating button with a M3 screw to carry the anode supply current from the lower compartment. The top plate also serves to mount the loading capacitor using a right angle bracket bolted to the top plate with countersunk screws to prevent puncturing of the PTFE sheet, and the bottom part is the amplifier box base plate. This has an effective area of 120cm^2. The formula to calculate the capacitance was given earlier.

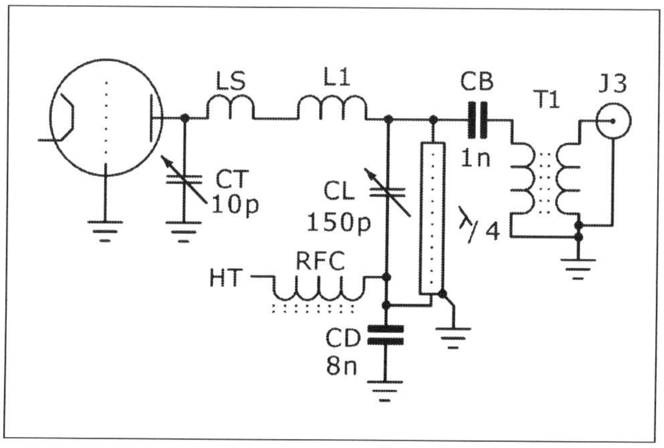

Fig 12.8. Complete anode network including the DC feed and decoupling capacitors

Fig 12.9. Anode RFC wound with URM-43 coax on a plastic former

PTFE has a dielectric constant of 2.12 and a dielectric strength of 60MV/m, so the 0.25mm sheet can safely withstand ≈10kV. The value of capacitance is approx 8700pF. This capacitor has a very low reactance at 50MHz (0.36Ω) and a Q of greater than 1000 (RS = 0.00036Ω). At 1kW output the current flowing to ground via the loading capacitor and the decoupling capacitor is 4.5A. The most significant loss resistance is the contact resistance between the variable capacitor moving plates and the grounding bush; this amounts to approx 10mΩ. The high voltage is fed via a small RFC of 5μH rated at 5A. The output transformer is wound on a Micrometals Inc T130-7 ferrite toroid (33mm diameter) with bifilar twisted 18SWG enamel copper wire (1.2mm diameter). The total measured loss in the network is approx 0.2dB, which is equivalent to 4% of the output power, ≈40W at 1kW output.

The RFC is bolted to the end wall of the anode compartment with 4mm screws and stand off pillars. 6mm cross-holes through the plastic tube retain the coax. The bifilar wound output transformer is to the right of the picture. The DC blocking capacitor is not the final type used. This picture was taken during the optimisation of the pi tank inductor at low power.

SINGLE VALVE VERSION

A single Gi7bt with 2kV-anode supply will deliver about 600W under CW operation at 50MHz. The air-cooled anode dissipation is 350W and with an efficiency of 60% it will require about 1kW DC input. The circuit values are:

Anode tuning capacitor	=	6pF
Anode loading capacitor	=	96pF
Anode tank inductor	=	950nH

RF CAPACITORS

In most high power amplifiers we need fixed value capacitors that can withstand not only high DC and RF voltages but are also able to handle considerable RF current. If the capacitor is lossy it will dissipate a large amount of power causing heat to be generated and the capacitor to possibly fail. Unfortunately some of the designs published over the years show unsuitable capacitor types and these can cause excessive power loss. **Fig 12.10** shows some high voltage and high current capacitors suitable for valve amplifiers.

The left-hand capacitor is a 3000pF coupling capacitor rated at 20kV AC, it can safely handle 100A at 30MHz. It was used in a 10kW HF transmitter and is 60mm diameter. The capacitors to the right are high RF

Fig 12.10. High voltage RF capacitors

current tuning/loading and DC blocking types specified for RF operation. The lower right disc shaped capacitor was used in the 6m amplifier. This is 10nF rated at 5kV AC and 30A RF current rating.

For VHF / UHF amplifiers that require a low loss DC block between the anode and the output network a PTFE *sandwich capacitor* is an excellent method. The value of this capacitor will be determined by the design. A diagram of a simple method using stock right angle material and plates is shown in **Fig 12.11**. This uses thin PTFE sheet as the dielectric. PTFE has a voltage withstand of approx 60MV/m. Hence, a thickness of 0.25mm will safely withstand ≈10kV and can handle a lot of RF current. This method is far superior to the doorknob capacitors often shown in published designs and is virtually loss-less. Many VHF amplifiers require a value of about 500pF; this can be attained with a capacitor of about 6cm^2 when 0.25mm PTFE sheet is used.

For high voltage DC decoupling where relatively small amounts of RF are flowing, another method is to use thin single-sided glass-fibre printed circuit board (0.8mm). The copper is etched away around the edge of a rectangular or square piece for ≈12mm and the board screwed or glued with super-glue to a flat metal panel. This can be a wall of the amplifier chassis. RFCs from here to the DC input connector and the rest of the amplifier complete the circuit.

VACUUM VARIABLE CAPACITORS

For very high voltage variable capacitors we can utilise a type known as 'vacuum variables'. As the name suggests, these have an evacuated envelope, normally glass or ceramic, with a variable capacitor contained within it. The breakdown voltage of a perfect vacuum is infinite, because it contains no molecules of gas. In practice obtaining a perfect vacuum is very difficult and as the moving part of the variable capacitor needs to exit the envelope this is a source of potential leakage allowing some air to be sucked in over time. However, vacuum variables can withstand very high voltages, up to 50kV for some types. Because the high vacuum allows a much higher operating voltage the gap between the fixed and moving plates can be made very small and the amount of a capacity we can achieve is greater than an air variable of the same physical volume. Normally the vacuum variable has a slow motion drive that allows fine setting of the capacitance value.

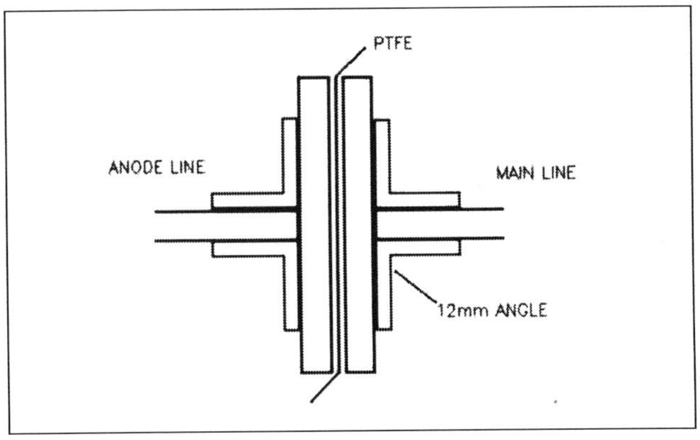

Fig 12.11. Homemade PTFE sandwich capacitor construction

CATHODE BIASING OF TRIODES

Grounded grid triodes, except the Class B zero bias types, will require a means of adjusting the bias. Because the grid is grounded to the amplifier metalwork for best stability the cathode terminal is now positive with respect to ground. To attain the correct grid-cathode voltage requires a circuit to hold the cathode at some voltage above ground. There are several methods we can use and depending on the anticipated peak anode current this will determine the best option.

RESISTOR BIASING

The simplest method, although far from being the best, is to insert a resistor between the cathode and ground. If this resistor is a variable type we can vary the cathode voltage with zero drive to obtain the required quiescent current. Suppose we need a quiescent current of 50mA and the required grid-cathode voltage is +20V for this condition. (We can obtain the required voltage by examining the valve constant current curves). Since R = V/I, for a voltage of 20V when 50mA is flowing we need a resistor of 400Ω. If we used a 390Ω wire-wound resistor we would be approximately correct. But valves have a tolerance and some valves will require slightly different bias voltages to obtain the correct quiescent current. Hence, part of the cathode resistor needs to be variable.

Now consider what happens when we apply RF

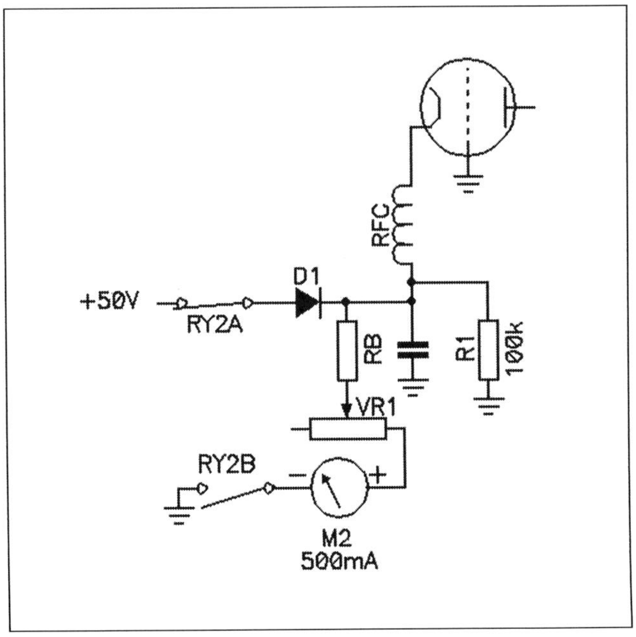

Fig 12.12. Cathode biasing with resistor

drive and the anode current increases. At 100mA the voltage developed will be +40V; at 200mA it will be +80V etc. Hence the bias will move from a linear type towards Class C with very little drive power. If the mode is CW or FM this will be OK but it will severely limit the output power possible. At around +50V the valve is into the cut-off region and very non-linear. So this method is really only useful when the valve draws little anode current. The cut-off bias is obtained from a +50V DC supply fed via the relay contact when in standby, diode D1 can be a 1N4007. On transmit RY2A is open and RY2B is closed.

ZENER BIASING

Let us now look at another popular method of obtaining a fixed bias voltage. This replaces the resistor with a zener diode.

This seems to be perfect until we examine the circuit more closely. Zener diodes have a characteristic known as 'dynamic Impedance' which means that they change their resistance with applied voltage. Below the zener voltage they are not conducting, but above the zener voltage they shunt current to ground to try and maintain a constant voltage. This works when the resistor feeding them is a fixed value. But in a valve amplifier the resistor value is anything but constant. The 'resistor' feeding the zener is the anode load resistance and as we have seen this varies over a wide range. At quiescent current levels it will be high but as the anode current rises it falls to a lower value. This means the zener diode needs to draw more current to hold the voltage constant. The zener voltage can vary by as much as 20% under these conditions and so the voltage varies with the drive level. Hence, it is also really not suitable when high peak currents are anticipated. A typical 1kW DC input amplifier can experience peak anode currents of up to 1.5A. Although the average anode current is only 0.75A the peaky nature causes the anode current to attain high peaks for short periods. Hence, the zener diode method is not the best. We also need to have some variation in the bias voltage to be able to set the quiescent current, this can be obtained by using several zener diodes connected in series and tapping to find the best voltage. This is messy. The other problem is the power dissipated in the zeners; they need to be high wattage types to safely carry the high peak anode current.

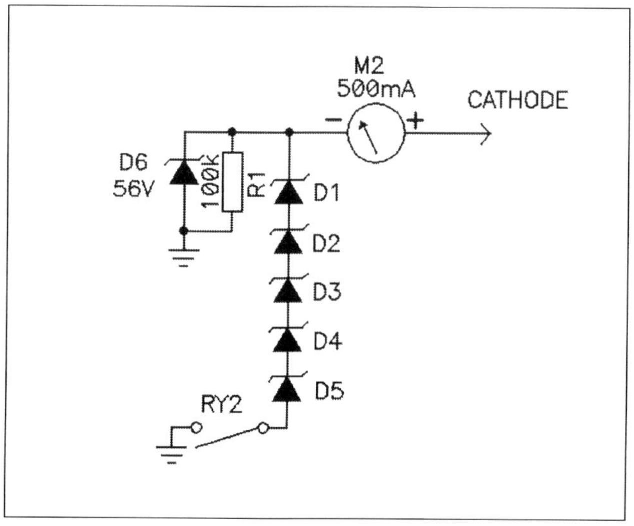

Fig 12.13. Cathode biasing with zener diodes

CONSTANT VOLTAGE SOURCE BIASING

We will now look at a simple shunt stabiliser for grounded grid triodes that can handle peak anode currents of 3A with ≈1% variation in the cathode voltage. This is shown in **Fig 12.14**. It uses a high current Darlington NPN transistor as an 'amplified zener' in a Constant-Voltage source. The reference voltage is obtained by a 500mW zener diode D1 and the voltage is applied to the base of TR1 with a potentiometer. This circuit is intelligent because it has feedback from the cathode to the transistor base and hence corrects any error occurring as the anode current varies. The transistor shown has a current gain of 1000 minimum and is rated at 60V collector-emitter voltage at 10A, so is adequately rated. It needs bolting to a heatsink to dissipate the power. At +20V cathode bias and a peak anode current of 1.5A it is dissipating 30W.

All the circuits shown incorporate some safety features. When the valve is in standby (not transmitting) the anode current needs to be cut off. The 56V zener connected in parallel clamps the cathode voltage to this value. The circuit was designed for the Gi7bt triode, which requires approx +40V cathode voltage to reduce the anode current to zero. In standby the relay contact is open and isolates the regulator, the 56V zener then cuts off the valve. The 100kΩ resistor is included as belt and braces in case the 56V zener fails open circuit. This can be a 0.5W type.

MECHANICAL CONSTRUCTION

For high-power amplifiers we need to devise a suitable mechanical construction. It is important to have close fitting covers with plenty of screws for the anode and grid compartments that prevent the leakage of RF, either between the compartments or out of the amplifier. The popular method is to make two boxes from folded up metal such as aluminium and to bolt these back to back. Because folding up boxes is a complex operation this is best entrusted to someone who has the necessary skill and bending facilities. The method I used was to form the boxes in two halves each one an L-shape with folded lips to attach the covers. A typical enclosure is shown in **Fig 12.15** which is the 6m high-power amplifier using two Gi7bt triodes.

The top, bottom and central plates are flat sheets of aluminium with holes to fasten to the L-shaped pieces.

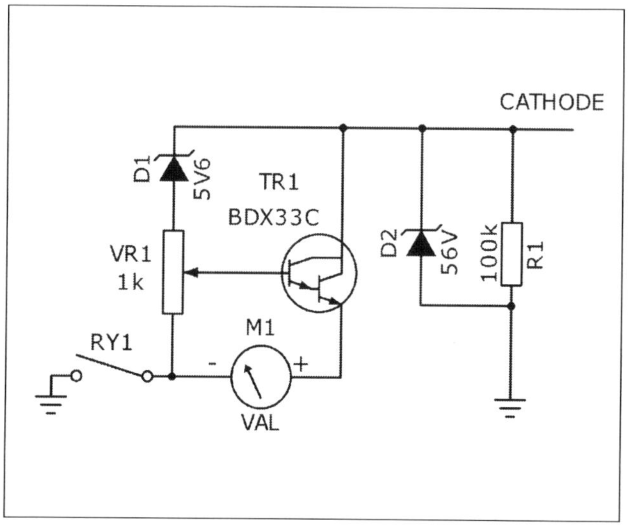

Fig 12.14. Cathode biasing with a shunt regulator

Fig 12.15. Anode compartment of the 6m amplifier with the anode cooling block placed to check fit

As these are all identical, apart from the holes needed in the central plate, these can be jig-drilled as a stack to ensure the holes line up. The top and bottom cover plates are fitted with screws into pressed inserts in the L-shaped pieces (PSM inserts or 'clinch nuts'). Self-tapping screws into soft material such as aluminium are a poor choice as they often strip after being removed and replaced a few times. Clinch nuts are made from steel and withstand repeated removal of the screws. By making the boxes in sections this makes the drilling of any holes simpler and the assembly easier; I detest 'key-hole surgery' working in a confined space. When all the holes have been drilled, the corners of the L-shaped pieces are joined by aluminium welding to form an RF-tight seal. Alternatively pieces of aluminium angle can be used to fix them together with screws. The top and bottom boxes are fastened to the central plate by screws passing through all three pieces. This forms a rigid enclosure with good RF shielding properties.

The holes in the central plate need opening up to suit the nylon insulating screws and insulator button for the central screw. They were left small to jig drill the central plate with the correct spacing from the decoupling plate. The loading capacitor is fixed to the decoupling

Fig 12.16. Amplifier metalwork dismantled to show the construction method. The two large holes are for the grid collets for the Gi7bt triodes that are equipped with finger-stock

Fig 12.17. High voltage decoupling capacitor plate with PTFE sheet

plate with the angle bracket shown alongside it. The angle bracket is fixed with countersunk screws with the heads under the top plate to prevent piercing the thin PTFE sheet. On this particular capacitor there were 3/16th inch (4.2mm) diameter holes moulded into the ceramic end plate, three M4 or 4BA nuts and bolts with fibre washer were used to attach the capacitor to the bracket. The fibre washers prevent the brittle ceramic from cracking, do not over-tighten!

USEFUL TIPS

When building an amplifier some constructors have difficulty with seemingly simple tasks. For example, we need to extend the shaft of a tuning capacitor that then passes through a bush on the front panel. Trying to determine the exact place on the front panel is tricky by measurement and one often finds that, no matter how much care is taken, the hole is still in the wrong place. This spoils the looks and should be avoided. Here is a dodge I learned many years ago which guarantees the hole is in the correct place every time.

The capacitor is mounted and then a piece of round rod the same diameter as the shaft is cut so that it just touches the front panel. The bush is slid on to this and the correct position is then marked with a felt-tip pen. After removing the panel the centre of the hole is found and this is centre popped and then the pilot hole drilled. Start with a small drill (3mm or 1/8 inch) and gradually open up the hole is several increments with larger drill

Fig 12.18. Marking out the correct place for the bush hole

Fig 12.19. Shaft bush mounted and the extended shaft in place

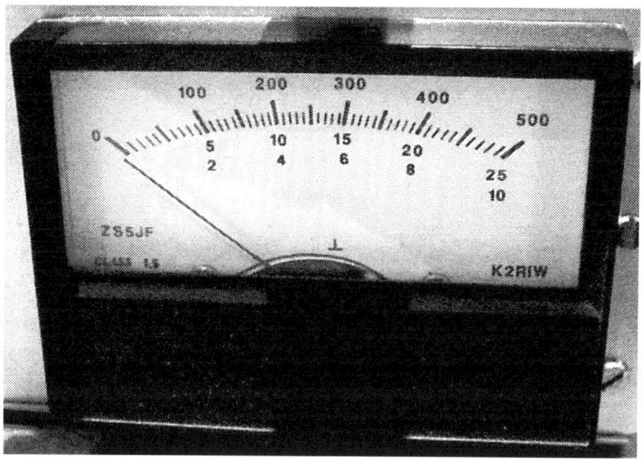

Fig 12.20. Modified meter for K2RIW 70cm amplifier

bits, this ensures a clean round hole and not a three-cornered one. Make the hole for the bush a little larger than normal to allow some adjustment, about 1 - 2mm is normally the best. **Fig 12.19** shows a bush mounted with the extended shaft. A large star washer on the inside prevents the bush moving when fully tightened, the outer washer is a flat type. By making the hole larger than required we can shift it a little off-centre to provide some friction, which stops the capacitor moving once set.

Another method is to machine the end of a piece of steel rod the same diameter as the shaft to a point like a centre punch. The rod is made so that the point just touches the front panel. Then a light tap with a small hammer on the outside of the panel forms a centre pop mark that accurately positions the shaft centre.

Note: this is the loading capacitor for the Gi7bt amplifier and the shaft extender must be an insulating material because the capacitor carries the full DC anode voltage. This shaft is made from paxolin rod 1/4-inch diameter to suit the shaft coupler and panel bush. The bush is rescued from an old potentiometer by carefully breaking it apart. The shaft coupler is a standard item or it can be made from 12mm diameter brass or aluminium rod if you have access to a lathe.

SCALING METERS

Often we need several meter scales to indicate different measurements. Because meters are expensive and we often have a limited area of front panel it would be nice if we could have one meter which can fulfil all our needs with switching. When building the K2RIW 70cm amplifier I had a suitable meter but it was scaled 0 to 1mA. It could be modified by rub on lettering but this is a messy process and it never really looks good, no matter how careful we are. I used a different technique that provided a professional finish. The meter was dismantled and the scale plate was carefully removed. This was photographed and a 2:1 positive was made. Then using some drawing film I drew the new scales using the positive under the drawing film to pick up the arc and start and finish points. The meter ended up with three new scales now drawn with drawing ink on the film. This was photographed and reduced to the final size and printed on Polish paper (as in Poland), a special white film. The scale plate was turned over and the new scale glued to the back, which now became the front. The finish was very good and worth the effort. I even added my callsign!

The scales are: top - 500mA anode current, middle - 25mA screen current and anode voltage 2.5kV, bottom - grid current 10mA and relative output power.

DUMMY VALVES

When using valves that require finger-stock to solder into a cavity or some other location it is useful to make a dummy valve to hold the finger-stock correctly aligned whilst the soldering is taking place. This needs to be a material that can withstand the high temperatures involved as well as something that the solder will not stick to. It is not a good idea to use a real valve because you may find the valve becomes a permanent fixture if the solder flows on to the valve. A good idea is to machine a dummy valve from a hard close-grained wood or aluminium stock bar.

A typical dummy valve is shown in **Fig 12.21**. This is a 2C39 type and can be used when soldering the anode and grid finger-stock into a cavity. It is also useful when assembling a cavity amplifier to get everything lined up correctly because a glass-seal or ceramic-seal valve can be damaged if the grid and anode are not aligned correctly.

Fig 12.21. Dummy 2C39 machined from aluminium round bar

Chapter 13

The pi network revisited

MANY amateur handbooks contain formulas to calculate the component values for the pi tank for RF power amplifiers. Unfortunately many of these are an over-simplification of the actual formula and consequently the values calculated are out, sometimes by a large margin. In many cases this is not too serious because the variable capacitors normally chosen for anode tuning and loading have a greater tuning range than required and often the amateur is unaware of the error when an amplifier is constructed.

The *RSGB Handbook* has been using the same formulas, which are too inaccurate for some applications, for many years. At HF this is not normally a problem for the reason already given, but at 50MHz and higher the errors become significant and may prevent the network from matching correctly. So to correct the situation let us give you the correct formula if you need to calculate a pi tank network with some degree of certainty. Frank Davis of Motorola Semiconductor Products Inc originally published the information contained below in 1968 in an application note *Matching Network Designs with Computer Solutions* (AN-267).

See **Fig 13.1**. Note that the normal DC blocking capacitor has been omitted from the diagram, as it serves no matching function. The reactance of the DC block is negligible compared with the anode load resistance R_A.

The anode of the valve(s) presents a resistance R_A to the network plus a shunt capacitance C_A due to the inter-electrode capacity. The load resistor R_L can be any value within reason; normally we select an antenna impedance of 50 or 75Ω. The anode capacitance needs to be subtracted from the calculated value of C1 to obtain the correct tuning capacitor value. At lower HF frequencies this is insignificant but at 21MHz and above it can be sufficient to prevent the network resonating correctly. If the anode capacitance is greater than the value of C1 required the network couldn't be brought to resonance unless the value of L is changed, which changes the loaded Q value (Q_L). For a higher loaded Q the value of C1 becomes larger and L becomes smaller.

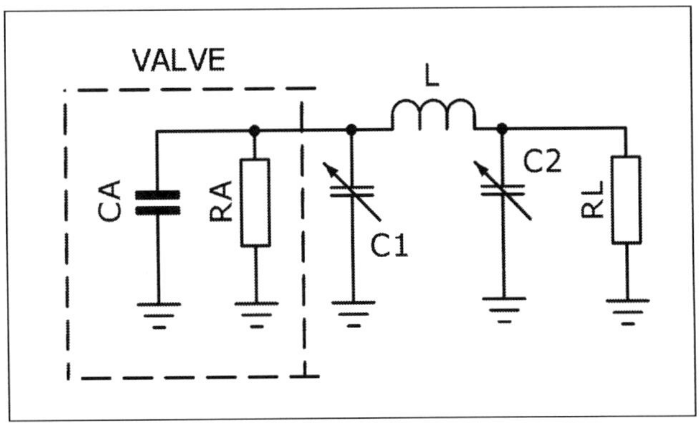

Fig 13.1. Pi tank network

The formula to calculate X_{C1}, X_{C2} and X_L are shown below, where R_A is greater than R_L.

$$X_{C1} = \frac{R_A}{Q_L}$$

$$X_{C2} = R_L \sqrt{\frac{R_A / R_L}{(Q_L^2 + 1) - (R_A / R_L)}}$$

$$X_L = \frac{Q_L R_A + (R_A R_L / X_{C2})}{(Q_L^2 + 1)}$$

The first step is to choose a loaded Q value. Normally for best efficiency we would select this to be low, but this causes problems if the network needs to match over a wide range of values for the anode load resistance and the antenna may be reactive off resonance. Choosing a higher loaded Q alleviates this problem and also gives better harmonic attenuation. The disadvantage of a higher loaded Q is the higher circulating RF currents in the network. Often a compromise has to be made for loaded Q and a value between of 10 to 14 serves to handle the various problems. Many handbooks advocate a lower value of loaded Q than a higher one. This in many cases is a false assumption, as we will see later. For many years the magic figure of 12 for the loaded Q has become entrenched in the amateur fraternity but it isn't always the best choice. Bill Orr, W6SAI, in his books advocates a higher Q_L than a lower one and the reasons will become apparent later.

Having chosen a loaded Q we can calculate the reactance for C1. This will need converting to a capacitor value using the standard formula. Having determined X_{C1} we can calculate X_{C2} and X_L. To make life easier we have included the tables for various loaded Q and anode load. These are all for a 50W antenna impedance at a frequency of 1MHz.

PI TANK NETWORK TABLES

The following tables give the calculated values for the network values when transforming from the value of R_A to a 50Ω resistive load with a fixed loaded Q value. The values are calculated for a frequency of 1MHz. The accepted maximum impedance ratio is normally 100:1 so the tables only go as high as 5kΩ when the load impedance is

50Ω.

To convert to a new frequency, divide all the component values by the new frequency. For example, for 14MHz divide all the values by 14. If the exact anode load is not listed, we can interpolate between two adjacent values.

$Q_L = 20$

Anode load Ω	C1 pF	C2 pF	L µH
5000	636.60	5522.29	43.13
4750	670.10	5712.62	41.08
4500	707.33	5916.91	39.03
4250	748.94	6137.20	36.98
4000	795.75	6375.94	34.93
3750	848.80	6636.13	32.87
3500	909.42	6921.52	30.80
3250	979.38	7236.85	28.73
3000	1061.00	7588.19	26.65
2750	1157.45	7983.50	24.56
2500	1273.20	8433.45	22.47
2250	1414.66	8952.73	20.37
2000	1581.50	9562.25	18.26
1750	1818.85	10293.03	16.13
1500	2122.00	11193.42	14.00
1250	2546.40	12344.13	11.84
1000	3183.00	13892.62	9.67
750	4244.00	16146.73	7.46
500	6366.00	19903.30	5.21

Table 1

$Q_L = 18$

Anode load Ω	C1 pF	C2 pF	L µH
5000	572.94	4774.50	47.74
4750	603.09	4952.66	45.48
4500	636.60	5143.38	43.22
4250	674.04	5348.51	40.95
4000	716.17	5570.25	38.68
3750	763.92	5811.33	36.40
3500	818.48	6075.16	34.12
3250	881.44	6366.00	31.83
3000	954.89	6689.35	29.53
2750	1041.71	7052.40	27.22
2500	1145.88	7464.79	24.90
2250	1273.20	7939.79	22.58
2000	1432.35	8496.28	20.24
1750	1693.97	9162.24	17.89
1500	1909.80	9981.29	15.52
1250	2291.76	11026.24	13.13
1000	2864.70	12430.01	10.72
750	3819.60	14470.11	8.28
500	5729.40	17864.54	5.78

Table 2

$Q_L = 16$

Anode load Ω	C1 pF	C2 pF	L µH
5000	509.28	3988.28	53.42
4750	536.08	4156.54	50.90
4500	565.86	4335.84	48.38
4250	599.15	4527.84	45.85
4000	636.60	4734.54	43.31
3750	679.04	4958.40	40.77
3500	727.54	5202.45	38.22
3250	783.50	5470.54	35.66
3000	848.80	5767.58	33.09
2750	925.96	6100.01	30.51
2500	1018.56	6476.44	27.92
2250	1131.73	6908.73	25.31
2000	1273.20	7413.72	22.70
1750	1455.08	8016.39	20.06
1500	1697.60	8755.66	17.41
1250	2037.12	9696.40	14.74
1000	2546.40	10957.11	12.04
750	3395.20	12784.94	9.29
500	5092.80	15810.22	6.49

Table 3

$Q_L = 14$

Anode load Ω	C1 pF	C2 pF	L µH
5000	445.62	3134.89	60.53
4750	469.07	3298.18	57.70
4500	495.13	3470.62	54.86
4250	524.25	3653.73	52.01
4000	557.02	3849.32	49.15
3750	594.16	4059.62	46.27
3500	636.60	4287.35	43.39
3250	685.57	4535.94	40.50
3000	742.70	4809.74	37.59
2750	810.22	5114.45	34.67
2500	891.24	5457.71	31.74
2250	990.26	5849.95	28.78
2000	1114.05	6306.03	25.82
1750	1273.20	6847.94	22.83
1500	1485.40	7509.90	19.82
1250	1782.48	8348.93	16.78
1000	2228.10	9469.09	13.71
750	2970.80	11087.32	10.59
500	4456.20	13764.41	7.401

Table 4

Q_L = 12

Anode load Ω	C1 pF	C2 pF	L µH
5000	381.9	2135.22	69.53
4750	402.06	2309.19	66.34
4500	407.9	2225.2	65.38
4250	449.36	2674.25	59.89
4000	477.45	2869.12	56.64
3750	509.28	3075.07	53.36
3500	545.65	3294.72	50.075
3250	587.63	3531.22	46.76
3000	636.60	3788.53	43.43
2750	694.47	4071.71	40.08
2500	763.92	4387.46	36.71
2250	848.80	4744.94	33.31
2000	954.90	5157.05	28.89
1750	1091.31	5642.85	26.45
1500	1273.2	6231.96	22.98
1250	1527.84	6973.60	19.47
1000	1909.80	7957.50	15.91
750	2546.40	9370.50	12.30
500	3819.60	11695.10	8.60

Table 5

Q_L = 10
No solution possible above ≈5000Ω

Anode load Ω	C1 pF	C2 pF	L µH
5000	318.30	318.30	79.57
4750	335.05	799.92	76.73
4500	353.66	1112.78	73.38
4250	374.47	1380.98	69.87
4000	397.87	1630.80	66.26
3750	424.40	1874.10	62.57
3500	454.71	2118.20	58.82
3250	489.69	2368.81	55.02
3000	530.50	2631.19	51.18
2750	578.72	2910.95	47.29
2500	636.6	3214.67	43.37
2250	707.33	3550.78	39.41
2000	795.75	3930.71	35.41
1750	909.43	4370.94	31.36
1500	1061.0	4896.72	27.27
1250	1273.20	5549.75	23.13
1000	1591.50	6405.66	18.93
750	2122.0	7621.5	14.64
500	3183.0	9601.90	10.255

Table 6

Q_L = 8
No solution possible above ≈3000Ω

Anode load Ω	C1 pF	C2 pF	L µH
3000	424.40	918.85	60.88
2750	462.98	1357.23	56.73
2500	509.28	1743.40	52.32
2250	565.86	2122.0	47.74
2000	636.60	2516.38	43.04
1750	727.54	2946.88	38.24
1500	848.80	3438.03	33.34
1250	1018.56	4026.21	28.35
1000	1273.20	4774.50	23.26
750	1697.60	5811.33	18.04
500	2546.40	7464.79	12.66

Table 7

Q_L = 6
No solution possible above ≈1750Ω

Anode load Ω	C1 pF	C2 pF	L µH
1750	545.65	760.88	46.96
1500	636.60	1537.53	41.83
1250	763.92	2205.25	35.98
1000	954.90	2934.58	29.77
750	1273.20	3854.80	23.26
500	1909.80	5230.20	16.44

Table 8

In the above tables, the anode capacitance (C_A) is assumed to be 0pF, which in a practical amplifier will not be the case. Hence, the anode capacitance needs to be subtracted from the value of C1. For low frequencies this is not important but when R_A is a high value and C_A is similarly significant this limits the ability to resonate the network and the value of L will need reducing to compensate.

The other factor not taken into account is the effect the anode RFC shunting the anode to ground. This will alter the real value of C1. Similarly when an output RFC is connected in shunt across C2 this also alters the real value to some degree. No allowance has been made for straps connecting the various pi tank components; these at low frequency are relatively insignificant but above ≈14MHz should be factored into the value of L.

However, the values given are a good starting point

for a new amplifier design and, provided the capacitors for C1 and C2 have sufficient range, can be expected to work satisfactorily.

The most important factor is the reactance of C1; this determines the value of loaded Q and the impedance presented to the anode by the tank network. If C1 is correctly adjusted than L and C2 could be found empirically.

Note the significance of the loaded Q. When this is low the range over which the valve can be matched to a 50Ω load is extremely limited. It is preferable to design for a higher loaded Q rather than a lower value; this also improves the harmonic attenuation. Doubling the loaded Q increases the harmonic attenuation by 6dB.

It is interesting to examine how the values of C1, C2 and L vary for different anode load values. At 2.5kΩ the value of C1 is twice the 5kΩ value, and at 500Ω it is 10 times the 5kΩ value, so it scales directly with anode load.

HF AMPLIFIER WORKED EXAMPLE

We will choose an anode load of 3kΩ at a frequency of 1.8, 3.5, 7 and 14MHz and a loaded Q of 12. From **Table 5** we find the 3kΩ row of values for C1, C2 and L.

C1 = 636pF
C2 = 3788.53pF
L = 43.43μH.

To convert to 1.8MHz we divide these constants by 1.8. This gives the values as:

C1 = 353.3pF
C2 = 2105pF
L = 24.13μH

To convert to 3.5MHz we divide the table values by 3.5. This gives the values as:

C1 = 181.7pF
C2 = 1082.4pF
L = 12.4μH

For 7MHz the 3.5MHz values are halved.

C1 = 90.85pF
C2 = 541.2pF
L = 6.2μH

For 14MHz we halve the 7MHz values.

C1 = 45.25pF
C2 = 270.6pF
L = 3.1μH.

Other bands can be found in the same manner. For example, 21MHz would require 1/3 of the 7MHz values etc.

SOME TRADE SECRETS ABOUT COMMERCIAL AMPLIFIERS

What is not obvious about some of the commercially manufactured linear amplifiers is some of the tricks the designers pull. An amateur who is building his first amplifier, in choosing the loaded Q will often select the same Q for all the bands. Commercial amplifier designers almost never do this. The reason is volume. For the low HF bands the volume the tank coil occupies is quite high if the loaded Q is low. Take a look at the tables again and you will see that as the loaded Q goes down the inductor value goes up. At 160m and 80m this can make a tank inductor occupy a lot of the available volume in the anode compartment. With a loaded Q of 8 and an anode load of 2kΩ at 1.8 MHz it requires an inductor of ≈24μH, with a loaded Q of 12 this is now only 16μH, a saving in volume of ≈33%. With a loaded Q of 14 it now requires an inductor of 14.3μH which is an approximate volume saving of almost 50%.

At 15m and 10m the opposite problem arises. The various straps and other wiring inductance eat into the tank inductor value and it ends up being lower in value because we have to remove some turns to compensate. This can make the coil impractical to make, if it is less than about 2-turns it may not be able to physically span the distance between the anode tuning capacitor and the loading capacitor. Hence, we need to increase the inductor value to make it physically bigger. To do this we need to use a lower loaded Q. If we take the table values for a 2kΩ anode load of the different loaded Qs we find that the inductance required at 29MHz is:

Q = 14	L = 0.89μH
Q = 12	L = 1.03μH
Q = 10	L = 1.22μH
Q = 8	L = 1.48μH.

This is actually a fortuitous situation because at the lower frequencies the resistive losses in the components are generally less than at the higher frequencies. Hence, we do not need such a high component Q at the lower frequencies and this allows the coil to be wound on a smaller former, occupying less volume but having higher losses. Not only this but with a lower loaded Q the circulating current in the tank network is lower and the resistive losses at the higher frequencies will be less. On average the circuit losses will be about the same on each band and only the valve(s) will determine the maximum output power across the frequency range. As the gain is expected to be less on the higher bands this simply means we need more drive power from the exciter as the frequency increases.

EFFECT ON LOADED Q WITH FREQUENCY

What isn't obvious from the various formulas and tables is how the loaded Q of a pi tank network varies with frequency. If the loaded Q had been selected as 12 and the circuit values determined for the top of the band the loaded Q at the bottom of the band will be higher with a fixed value inductor. This can be seen by the value of C1. If you remember that the loaded Q is determined by the value of C1, increasing C1 lowers the reactance of XC1 and raises the loaded Q value. When we change the operating frequency to the bottom of the band we need to increase the value of C1 (lower the value of XC1) to bring the network to resonance. Hence, the loaded Q goes up. For bands such as 10m the effect isn't so apparent. If the loaded Q had been selected as 12 at 29.7MHz the loaded Q increases to about 14 at 28MHz, which is not too serious. But for bands like 80m where the percentage change in frequency is much larger the loaded Q can be become much higher. For the loaded Q of 12 if set up at 3.8MHz the Q will increase to about 16 at 3.5MHz. Looking at it the other way round, if the loaded Q was set to 12 at 3.5MHz the Q at 3.8MHz will fall to about 8 at 3.8MHz because the value of C1 is less at the top of the band. The best option is to pick a frequency at mid band for the calculation, then at the top of the band the loaded Q will decrease a little and at the bottom of the band it will increase by the same percentage.

WORKED EXAMPLE FOR A VHF AMPLIFIER

At VHF the anode capacity and any stray capacity become significant and need to be considered. Because the tuning range required is small we can usefully increase the loaded Q value to make the inductor value required more practical. However, we need to consider the minimum tuning capacitance to resonate the network. If we run through the values for a 50MHz amplifier we will get an insight of the problems. We will design an amplifier that covers the full band up to 52MHz.

We will select a 6146 valve that is rated at up to 60MHz as an RF amplifier. From the manufacturer's data we get the parameters of anode capacity of 8.5pF, when strays are included this will be about 10pF. The anode load is given as 3600Ω and the nearest value in the tables is 3500Ω, which will be suitable. We will initially choose a loaded Q of 12 as a first attempt. Finding the table values we have the constants at 1MHz:

C1 = 545.65pF
C2 = 3294.72pF
L = 50.075H

Before we go any further we need to establish if the anode capacitance is significant, so we find the value for C1 by dividing by 52, this being the worst case.

C1 = 545.65/52 = 10.5pF

Since the effective anode capacity is 10pF this scheme will not work because the anode tuning capacity required is virtually the same as the valve capacitance. So we need to choose a higher loaded Q. We will select the table for Q_L = 14 and try again.

C1 = 636.6pF
C2 = 4287.35pF
L = 43.39μH

C1 at 52MHz is now 636/52 = 12.2pF which when we subtract the anode capacity requires an anode tuning capacitor of ≈2.2pF. Although a small value this can be made to work if a low residual value capacitor is available. We will assume that a capacitor can be obtained or fashioned from a homemade disc variable. Now we can calculate the values for C1, C2 and L for the lowest frequency of 50MHz:

C1 = 636/50 = 12.72pF (actual value = 12.72 - 10 = 2.72pF)
C2 = 4287.35/50 = 85.7pF
L = 43.39/50 = 0.86μH

If we had tried to use a loaded Q of 10 the value of C1 required would be only ≈8pF and hence the anode tuning capacitor would need to be -2pF, which is an impossible value. One way to overcome this is to dispense with C1 altogether and use the anode capacity to resonate the network. A way of achieving this is to make the inductor variable in value by squeezing the turns tighter together to increase the inductance. A design such as this appears in the *ARRL Handbook*, using a leadscrew to alter the turns spacing of a copper tube wound inductor.

For higher loaded Qs the circuit values are approx as follows:

Q_L = 15
C1 = 3.64pF (actual)
C2 = 95pF
L = 0.81μH

Q_L = 16
C1 = 4.5pF (actual)
C2 = 104pF
L = 0.76μH

Q_L = 17
C1 = 5.5pF (actual)
C2 = 112pF
L = 0.72μH

Q_L = 18
C1 = 6.4pF (actual)
C2 = 121pF
L = 0.68μH

At 50MHz it is preferable to increase the loaded Q to attain better harmonic rejection to the signal appearing at 100MHz, which falls in the FM broadcast band. It also makes the anode tuning capacitor a more practical value.

From this exercise we can see that the valve anode capacity has a very distinct bearing on whether we can obtain resonance. Often the loaded Q needs to be increased beyond the normally accepted values to make a practical VHF amplifier. If a valve with a lower anode capacity is selected the design is a lot easier.

FORMULAS TO CALCULATE ANODE LOAD

Over the years there have been many different formula to calculate the anode load resistance. Many of these make assumptions that may not be valid. For many years the *ARRL Handbook* has used one formula that does not stand up to closer examination. Similarly, the *RSGB Handbook* has also carried a formula, since about 1961, which also has gross errors.

The main problem is calculating the actual peak RF voltage swing at the anode and deriving the rms voltage. In a Class-C amplifier the anode RF voltage is grossly distorted, in a Class-B amplifier not as much and in a Class-A amplifier hardly at all. The various formulas attempt to get around these problems with 'fudge factors', none of which actually give the correct result. Although the errors may be quite small they are still evident when constructing a VHF amplifier. At HF the errors are not as obvious because the anode network normally has plenty of range to counter small errors.

The various formulas are given below and we will comment on them in due course:

$$R_A = \frac{(0.57 \times V_a)^2}{P_o} \quad (1)$$

$$R_A = 464 \times \frac{V_a}{I_a} \quad (2)$$

$$R_A = \frac{V_a}{1.3 \times I_a} \quad (3) \quad \text{Class-A operation}$$

$$R_A = \frac{V_a}{1.57 \times I_a} \quad (4) \quad \text{Class-B operation}$$

$$R_A = \frac{V_a}{2 \times I_a} \quad (5) \quad \text{Class-C operation}$$

where:
R_A is the effective anode load resistance in ohms
V_a is the anode DC supply voltage in volts
I_a is the indicated anode DC current in amps - note in equation 2 the value is used as mA.
P_o is the RF power output in watts.

Equations 1 and 2 are from the *RSGB Handbook*, 3 to 5 are from the *ARRL Handbook*. None of these give the correct value for anode load. The reason we can say this with certainty is that Eimac and others in their valve data give the anode load values for different operating conditions and different anode voltages. Using any of the formulas above always returns an error, in some cases a large error. Let us look more closely at the equations.

Equation 1 appeared in the *RSGB Handbook* at least as early 1961, the oldest copy I possess; it also appears in later copies. In the text the assumption is made that the anode voltage will only swing over a range of 80% of the anode supply voltage. This is a reasonable assumption because a valve cannot operate unless it has a certain anode-cathode voltage. Hence, the anode voltage cannot swing too close to ground.

The other assumption made is that the anode efficiency is 70%. Hence, it is only suitable for valves operated in Class-C where 60 to 75% efficiency is often the norm, but for linear amplifiers where the efficiency is often a lot less this formula falls down.

However, if we consider the valve as a 'black box' and insert an ammeter in series with the high voltage supply we get a different answer. Suppose the anode supply is 1kV and the valve is drawing 250mA. Using Ohm's law we can calculate what resistor value is required to draw 250mA when fed with 1kV. It is 4kΩ. This is the static condition and the valve behaves as a variable resistance as we change the grid-cathode voltage. However, under dynamic conditions this no longer applies. To understand the difference we will look at a typical valve and run through the calculations to find the *effective anode load resistance*.

If we examine the Eimac data for the 3-500Z triode we can list the various parameters. With an anode voltage of 3kV and an indicated average anode current of 370mA it is drawing ≈1.1kW DC input. The RF output power is given as 750W under single tone operation and hence the efficiency is ≈67%. Applying the various formulas gives the following answers:

1. 3898Ω
2. 3762Ω
3. not used for the 3-500Z
4. 5164Ω
5. not used for the 3-500Z

Eimac gives the correct value of the anode load as 5kΩ under these operating conditions.

Equation 1 contains an error in the term 0.57. Because we are only using 80% of the anode voltage swing this factor should be $(0.8)^2$, which is 0.64. If we redo the calculation using this new constant we get closer to the correct result, 4915Ω. It the constant is raised slightly to 0.645 we obtain a more correct result of 4992Ω. However, as we will see later the anode voltage swing of the 3-500Z is somewhat more than 80% in practice.

Similarly if we use equation 2 we get similar errors. This gives a result that is in error by approx 25%. The *ARRL Handbook* in the same time period carried the same formula but the constant used was 500. A better approximation is about 620. This gives the result as 5027Ω.

The only formula that gives an answer close to the correct one is equation 4. But this formula only applies to valves operated in true Class-B and as one valve manufacturer pointed out in the 1970s hardly any amplifier is actually operated in Class-B because we obtain better IMD performance and gain in either Class-AB1 or AB2. The constant 1.57 if increased to 1.62 gives the correct result for the 3-500Z data.

Similarly when the valve is a tetrode or pentode we need to consider the screen voltage. The anode voltage cannot safely swing down to much below the screen potential because this causes excessive screen grid current to flow and this limits the anode voltage swing to even less than a triode. If the screen grid voltage is, say, +750V then the lowest we can safely swing is about +700V. When the anode voltage is less than the screen grid potential the screen grid tries to take over the anode's role and it draws a high current.

If we go back to our black-box valve and use the anode voltage and current of 1kV and 250mA under DC conditions the valve behaves as 4kΩ. But using equation 2 with the corrections it behaves like a resistance of 2200Ω. If we assume this is a Class-C amplifier delivering an efficiency of 66.66% then the DC input power is 250W and hence we can expect 166.66W output. Using equation 1 corrected we obtain a value of 2496Ω and using equation 5 we obtain 2kΩ. So we have a spread of about 20% over the results and it is probably acceptable to pick a value somewhere in the middle as being close to correct, in this case about 2250Ω. Hence, for Class-C operation equation 2 gives the closest answer to what is probably the correct one.

So we can conclude by this exercise that there

must be an awful lot of amplifiers built over the years with incorrect equations and the performance could be improved by changing the tank network values. Not only are the tank network components incorrectly calculated, but the anode load is also in error. Because the correct anode load value is critical to establish the value of C1 with the chosen loaded Q all the other values hinge on this being correct in the first place. The fact that the amplifiers work is probably due to having more tuning range in the tank variable capacitors than the formulae actually requires, but they are not attaining the loaded Q the constructors believe they are. In many cases the constructor will have had to alter the tap points on the tank inductor to get close to acceptable efficiency figures and required power output. The component that normally needs altering is the inductor in the pi tank network; C1 and C2 normally have sufficient tuning range to compensate.

Eimac, in their document *The Care and Feeding of Power Grid Tubes*, give the definitive formula. The resonant anode load is equal to "the anode peak RF voltage divided by the anode peak RF current". The problem is to ascertain both of these values with any certainty. Similarly Eimac recommends that the value of the RFC reactance feeding the anode supply needs to be approx 100 times larger than the value of the series inductor reactance in the Pi tank network. In many Pi tank networks this criteria is not met and consequently substantial RF current flows to ground in the RFC causing a large power loss and degradation in efficiency.

OBTAINING THE ANODE LOAD VALUE FROM FIRST PRINCIPLES

Very often no value for anode load is given for a particular valve when operated under different conditions than the manufacturers test circuit. Sometimes we use a higher or lower anode DC supply voltage or run the valve at less than its full rated power. In order to derive the effective anode load resistance we need to go back to first principles. Assuming the signal at the anode is a quasi-sinewave then we can apply standard formulas to derive the effective load resistance.

If the anode voltage is swinging about an imaginary point on the anode voltage scale we can derive the average anode voltage. Assume the anode DC supply is 1kV. Hence, the anode can only swing down from this value to some lower voltage. Let us assume that when the valve is fully driven it swings down to 250V. This is a peak voltage swing of (1000 - 250) = 750V. Hence, the anode voltage can be visualised as sitting somewhere between 1kV and 250V at an imaginary DC value of half the difference, at 500V DC. Hence, it appears to be swinging about 500V DC up to 1kV and down to 250V, just like a Class-A amplifier would be if the anode load were a fixed value resistor. The total signal voltage is still 750V pk.

Since the anode is connected via the DC blocking capacitor to the pi tank network input this only sees the AC variation. To convert to rms we divide the peak value by 1.414; or multiply by 0.707, which gives a value of 530V rms.

The valve anode has some capacitance to ground. When the pi tank network is adjusted for maximum power transfer the reactance of this capacitance disappears. But the actual capacitance does not disappear, it is still sitting there across the anode to ground and because a varying voltage is across the anode to ground this capacitor has a RF current flowing in it. The value of the current is determined by the reactance at the operating frequency. Similarly the inductor used to negate the anode capacitance reactance also has an identical current flowing, but this one is 180° out of phase with the capacitor current. As far as the valve anode is concerned the two currents being equal but opposite phase cancel out and no net current flows. But the capacitor current is always there. Part of the pi tank inductance is used to cancel the anode capacitive reactance and the remainder is used for the impedance transformation.

The standard pi tank formulas assume both the source and load are pure resistances. In a valve amplifier this is never the case, except when the antenna looks like a pure resistance with a VSWR of 1:1, which only a dummy load gives. In practice no such antenna exists, the antenna will always have some reactive component and resistive portion that varies with frequency. At one frequency it may be capacitive and at another it may be inductive. If the antenna presents an inductive load to the pi tank network we correct this be adding a little more capacitive reactance across the output terminals. In simple terms we change the loading capacitor value slightly. To add more capacitive reactance we need to lower the loading capacitor value. The same argument holds when the antenna is capacitive, we also change the loading capacitor value to a slightly greater value to compensate.

Now consider the case when we add the anode tuning capacitor across the input to the Pi tank network. This also has some reactance and hence RF current must be flowing to ground via this capacitor. Both of these capacitors need to be charged and discharged on each half cycle of the RF voltage waveform, this absorbs power. If the anode voltage swing is high and the capacitance also high this is a large RF current flowing.

Fortunately in a pure reactance no power is dissipated because the voltage and current are 90° out of phase. In a practical circuit this is not the case and some power will be dissipated because the component has an associated loss resistance, which lowers the Q, and hence power can be dissipated. Normally with high Q air variables this factor is small and can normally be ignored. The inductor is normally the component that has the lowest Q and hence the greatest loss resistance. For high power amplifiers the inductor Q needs to be as high as possible.

The anode tuning capacitor is normally on the *cold side* of the DC blocking capacitor so it only experiences the RF voltage. The reactance of the DC blocking capacitor is normally so low that we can ignore its effect on the network. So the RF voltage we see when looking across the anode tuning capacitor is the same value as that appearing at the anode. As we have chosen a value of 530V rms we can calculate the RF current flowing to ground in the anode tuning capacitor if we know its reactance. This capacitor in a HF amplifier is normally many times the anode capacitance so it tends to swamp the affect the anode capacitance has. The peak voltage appearing across the anode tuning capacitance is given by the formula:

$$V = \sqrt{(2P \times R_A)}$$

where
V is the peak RF voltage in Volts
P is the power output in Watts.
R_A is the anode load impedance.

Assuming the pi tank network has negligible loss, which is normally the case when high Q components are used, then the anode load can be calculated from the measured power output at the far end of the pi tank network with good accuracy, if we know the transformation ratio.

The formula can be used to ascertain the peak RF voltage occurring at the anode for a particular anode load. For the Eimac 3-500Z operating with 3kV anode voltage the anode load is given as 5kΩ, hence the effective peak RF voltage at 750W output is 2738V which is equal to ≈91% of the DC anode supply. Hence, the anode swings from 3kV down to 262V and this is a rms value of 1936V.

$$\text{Since } R = \frac{V^2}{P}$$

Note that here we are now using P and not 2P as in the previous equation because now we are using the rms voltage and not the peak voltage. This is a common mistake made by some amateurs and it gives a large error if incorrectly used.

We know P by measuring the power with an accurate power meter into a dummy load, so then R can be ascertained. With a 1936V rms RF signal and a power output of 750W, R = 4997Ω, which agrees with the Eimac figure of 5kΩ. The Eimac figure is probably rounded up for convenience to 5kΩ, the value of 4997Ω represents an error of 0.06%.

> What isn't obvious from this is the effect the unloaded Q has on the anode RF voltage. When tuning up a pi tank network we normally begin by setting the loading capacitor to be fully meshed, that is maximum capacity. Then we apply a little drive power and swing the anode tuning capacitor until we find the largest dip in the anode current. Under this condition little RF power is being fed to the antenna and the anode load resistance is much higher than normal, because the valve is lightly loaded. Under this condition the peak RF voltage on the anode is a maximum. By examining the formula above we can see that as R is high and P is low then the anode RF voltage must be higher than normal. If the voltage rating of the anode tuning capacitor is insufficient it will arc over.

The other standard formula, often used for solid-state amplifiers, is:

$$R_L = \frac{(V_{CC} - V_{sat})^2}{2P}$$

Assuming the saturated anode voltage is 262V and the supply is 3kV this gives a value for R_L of:

$$R_L = \frac{(3000 - 262)^2}{1500} = 4997\Omega$$

which also agrees with the 5kΩ value given by Eimac.

But we also need to look at the DC conditions when operated at 750W output. The data sheet gives an anode supply voltage of 3kV and an average anode current of 370mA. Using Ohm's law we can calculate the DC resistance the valve presents to the power supply.

Since R = V / I,
R = 3000 / 0.37 = 8108Ω.

So the power supply regards the valve as a fixed resistive load of ≈8.1kΩ but the anode appears as a resistance of 5kΩ to the pi tank network. How do we explain this?

The answer is multi-facetted. Part of the reason the anode looks like a lower impedance is because there is some reactance present in the anode capacitance and also the anode tuning capacitance. If we take the operating frequency to be 2MHz, which is the frequency used by Eimac to extract the results from the test amplifier, we can calculate the reactance of the anode capacity. The data sheet gives an anode capacitance of 4.1pF, which is pretty insignificant. At 2MHz this is a reactance of approx 19.4kΩ so it isn't having any significant affect across the 5kΩ anode load.

The real reason is that the indicated anode current isn't in fact a fixed value, nor does the anode voltage swing all the way down to ground. The anode current is also swinging around an average value, in this case 370mA, which the meter indicates. On one half of the input signal waveform the anode current swings up to maximum, this corresponds to when the anode voltage falls to a minimum. On the other half cycle the anode current swings down towards zero current and this coincides with when the anode voltage approaches the maximum. So the anode current meter only tells us the average anode current flowing and not the peak values. (The average current of a true sinusoidal waveform is 2/π = 0.6366.) The minimum anode current will be determined by the value of quiescent anode current when no signal is applied. In the Eimac data sheet this is given as 160mA.

So the anode is swinging between 160mA and some higher value with an average value of 370mA. Hence, the total anode current swing is 2 x (370 - 160) = 420mA as a rough approximation. This is a rms value of 297mA approx. Now if we use the rms value for the anode voltage of 1936V and the rms anode current we get an effective load resistance of:

R = V / I   1936 / 0.297 = 6518Ω.

The reason this is higher than the actual value of 5kΩ is because we made the assumption that the anode current swung an equal amount either side of the average indicated current. In a Class-B amplifier the upward swing of the anode current is greater than the downward swing, so the actual true average current is shifted up a little. Because this is not a true sinewave the meter takes an average of the distorted waveform. If we rearrange the formula we can find the effective anode current rms value.

Since I = V / R and we know R = 5kΩ and V rms = 1936V, then I rms = 0.387A. This is a value 30% larger than the previously calculated value assuming the anode current swung an equal amount about the average indicated current. In fact the anode current does not swing down to the idle value and no more. Because grid current is flowing the anode current gets driven below the quiescent idle current value by some amount.

A very rough approximation is to take the indicated anode current value under single tone drive conditions and to double it to find the peak value. If the anode current were a true sinewave this is a valid assumption. So let us assume the anode current is a true sinewave and see what error this gives.

Since I_a indicated is 370mA and I_q is 160mA then peak anode current is (2 x 370mA) = 740mA. The effective anode current is now (740 - 160) = 580mA peak and the average value is 580 x 0.637 = 369mA. This gives an anode load value of 5240Ω, which is fairly close to the real value of 5kΩ. Clearly there is a degree of fudge factor still involved but the answer is probably close enough to design with. This represents an error of about 5%, which is tolerable.

Let us explore the anode load resistance more. We know that the anode voltage swings about a nominal value by a certain amount. When the anode current is a maximum the anode voltage is a minimum. Let us take the values we obtained earlier and calculate the apparent anode load resistance. When the anode current swings up to maximum the peak current is 370 + (387 / 2) = 565mA approx. On the anode current minimum crest it swings down to 370 - (387 / 2) = 176mA approx.

Now using Ohm's law we can find the effective anode load under these two conditions. For the maximum anode current swing the anode voltage is 262V and the value of R is:

$$R = 262 / 0.565 = 463\Omega$$

When the anode current swings down to the minimum condition the anode voltage swings up to almost 3kV, let us assume it gets to 2.9kV, then R is:

$$R = 2900 / 0.176 = 16.48\text{k}\Omega.$$

Neither of these is close to 5kΩ but if we take the average value it works out to be ≈8kΩ, which is what the anode voltage and average current gives us.

There is unfortunately only one accurate method of determining the actual anode effective load resistance and that is to build a test amplifier. This is how Eimac and many other valve manufacturers do it. Having built the amplifier the valve is loaded up with an empirically designed or estimated values for the pi tank network and adjusted for resonance to obtain some data points, anode voltage, anode current and power output. Then the DC supplies are switched off and the valve heaters left powered up, so we get a true value for the anode capacity.

If the anode capacity is very low, such as the 3-500Z valve, we can switch the heater off and assume the anode capacity remains constant. This is a valid assumption because Eimac and many others measure the capacity of a cold valve at a very low frequency, typically 100kHz, in a shielded test fixture to get an accurate measurement.

Next a non-inductive resistor with very short leads is connected between the anode and ground with a value calculated from the various formulae to be the approx anode load resistance. A Vector Network Analyser (VNA) then 'looks into' the 50Ω output back towards the valve as a '1-port network'. The anode tuning and loading capacitors are left at the settings for best power output and the VNA measures the impedance appearing at the output connector. If the anode load resistor selected is correct then the amplifier looks like a perfect match to 50Ω at that frequency. If it doesn't, the VNA can extract the correct value to use by measuring the magnitude of the reflected signal and the phase angle. When the resistor value is exactly the same as the anode produced when powered up we have the correct value of the anode load.

This method takes into account all the stray capacitive and inductive components connected to the anode.

To measure the anode voltage swing when the valve manufacturers build a test amplifier they include a voltage divider made from high value non-inductive resistors. If we connect a 900kΩ resistor to the anode and a 100kΩ resistor in series between the anode resistor and ground we can connect a cathode ray oscilloscope or a rms reading RF broadband sampling voltmeter across the bottom 100kΩ. With the potential divider values chosen the signal is one-tenth that appearing at the anode. (A convenient method is to use 10 100kΩ/2W non-inductive resistors [carbon composition] connected in series with very short leads. A normal 2W resistor is only rated at about 500V DC so we need to series quite a few to get a safe working condition.) The DC and AC conditions can then be measured without drive and with different drive levels. From this we can ascertain the peak to peak or rms RF voltage occurring.

If you make this measurement you will be surprised to find that the anode voltage as well as swinging down to some low value also swings above the anode DC voltage by some amount. In the unloaded condition this can be as much as the anode DC supply voltage, so driving an amplifier to maximum anode current when it is unloaded will probably cause the anode tuning capacitor to arc over and should be avoided.

Eimac give the following diagram (**Fig 13.2**) in their document for the anode voltage swing at resonance, which shows the anode RF voltage swings as much above the DC supply as it swings below it.

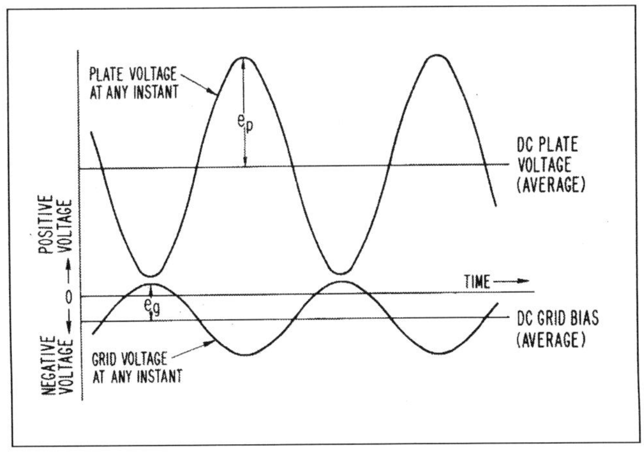

Fig 13.2. Anode RF voltage swing (courtesy of Eimac)

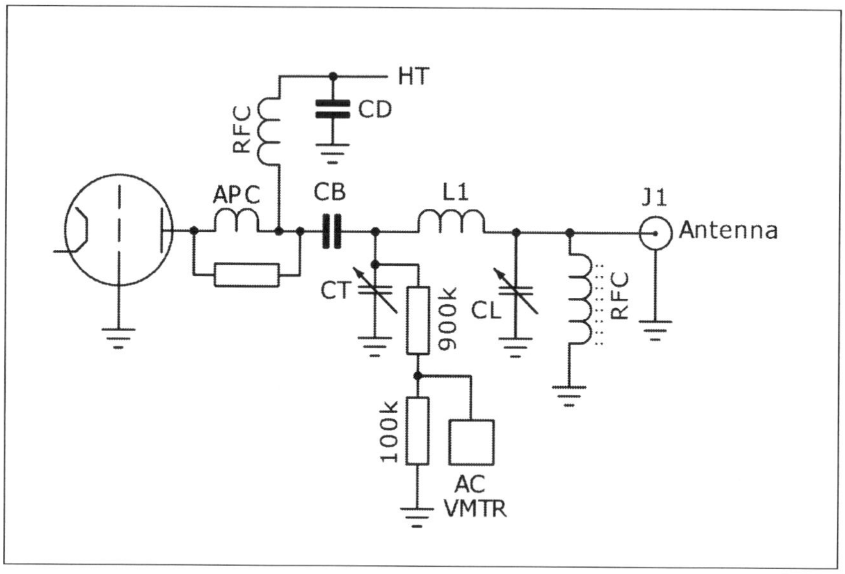

Fig 13.3. Method of measuring anode RF voltage

transformation ratio of 100:1 for the pi network when 50Ω antennas are used. This is effectively what happens when we use the amplifier for SSB. At very small drive levels the anode load resistance approaches infinity for the correct transformation match to the 50Ω load. The extra anode dissipation is not normally a problem because the valve chosen will be capable of dissipating far more power when operated at full output than when the drive is reduced. So no damaging effects are normally encountered at lower output levels because the anode has excess dissipation available.

If the pi tank network has insufficient tuning range then we cannot get the correct power output and hence the efficiency suffers. This translates into more power being dissipated in the anode than would occur if the matching were correct.

Similarly anode current can be measured with a current probe or by inserting a low value resistor in the cathode ground return to the power supply. This gives us the rms or peak anode current flowing. Knowing these parameters allows us to calculate the anode load with more accuracy using simple formula like Ohm's law. (Often the RF voltage measurements are made with 100:1 dividers fitted to RF probes with rms responding instruments. These are connected across the anode tuning capacitor and the loading capacitor to establish the true rms voltages present.)

EFFECT OF REDUCING DRIVE POWER

To control the RF power output we often simply vary the drive level to suit the operating conditions. When we need more power output we increase the drive level, which gives more anode current. But what happens to the anode load when we change the drive level?

Consider the case for an amplifier capable of 2kW DC input with a fixed anode voltage of 2800V. Now if we calculate the anode load resistance for the two cases we see the large change in impedance.

At 2kW input the anode load resistance is approx 2500Ω and the pi tank network would be designed for this value. At 1kW input the anode load resistance is now 5000Ω. If the amplifier were tuned up at the 2kW input level and then the drive reduced to 1kW input we have a large mismatch to the network because the transformation ratio required has doubled. For operational modes such as RTTY where the amplifier cannot safely be run at the 2kW input level we have to reduce the input power. Often if the input is reduced to about 50% the amplifier can be run safely at full carrier for short periods. So if the amplifier is needed to be tuned up at this lower power the anode tuning network is very different and it may be that insufficient tuning range exists in the variable capacitors. In many cases it is possible to find some setting during tune up where the power output is acceptable, but the efficiency is poor because we do not have the optimum value for the inductor.

If we now reduce the drive more to obtain a DC input power of 500W the anode load resistance doubles again to 10kΩ. As we have seen this beyond the normal

If the tables listed earlier are examined we see a serious problem with the inductor value. Let us take the anode load value of 2500Ω at 1MHz for a loaded Q of 12, this gives a value for L of 36.71µH When the anode load changes to 5000Ω to maintain our desired loaded Q of 12 L needs to be changed to 69.5µH This is almost a 2:1 variation in the inductor value. Because the inductor in the tank network is often a fixed value for each band this totally destroys the loaded Q. If the tables are studied you will see that as the inductor value goes down the loaded Q goes up. For the case where the inductor is half the required value the actual loaded Q is approx doubled to 24 in our example. The only way to correct this situation is to make the inductor variable and set it to the correct value for different power levels. This makes the design of the tank network difficult unless a roller inductor ('roller coaster') is used. A method used by one commercial amplifier manufacturer is to use extra tapping points on the main inductor, this allows the operator to select the correct inductance value for the input power being used.

The other thing that occurs is the value of C1 and C2 need to be larger as the anode load falls. In most HF amplifiers the anode tuning and loading capacitors have a maximum value for each band. If the capacitors cannot be increased to the correct value the network ceases to be resonant and extra power is dissipated in the anode because of the mismatch.

To illustrate this problem here are the correct values for the different anode load values at 1MHz. We will maintain a loaded Q of 12 for each power output level.

P = 2kW R = 2500Ω
C1 = 763pF
C2 = 4387pF
L = 36.71µH

P = 1kW R = 5000Ω
C1 = 382pF
C2 = 2135pF
L = 69.53µH

A simple way of getting around this problem is to alter the anode DC supply voltage. If the anode supply is reduced to 50% when the 1kW operation is required we need to double the anode current to obtain 1kW input. This lowers the anode load resistance to ≈2500Ω and so the same pi network values can be used without any problem. It is rather surprising that hardly any commercial amplifier manufacturer uses this easy solution. As the vast majority use a voltage doubler type of power supply this could be achieved very easily by switching the rectifier from a doubler to a bridge configuration which only requires one high voltage contact on a suitable relay.

MATCHING TO A GROUNDED GRID RF POWER AMPLIFIER CATHODE

When a valve is operated in grounded grid the cathode becomes the input terminal. This presents a much lower impedance than a grid driven valve. The exact impedance value is highly dependent on the value of grid current flowing. Eimac in the data sheet for the 3-500Z grounded grid zero biased triode gives some values for HF operation at 2MHz. With an anode voltage of 3kV and a drive power of 30W the grid resistance is 115Ω. We have to consider the input capacitance to calculate the exact impedance, for the 3-500Z this is given as 7.4pF. Hence, at HF this is relatively insignificant and can usually be ignored.

The driver stage at 50Ω source impedance needs to be matched to the cathode with low loss. Because it is inconvenient to also have to retune the input network as we move about the band this is often a fixed tuned network with sufficient bandwidth to cover the whole band with low VSWR. Therefore we need to use a low value of loaded Q. Most handbooks recommend a Q of less than 6. However, Eimac in the application note for the 3-500Z recommend a loaded Q of at least 5 and preferably between 12 and 15 for best intermodulation performance. The higher the loaded Q the more like a true sinewave the driving signal will be.

Q_L can be expressed as the bandwidth at the -3dB points in a band-pass filter, which is what the 3-pole pi input network is in effect ,although at first sight it appears to be a low-pass filter. If the band-pass filter is wider than the band to be covered, the input VSWR and losses will be low. For an amplifier input network at 3.5MHz the bandwidth needs to be at least 300kHz to cover the UK allocation of 3.5 to 3.8MHz. For the American allocation of 3.5 to 4.0MHz it needs to be 500kHz wide. If the network is centred at mid-point of the band this is ±150kHz for the UK and ±250kHz for the US allocation.

$$Q_L = \frac{f}{BW} \quad \text{and} \quad BW = \frac{f}{Q_L}$$

Hence, for a bandwidth of 300kHz at 3.65MHz the value is 12.16. A loaded Q of less than this will therefore satisfy the requirements. Because the bandwidth is measured at the -3dB points we need to increase the value to allow full drive power at the band edges, for 3.5MHz a value of 450kHz would be about correct for the UK allocation and approx 700kHz for the US allocation. Hence, for the UK allocation we choose a Q of 3.65/0.45 = 8.11 and a value of 8 would be suitable. For the US allocation the value would about 3.75/0.7 = 5.35 and a value of 5 would be appropriate. This is lower than the value recommended by Eimac although it will probably be adequate.

The formulas to calculate the circuit components were given earlier and can be used.

For a single 3-500Z the approx circuit values are:

1.8MHz	C1 = 6143pF	C2 = 9232pF	L = 2.08μH
3.5MHz	C1 = 3155pF	C2 = 4748pF	L = 1.06μH
7.0MHz	C1 = 1574pF	C2 = 2374pF	L = 0.53μH

Where C1 is the capacitor connected from the cathode to ground and C2 is across the 50Ω input. The other bands can be found by dividing the values by the ratio of the new frequency versus the old frequency. Hence, 14MHz will use values half of that of 7MHz etc.

The problem with the pi input network is that the value of the inductor becomes very small for the higher bands and a better option is the L-match or T-match for these.

If two 3-500Z valves are used in parallel, the cathode impedance falls to about 57Ω and as this is very close to the 50Ω driver source impedance many amateurs dispense with the input network and simply feed the cathodes via a DC blocking capacitor. This works to some extent but the 'flywheel effect' of the resonant network is then lost and higher intermodulation distortion occurs.

One of the things not apparent from this is the way the cathode reactance changes with increasing frequency. At high frequency the inductive parameters dominate and it appears as a net inductive component. We can obtain an insight of how this modifies the cathode impedance by examining the input network used by the Kenwood TL-922 HF amplifier which has two 3-500Z triodes connected in parallel. For the bands from 1.8MHz to 14MHz the input network is a pi type. On 21MHz and 28MHz it changes to an L-match with the shunt capacitor on the 50Ω end. Because in an L-match network the shunt capacitor is always across the higher impedance end we can see that now the network is working as a step-down transformer and hence the cathodes are a lower impedance than 50Ω.

Modeling the input networks using analytical software we can determine the approx cathode impedance for the different bands. Although the inductor values are not given on the schematic the shunt capacitors are and by iterating through values we can find the best input match to 50Ω. From this we can then extract the relevant circuit values.

For example, the capacitor values for 1.8MHz are the same at both ends of the pi network, hence we can see that the cathode is the same value as the source impedance of 50Ω. This agrees with the values discussed earlier and the two valves in parallel approximate a 50Ω load. The same condition also occurs for 3.5MHz but for 7MHz and 14MHz the capacitor values are different. The 50Ω source capacitor is a lower reactance than the cathode reactance and hence it is acting as a step-up transformation on these bands. On 21MHz and 28MHz the input is now an L-match with a step-down function.

The approximate cathode impedance and series inductor were determined to be:

160m	Z = 52.2Ω	L = 4.27μH
80m	Z = 61.6Ω	L = 2.38μH
40m	Z = 108Ω	L = 1.56μH
20m	Z = 165Ω	L = 0.98μH
15m	Z = 25.6Ω	L = 0.22μH
10m	Z = 24.5Ω	L = 0.17μH

Note how the cathode impedance changes dramatically between 80m and 20m, reaching a peak at 20m, which is probably due to the valves' inductance and other stray inductance in circuit canceling the grid-cathode capacitance and going through a resonant point. In addition to the cathode-grid capacity of approx 15pF there are also other small value capacitors used for cathode RF voltage metering which add an additional 45pF in shunt with the cathode to ground. So in total the effective cathode capacitance is approx 60pF.

This analysis is of course academic because there are lots of other components connected to the cathode that can alter the impedance. The filament choke, which is the normal bi-filar wound type on a ferrite rod, and the power supply feed choke are both effectively in shunt with the cathodes. These values are given as 32μH and 80μH which when connected in parallel is an equivalent inductance of ≈24μH. If this value of inductance is connected in the analysis circuit we can see the effect it has. In fact the effect is negligible and it can be ignored.

Another factor we can ascertain from the extracted circuit values is the loaded Q used. This is very low, typically 1 to 3 across the range and it is a slight puzzle why these values were chosen.

LOADED Q REVISITED

The obsession with a low loaded Q for an anode network may get you into trouble. The often stated reason for choosing a lower loaded Q is given as the *network efficiency*. Note this is not the overall efficiency of the amplifier, it is simply the measure of power loss by the network transforming from the valve anode resistance to the antenna impedance. A linear amplifier using Class-AB1 will have at best an anode efficiency of approx 45%. If your amplifier gives higher efficiency figures then it is not being operated in a linear manner and certainly not as a Class-AB1 stage. In Class-AB2 the efficiency is theoretically only 50%, for Class-B it is theoretically only 66%. Note that it is very unlikely we can get close to the theoretical values with practical valves and components and still maintain good linearity.

The network efficiency concerns only the three components forming the pi circuit: C1, C2 and L.

The network efficiency is a measure of how much power is dissipated (lost as heat) in the network components. The air variables normally have a very high Q, typically well in excess of 1200 for good quality types. The component in the Pi network with the lowest Q is normally the series inductor. Any RFCs across the network for anode DC feed or safety reasons may dissipate additional power if they are lossy.

The formula to give the network efficiency η is:

$$\eta = 100 \times \left(1 - \frac{Q_L}{Q_U}\right) \%$$

where Q_L is the loaded Q and Q_U is the unloaded Q.

If we take a couple of examples we will see that the loaded Q isn't such an issue with a low loss inductor (high Q).

Assume the pi network inductor has a Q of 500, which is quite reasonable for an HF amplifier where an air-wound inductor is commonly used. Also assume that the network unloaded Q is approx 80% of the inductor Q. The network unloaded Q condition exists when the loading capacitor is fully meshed and the amplifier is supplying little power to the antenna, that is the anode is presenting a high value of load resistance. To tune up a pi network stage we swing the anode tuning capacitor C1 through its travel until we find the largest dip in the anode current as shown on the meter. As the loading capacitor C2 is unmeshed more power is fed to the antenna and the anode current dip at resonance is less. When the amplifier is loaded to maximum output the anode current dip is quite small, about 10% at most. In this condition the network is now showing its loaded Q value.

Let us choose a loaded Q value of 10 and the unloaded Q is 400 as already stated. For this condition the network efficiency is 97.5%. For every 100W being sent to the antenna the network is dissipating 2.5W. The vast majority of this is being dissipated in the inductor but a small amount will be also be dissipated in the two capacitors because they do not have infinite Q values. This is hardly anything to get excited about; the inductor can most probably dissipate far more without overheating. HF tank inductors are large and have a large surface area to dissipate the heat. If the anode compartment is the type with forced air cooling for the valves then some of the air is blown around the inductors to keep them cool. If you compare the physical size of the inductor to a 5W resistor the inductor has a far greater surface area and hence can dissipate a lot more power safely. In several published designs in the older *ARRL Handbooks* amplifiers running 1kW output have used pi tank inductors officially rated at 500W without any ill effects, so there is a quite a bit of leeway.

Now we select a loaded Q of 20, which most amateurs regard as being too high. Under this condition the efficiency is 95%. For every 100W sent to the antenna the network is dissipating 5W. We can convert this loss percentage into dB. A loss of 1dB is in round numbers 20%. So the 5% loss in the network amounts to 0.25dB. The average amateur station operating on HF with normal coax has more loss in the coax than the pi tank inductor.

Let us now select a better inductor for the pi tank. Assume it now has a Q of 750, which is quite realisable with the correct winding technique. If we also assume the unloaded Q is 80% of the inductor Q, i.e. 600, then the efficiency figure gets better. For a loaded Q of 10 the efficiency is now 98.33%. For a loaded Q of 20 it is 96.66%. So we can use high loaded Qs provided the series inductor Q is adequate. The myth about low loaded Q goes back to the days when people did not know how to wind inductors for lowest loss (highest Q). If the series inductor has a poor Q, the network losses are directly attributable to this and not the loaded Q factor. The same argument goes for the tuning capacitors, if these are unsuitable types and have poor Q they add to the network losses. Many HF amplifiers have padding capacitors for the lower bands because the anode tuning and loading capacitors have limited maximum values. Often these capacitors are unsuitable types and they dissipate a fair amount of power. Band switches also have some contact resistance and resistive losses in the various tapping wires used on the inductor. Although these DC resistive losses are quite small they all add as if they are connected in series and can reach appreciable values when the RF skin loss resistance is also considered. It must not be forgotten that the total resistive loss is the DC ohmic loss and the RF skin resistance loss added together.

Although the inductor may have a very high Q when measured out of circuit mounting in close proximity to shielded compartment walls does have a de-Q-ing effect. For optimum Q the metallic wall should be at least two coil diameters away from the inductor. In many amplifiers this is not achieved and the unloaded Q is not as high as possible due to space or volume limitations. Also the shield material has a large affect on the de-Q-ing, mild steel being far worse than aluminium.

The disadvantage of a low loaded Q is that the harmonic suppression is worse than a higher loaded Q network. Against this must be weighed the overall efficiency of the amplifier. If the valve is run in Class-AB1 and we require lower IMD products then we cannot expect any more than about 45% for the valve and then the small percentage loss in the output network becomes insignificant.

If a typical UHF cavity amplifier is considered the unloaded Q can reach values as high as 2000. In this type of amplifier loaded Q values of 25 or higher are very common. The low loss cavity has almost no resistive losses, especially if it made from a high conductance material such as copper and silver-plated to reduce the skin resistance losses. Cavity resonators used as antenna duplexers and filters have unloaded Q values of up to 5000 and the same can be achieved for valve cavity amplifiers if the construction technique is good.

Measurement on practical pi networks in HF amplifiers transforming over a large impedance ratio show that the actual insertion loss is very low. For a typical HF amplifier pi network for a 1kW amplifier with the correct choice of components and construction techniques the sort of loss we measure is about 0.1 to 0.2dB. This is borne out by valve manufacturers measurements on test amplifier circuits where the available anode power and useful load power are often given. At low frequency the losses are negligible if the components used have high Q. What does absorb and dissipate significant power are the other components in the anode network. If we list the various components in descending order of power dissipated we end up with a list that looks like this:

1	Valve(s)	≈ 55% power dissipated
2	Anode choke	≈ 10 - 15% power dissipated
3	Antenna safety choke	≈ 5% power dissipated
4	Pi tank inductor	≈ 2 - 5% power dissipated
5	Tuning capacitors	≈ 0.5 - 2% power dissipated

The power loss needs to be put into the correct perspective. If the amplifier is supplying 1kW to the load and 10% is dissipated in all the loss-making networks, that is 100W and the total attenuation is only 0.5dB.

Chapter 14

Water cooling for Gi7bt triodes

HERE is a blow-by-blow description of how to make a water-cooling block for the Gi7bt Russian triode valve. The pictures show a dual-valve block but a single or a triple valve block follows the same method. Water cooling blocks for other valves follow the same method, only the dimensions will vary.

NORMAL AIR COOLER

The Gi7b series is supplied with an air cooler that bolts to the anode. To remove this requires a peg-spanner. **Fig 17.1** shows the peg spanner and the air cooler removed.

MATERIAL REQUIRED

The block is a piece of 12 to 15mm thick aluminium approx. 100mm wide and 50mm long. The water pipe stubs are made from aluminium tubing approx 5mm internal bore and about 6.5 to 7.5mm outer diameter. I had a piece of 7.2mm OD thick walled tube that was left over from an antenna project.

STEP 1

Mark out the block for the valve centres and the water cooling passages carefully. Ensure the water passages will not be broken into when the top of the block is counter bored with the 15mm recess for the peg nut. Centre punch these lightly and spot with a pilot drill. Check the pilot drill is in the correct place. Mount the block in a machine vice so that it is vertical to the drill table. Do *not* try and hold it by hand, the cutting forces can whip the block out of your fingers and you may get severely cut by the rotating drill!

Set the drill to a low speed - about 300RPM is good - to obtain a clean round hole. Drill through the block using a drill smaller than the final size. For a finished size of 6.5mm start with a 4mm drill and then open up the holes in small increments, the second drill would be about 5mm and finally the 6.5mm drill. Ensure the drills are sharp and correctly ground. Use plenty of cutting fluid so the drill does not jam in the block. I use *Tapmatic™*, but methylated spirits, paraffin or kerosene are also good. If you have none of these then engine oil of SAE-30 is also suitable. The important thing is not to try to drill the holes dry; this will give a ragged finish. The water passage holes need to be drilled in easy stages, do not try and push the drill through in one go. Remove the drill every 10mm or so and clear away the swarf with a brush and apply more cutting oil to the drill flutes. This will give a clean, round hole that exits in the correct place on the other side of the block. De-burr the holes on each side of the block.

Note: A 4mm drill, unless it is a long series type, will probably be too short to go right through the block for the water passages. Hence, drill as far as possible until the flutes are full depth in the block. This will only leave about 10mm for the final size drill to clear. Now drill the holes for the anode studs by moving the block in the machine vice so that the block is parallel with the table. Start with a 4mm drill and open out to 6.5mm. Lightly countersink the holes on each side to remove any ragged edges. Sand the block top and bottom faces on a piece of 200 grit wet and dry paper on a flat surface to remove any burrs.

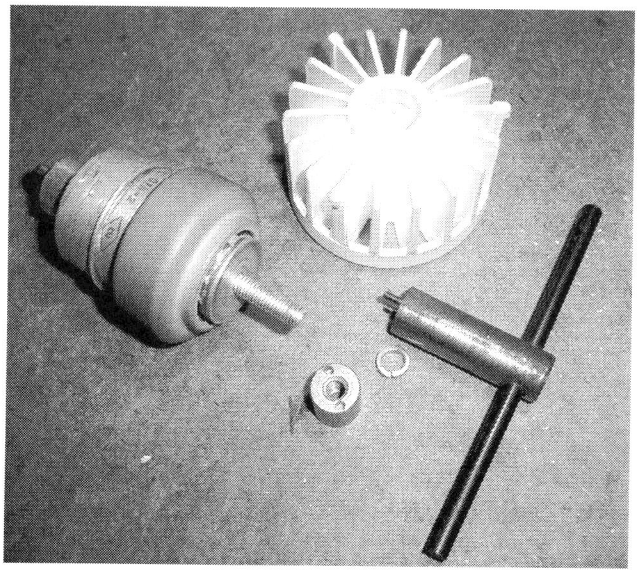

Fig 14.1. Air cooler and peg spanner

Fig 14.2. Drilling the water passages cross holes

Fig 14.3. Partially complete block with anode and water holes before the counter boring

The anode studs are too short to pass through a block 15mm thick and leave enough threads for the nut. The studs are made from a soft copper alloy, which will not tolerate over tightening. The peg nuts need a 15mm diameter counter bore to allow enough threads to be used. For a block 15mm thick this needs a counter bore of approx 5mm deep on the top face of the block. Carefully align the counter bore drill with the hole already drilled. If a milling machine is available this can be done with the table lead screws. If only a pillar drill is available, careful setting up will achieve equal results. I am fortunate to have a vertical milling machine as part of the arsenal.

Finally when all drilling and machining is complete de-burr all the holes to remove sharp edges and give the block a good wash in hot soapy water to remove any bits of swarf, cutting oil and finger marks. However, before you do so, read the next section about the water-cooling stubs to find out the correct size for the cross-holes.

Fig 14.4. Counter boring the peg nut holes

STEP 2

To make the water cooling stub pipes we need to cut the pieces of tubing square and straight. If possible part the tube off in a lathe. The exact length of the stubs is not too critical, 20 to 25mm would work fine. If a lathe is not available then a small hacksaw and a steady hand will work just as well. Alternatively a small pipe cutter will make a clean cut and square ends, although the tube will

Fig 14.5. Parting off a stub pipe using the tail stock drill as a length stop

tend to close up somewhat. This can be corrected by passing a drill through the tube the same as the internal diameter.

Chamfer both ends of the stub pipes by putting the stub into the drill chuck and applying a fine file. The end that goes into the block needs a longer chamfer if the intention is to fit this as a force fit. More of this later.

If a lathe is available it is a good idea to machine a small undercut with a parting tool about 2mm wide and 0.5mm deep to give additional purchase for the vinyl tubing. If only a pillar drill is available it is possible to do this with a thin flat needle file and a steady hand by spinning the piece of tube in the drill chuck and applying the file edge on to the tube.

Fig 14.6. Machining an undercut to assist the vinyl tube sealing. The drill chuck carries a drill the correct size for the tube ID to prevent the stub distorting under the parting tool pressure

STEP 3

We need to choose between whether we will press fit the stubs into the block or glue them in place with an epoxy resin. For most people a press fit is not too easy if the correct drill size is not available. In my case it so happened that I didn't have the correct drill to make the holes in the block suitable for a press fit as had been intended. As the work was being done on a Sunday when tool merchants aren't open it was decided to use two-part clear

Fig 14.7. A set of water pipe stubs ready for pressing into the block

epoxy resin (Araldite™ or similar).

For a press fit we need to make the hole for the stubs about 0.05 to 0.1mm smaller than the pipe OD. This extra diameter only needs to be about 5mm deep, so drill the holes at each side of the block to a suitable size. Then press in the stubs keeping them perpendicular to the faces. Using the drill press with a drill the same size as the internal diameter of the stubs ensures the stubs are perpendicular.

Note: Most jobbers' drills tend to cut slightly over size, especially in soft material like aluminium. So it safer to work up to the final size in small increment leaving the final drill to only remove about 0.2mm in diameter. If the hole is over size the stub will be too loose and then you will have to use two-part epoxy to fasten them in place.

Fig 14.8. Using a drill press and pilot drill to align the stub before pressing into the block

STEP 4

If using epoxy to fasten the stubs in place, we need to ensure the holes in the block are large enough for an adequate film of adhesive around the stub to make a proper seal. We also need to provide a deep chamfer to enable the epoxy to get a good hold of the two parts. For a two-part epoxy adhesive we need at least 0.1mm diameter larger than the stub pipes. This will then give secure sealing and good mechanical strength.

Fig 14.9. The stub pressed fully into the block

Thoroughly clean all the parts and abrade the ends of the stubs that will fit into the block with a coarse abrasive such as 60 grit to form a good key for the epoxy. Wash the parts in thinners or alcohol to remove finger grease and place to one side to dry. Mix up some two-part epoxy and smear each stub end with it. Place the stub into the block and rotate it a few times to spread the epoxy around the hole. Fit only one side at a time. Fit the stubs and place the block vertically and allow the epoxy to set thoroughly. If using a quick curing epoxy this can be as short as 10 minutes. When the one side is set, mix up some more epoxy, up end the block and do the second side. Although quick curing epoxy cures in about 10 minutes it only attains full strength after about 12 hours, so resist the temptation to handle the block until it has cured thoroughly. Placing the block in a warm place speeds up the curing, but not more than about +60°C should be used.

When both sides are thoroughly cured check if the water passages are clear. If epoxy resin has seeped it may block the stubs. Pass a drill through each stub the same size as the internal diameter to clear this out. Finally blow through with compressed air to remove any debris and wash again in hot soapy water and then dry.

Fig 14.10. A countersink cutter making a deep chamfer for the epoxy stubs. The block has been drilled to suit the epoxy fastening method with a drill 0.2mm larger than the stubs for a depth of 5mm

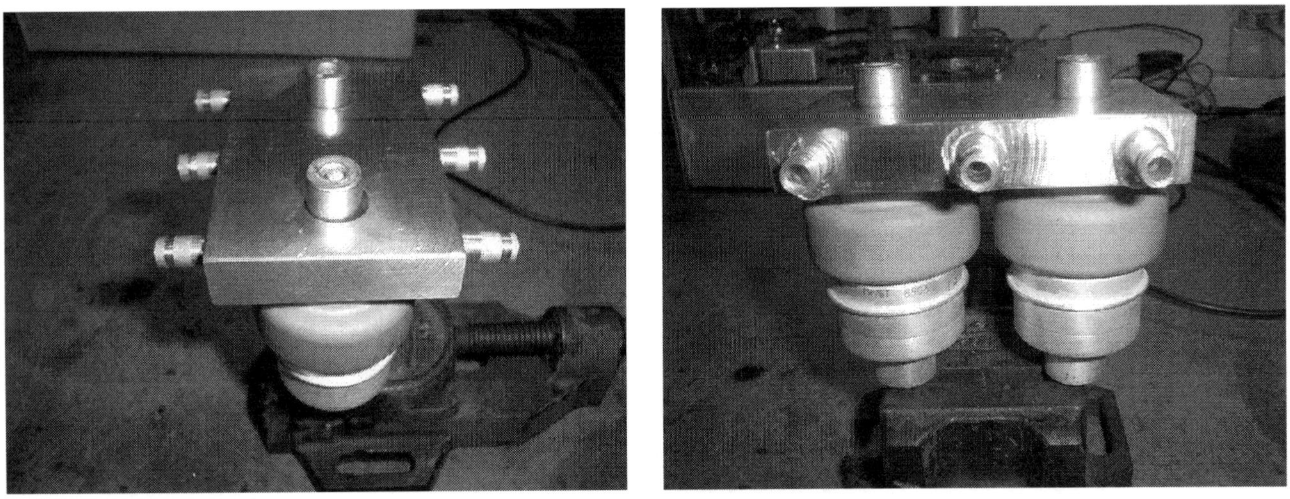

Fig 14.11 (a) and (b). Two views of the finished water cooling block, ready for a 1kW amplifier

ALTERNATIVE CONSTRUCTION METHOD

If you are fortunate to have access to a milling machine, an alternative method is to make the block in two pieces with an annular groove machined for the water to flow in. The basis of a design is shown in **Figs 14.12** and **14.13**.

The bottom portion of the block is made from a piece of aluminium 10mm thick. The two anode stud holes are drilled as for the first design. An annular groove 6mm wide and 6mm deep is machined around each anode stud hole and the water pipes enter through the side walls as before. The top is closed with a piece of 3mm thick aluminium plate with a gasket made from thin cork or paper gasket material. M3 screws tapped into the bottom piece serve to clamp the top plate to form a watertight seal.

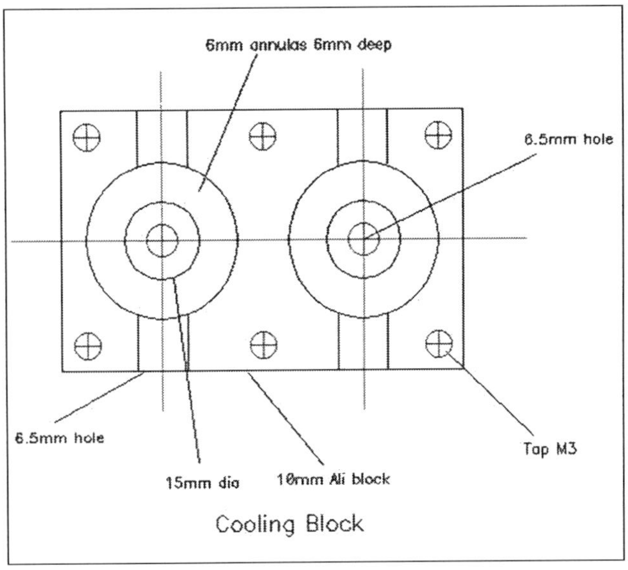

Fig 14.12. Bottom portion of cooling block

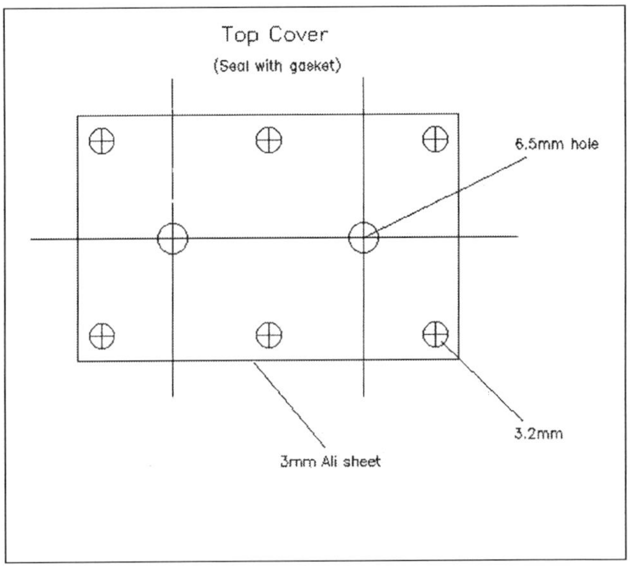

Fig 14.13. Cooling block top cover

Glossary

THE various terminologies and their abbreviations, where applicable in parenthesis, are shown here. Although not exhaustive, the most commonly used ones are included.

Absolute Zero The lowest temperature possible to attain. On the Kelvin scale this is 0K, this is the same as -273.2° Celsius. At absolute zero the random electron motion in atoms ceases and a conductor emits no electrical noise. This causes a normal conductor to exhibit zero resistance, a super-conductor.

Ampere (A) The unit to quantify the flow of electric current.

Amplifier A device that increases the amplitude of the input signal in either voltage, current and power.

Amplitude The 'size' of a particular item, for example voltage or current.

Amplitude Modulation (AM) A type of modulation method where a carrier is modulated in amplitude and two additional side bands are formed. See also DSB and SSB.

AM to PM Conversion The conversion of amplitude modulation into phase modulation. Normally caused by non-linearity in an amplifier. Frequency multipliers cause high levels of AM-PM conversion and this generates high levels of phase noise close to the carrier if the modulating signal is random noise. See also Phase Noise.

Anode The positive terminal of a valve or a semiconductor diode or other item.

Antenna A device that either radiates a radio frequency signal or receives an incoming signal from a distant transmitter.

Antenna Coupler A name given to a device which connects a transmitter to an antenna feedline, often with impedance matching and/or balanced to unbalanced networks to suit the two items. Also commonly known as an Antenna Tuning Unit (ATU).

Antenna Tuning Unit (ATU) See Antenna Coupler.

Attenuation The reduction in signal level due to a loss, often performed by special test equipment items known as attenuators. The unit is normally the dB. It can also be expressed as a power ratio or voltage ratio in a linear format. Radio waves travelling through space suffer attenuation, known as Free Space Attenuation or Path Loss Attenuation.

Attenuator A device designed to produce signal level reduction, the opposite of an amplifier.

Automatic Gain Control (AGC) A technique used in receivers to maintain a constant output level by reducing the gain of the amplifiers with increasing input signal level.

Band Pass Filter (BPF) A filter designed to pass only a limited portion of spectrum located at the centre frequency of the filter. See also LPF and HPF.

Bandwidth (BW) The portion of spectrum that a filter or other device is able to pass with minimum loss. Often the -3dB point on the response curve is taken as the criteria. See also HPBW.

Barkhausen Effect The inherent noise generated internally in an amplifier device which when fed back to the input causes oscillation to begin. Named after the German scientist Heinrich Georg Barkhausen. See also Shot Noise.

Bipolar In semiconductors, a type of transistor e.g. bipolar junction transistor (BJT).

Boltzman's Constant (K) 1.37×10^{-23} joules per degree Kelvin.

Broad Band Noise In amplifiers and oscillators, the level of noise power versus frequency. Often expressed in absolute power terms as dBm/Hz.

Carrier A radio frequency signal without any modulation, theoretically occupying zero bandwidth.

Cathode The opposite terminal to an anode in a diode. Also the ground return or common terminal in thermionic devices e.g. valves.

Capacitor A component possessing capacitance. See Farad. Circuit reference C. (Old terminology used the name Condenser, still used in automotive parlance).

Choke A common alternative name for an inductor. See also Radio Frequency Choke.

Clipping A deliberate or undesired reduction of signal amplitude to a constant level, often using diodes.

Compression The reduction in signal amplitude due to non-linear characteristics of an amplifier.

Coupler A special connector used to interface between two RF connectors. A 'Female Back-Back' coupler for example, allows two male plugs to be joined together in series. It hence consists of two female sockets connected together. Often available in both 'sexes' to suit the requirements. Also used (incorrectly) to describe a network to connect a transmitter to an antenna, often with impedance conversion contained, to match the different impedances in use. The correct name is Antenna Coupler.

Coupling Factor In transformers a measure of how well the two windings are able to transfer power, normally expressed in percentage terms. Also used In directional couplers to denote how much of the main line signal is coupled out to the measurement port, expressed in dB.

Cut Off Frequency In filters the frequency at which the attenuation attains some value, normally -3dB. Also used at microwave to denote the frequency where a wave-guide no longer allows signals to propagate along it.

Cycle A periodic voltage or current that describes the variation over a period of time e.g. cycles per second (c/s). A pendulum has a period of time to complete a cycle when swinging backwards and forwards.

Decibel (dB) The expression of gain or loss in a logarithmic format. The decibel is one-tenth of a Bel. Invented by the Bell Telephone Corporation to denote the signal loss occurring in telephone circuits. Today the Bel is rarely used.

dBc The relative amplitude of harmonics or distortion products referred to the carrier amplitude.

dBd A convention for expressing the 'gain' of an antenna in decibel form relative to a half-wave dipole for the same frequency. To convert to dBi add 2.15 to the dBd figure.

dBi A convention for expressing the 'gain' of an antenna in decibel form relative to a pure isotropic radiating source. To convert to dBd subtract 2.15 from the dBi value.

dBm A convention for expressing power in a logarithmic format with 1mW being the reference point. 1mW = 0dBm. Hence 10mW is written as +10dBm. A power of one-tenth of a mW is written as -10dBm etc. To convert to dBW subtract 30 from the dBm value. Hence, +30dBm = 0dBW = 1W.

dBW A convention for expressing power in a logarithmic form with 1W being the reference point. 1W = 0dBW, hence +10dBW = 10W, -10dBW = 100mW etc. To convert to dBm add 30 to the dBW value. Hence 0dBW = +30dBm = 1W.

Deviation The moving of the carrier frequency in a FM transmitter about the nominal carrier frequency by the modulating signal. Deviation is normally expressed in kHz plus and minus. For an amateur FM transmission this is limited to quite small amounts, typically ±3kHz maximum. Broadcast FM transmitters use up to ±75kHz.

Diode A two electrode device capable of rectifying an AC signal to convert it into a DC form.

Dipole In antennas, a radiating element consisting of two metallic rods, normally an electrical length of half a wavelength, hence known as a *half-wave dipole* ($\lambda/2$).

Directional Coupler An item of test equipment to obtain a sample of the signal flowing in a transmission line for measurement purposes. Sometimes (incorrectly) reduced to Coupler.

Directivity In antennas, a measure of how the field strength is confined into a particular direction (see also gain). Also used in directional couplers to denote the sensitivity to the forward and reflected signals, expressed in dB.

Distortion The linearity of an amplifier affects the distortion generated. In a perfectly linear amplifier the distortion would be zero. The qualification of distortion depends on the type of equipment. In audio amplifiers the measurement is expressed in percentage as Total Harmonic Distortion (THD). In RF amplifiers the distortion is more often expressed as the levels of spurious product, harmonics and side bands, developed relative to the amplitude of the carrier. In SSB transmitters the measurement is often reduced to the Inter-Modulation Distortion (IMD) products and are measured using a Spectrum Analyser in dBc.

Double Side Band (DSB) A modulation method similar to AM where the carrier is eliminated leaving just the LSB and USB products.

Dynamic Resistance (R_D) A feature of parallel resonant networks. At resonance the network behaves as a very high value resistance with no reactive components. The unit is Ohms (Ω). Forms the basis of band pass filters.

Electronic Counter-Measures (ECM) Either a passive or active means to cause disruption or confusion to an enemy force. During WWII aluminium strips dropped from aircraft called 'Window' or 'Chaff' were used to blind the German radar systems. Other techniques employ high power transmitters to obliterate enemy transmissions. See also Jammer.

Electronic Counter Counter-Measures (ECCM) The avoidance of jamming by electronic means. See also Jammer.

Electrode Either a part of a thermionic device such as a valve or a plate in a battery.

Electro-Motive Force (EMF) The voltage supplied by a signal generator when the output is open circuit. See also Potential Difference. EMF is also the voltage we measure across an open circuit cell or battery.

Electro-Magnetic Compatibility (EMC) The measure of how well different radio and other electronic systems can coexist without interfering with each other. See also RFI.

Farad (F) The unit of capacitance. Named after the British scientist Michael Faraday. Because the Farad is impracticably large, smaller units such as the (μF), nanofarad (nF) and picofarad (pF) are often used.

Field Effect Transistor (FET) A type of semiconductor transistor, often also known as a Junction Field Effect Transistor (JFET).

Frequency (f) The number of AC complete voltage peaks and troughs (cycles) occurring in a unit of time. The unit of frequency is the Hertz (Hz); originally cycles per second (c/s) was used. Named after the German scientist Heinrich Hertz. Because the Hertz is not suitable for most measurements we use multiples to define higher frequencies. Audible signals cover about 10Hz to 15kHz, radio signals cover from about 30kHz to megahertz and gigahertz. The full radio spectrum covers from about 30kHz to about 1000GHz.

Frequency Modulation (FM) A modulation method where the carrier frequency is varied about the nominal with a modulating signal, such as speech or music, to convey intelligence. Used for broadcasting, commercial radio and amateurs. FM can be classified into Wide Band (WBFM) or Narrow Band (NBFM).

Gain (Ga) The numeric increase of voltage, current or power due to a device e.g. transistor, valve etc operated as an amplifier. When expressed as a voltage or power ratio gain has no units. When expressed in a logarithmic form the gain is expressed in dB. Also used for antennas to express the directivity in a particular direction.

Gain-Bandwidth-Product (GBP) The numerical summation of gain in power ratio terms and frequency spanned. For example, an amplifier has a gain of 10 across a bandwidth of 10MHz; the GBP is hence 100MHz. If the amplifier bandwidth is reduced to 1MHz the GBP remains the same but the gain across the 1MHz spectrum is now 100.

Gain Compression See Compression.

Gain Slope The curve shown on a graph of output power versus input power for an amplifier. If the amplifier is linear, the curve will be a perfectly straight line. With non-linearity the slope deviates from the straight-line condition and curves towards an arc. Also used as an expression where the gain changes with frequency, in this case the normal units are dB/decade or dB/octave of bandwidth. Gain sloping is used to flatten the gain of an amplifier across a wide bandwidth.

Half Power Bandwidth (HPBW) The bandwidth at the -3dB points on a filter or a modulated signal. The HPBW is normally used to define how much spectrum a signal occupies.

Henry (H) The unit of inductance. Smaller units such as the milli-Henry (mH), micro-Henry (μH) and nano-Henry (nH) are used at higher frequency.

Heptode A 6-electrode valve often used for receiver frequency mixers.

High Pass Filter (HPF) A filter designed to only allow signals above a certain frequency to pass. See also BPF and LPF.

Impedance (Z) The complex termination presented when reactance and resistance are both present. The unit of measurement is the Ohm (Ω).

Independent Side Band (ISB) A technique used commercially for long distance HF transmissions to convey two separate messages, perhaps speech and data. One side band carries one message and the other side band carries the other. Hence, one transmitter can send two different messages simultaneously. An alternative scheme uses both side bands to convey the same message to combat selective fading over a long signal path. The receiver selects the best quality signal and can change very rapidly between side bands if fading occurs.

Inductor A component possessing inductance. See Henry. Circuit reference L.

Insulator A material with a high resistivity which passes insignificant current.

Insulation A covering of an insulating material, e.g. PVC, used to prevent short circuits on items such as wires etc.

Inter Modulation Distortion (IMD) Normally the odd order distortion products caused by non-linearity of the amplifier device. Third Order products are known as IMD^3 and fifth order products as IMD^5. See also Distortion.

Inter-Series Adapter A special RF connector that allows one type of RF series male or female connector to be mated to another series type. For example, PL-259 to BNC male, N-type female to SMA male etc.

Jammer A transmitter used to deliberately disrupt radio communication traffic. When such an item is used it is known as jamming. Extensively used for ECM (Electronic Counter Measures).

jansky (Jy) In radio astronomy the unit used to quantify the RF power output of a distant radio star or other object. Named after Karl Jansky, the first person to observe radio signals from outer space in 1930. The correct way of writing it is jansky, i.e. with lower case first letter; the shortened unit is written Jy.

Joule (J) The practical unit of energy, equal to 1 Newton-metre (Nm).

Junction The common point where two or more conductors are joined. Also used in semiconductors to describe an internal part of the device e.g. base-emitter junction.

Kelvin (K) The unit for temperature measurement on the absolute scale, named after the British scientist Lord Kelvin. Absolute zero is -273.2° Celsius.

kilowatt (kW) A power level of 1000 Watts.

Lambda (λ) The Greek symbol used to denote wavelength.

Lecher Line (pronounced "lekker"). An early item of test equipment that was used as a means of measuring frequency, especially at UHF and above. It consists of two parallel wires fed at one end with RF power and the far end is open circuit. A movable indicator, e.g. a neon lamp or torch bulb, bridged across the lines showed the peaks and troughs of the voltage. By establishing the physical distance between adjacent peaks the wavelength could be established. Before accurate wave meters were made this was the accepted way to measure frequency. Also a means of establishing the VSWR on a transmission line.

Linear The description of how the amplitude of a signal is amplified, with a constant gain the output versus input power should be a fixed ratio. Today often (incorrectly) used to identify an RF power amplifier, i.e. Linear Amplifier.

Logarithmic The expression of gains or losses from a ratio form into a decibel form.

Logarithmic Amplifier A special type of amplifier designed to have a non-linear gain function that obeys a logarithmic formula using deliberate compression to limit the amplitude of the signal. Used in radar receivers and test equipment to obtain an accurate measurement of the signal-input power, for example spectrum analysers and measuring receivers.

Lower Side Band (LSB) The lower of the side bands produced by an AM transmitter. Also used to denote the side band transmitted when an SSB transmitter is used. See also SSB and USB.

Low Pass Filter (LPF) A filter designed to only allow signals up to a certain frequency to pass. See also BPF and HPF.

Magnetron A microwave power generator that is

self-excited. Used to generate high power for radar transmitters and also used in microwave ovens to heat food. Today the cavity magnetron is the most common type used. Developed by Randall and Boot at Birmingham University during WWII.

Megawatt (MW) A power level of one million Watts.

Microwatt (μW) A power level of one-millionth part of a Watt.

Milliwatt (mW) A power level of one-thousandth part of a Watt.

Mixer A device used to convert from one frequency to another using heterodyning. (Old name was *frequency converter* or *frequency changer*).

Multiplier In RF circuits, a non-linear amplifier adjusted to generate harmonics of the input frequency.

Noise Factor (F) The numeric value as a ratio for the amount of extra noise an amplifier introduces during the process of amplification. Noise Factor has no units. An ideal amplifier would add no extra noise and would have a Noise Factor of 1, hence Noise Factor is always greater than 1. See also Noise Figure and Noise Temperature.

Noise Figure (NF) The expression of Noise Factor in a logarithmic form. The unit is the dB. A Noise Factor of 1 is equal to 0dB.

Noise Pedestal In oscillator phase noise measurements, the apparent shape of the carrier sitting on top of a plateau of noise with the shape of an upturned wine glass. The noise falls away to lower levels at frequencies far removed from the carrier to the broad band noise level.

Noise Temperature (NT) The equivalent temperature on the Kelvin scale that a resistor would need to be to generate the same magnitude of noise as the Noise Factor or Noise Figure value. When the Noise Figure is very low it is usually more convenient to express it in Noise Temperature form. For example, NF of 0dB is the same as 0K Noise Temperature. 0.02dB ≡ 1.2K, 0.25dB ≡17.5K, 1dB ≡ 75K etc.

Non-Linear The opposite of linear. In a non-linear amplifier the product of output power to input power is not a fixed ratio. The non-linearity is a function of the amplifier device and usually the input drive level. At high input drive levels the amplifier compresses the signal and generates harmonics.

Occupied Bandwidth The bandwidth of a modulated carrier required to pass the signal.

Ohm (Ω) The unit of resistance. Defined as the resistance of a column of mercury at 0° Celsius of 14.4521g mass, of uniform cross-sectional area and 106.300cm in length. Named after the German scientist Dr G S Ohm.

One-dB Compression Point (-1dB) The power level at the output of an amplifier where the gain has fallen by 1dB from the small signal level gain due to gain compression (P_{1dB}). Also quoted as the gain where the output level has fallen by 1dB due to compression effects (G_{1dB}).

Output Third Order Intercept Point (OIP^3) The theoretical power level at which the fundamental linear power curve and the third-order non-linearity generated power would cross on the graph of power output versus input power.

Q factor (Q) The ratio between the resistance (R) and reactance (X) of an inductor or capacitor. Q = X/R.

Parallel Resonance The resonance in a circuit where the network exhibits maximum resistance to current flow and hence the highest voltage amplitude. See also Series Resonance.

Pass Band In filters, the band where the filter has reached the minimum insertion loss, normally the -3dB point. For a BPF the Pass Band is the same as the Bandwidth. In LPF and HPF the portion of spectrum from the cut-off frequency higher or lower (as appropriate) which suffers minimum insertion loss. For the LPF the pass band is normally from DC to the cut-off point, in an HPF it from the cut-off point to all higher frequencies.

Pentode A 5-electrode valve.

Percent Bandwidth In a filter, antenna or amplifier the ratio of bandwidth to frequency. For example, an FM broadcast receiver covers 88 to 108MHz, the centre frequency is hence 98MHz and the span is 20MHz. The percent bandwidth is therefore ≈20%. An antenna for the European 2m band needs to cover 144 to 146MHz, a bandwidth of 2MHz. The percent bandwidth is 2/145 = 1.4%.

Phase Modulation (PM) Similar to an FM transmission, but with PM the carrier frequency is not moved. The phase of the carrier is altered by a small amount, typically not exceeding 90°. PM transmissions may be received on a conventional FM receiver with little distortion.

Phase Noise Unwanted broad-band noise appearing close to the carrier frequency. Phase noise normally decreases in amplitude further away from the carrier. Most often caused by frequency synthesisers, although any oscillator has phase noise. Phase noise is expressed as dB below the carrier level at a certain offset from the carrier. For example, -100dBc at 10kHz offset.

Potential Difference (PD) In a signal generator, when the output is terminated in the characteristic impedance, the output voltage is called the PD. In this condition the voltage across the load termination is half the EMF value. In the case of a cell or battery it is the voltage developed across the load. PD will always be less than the EMF due to internal resistance in the cell or battery.

Radio Frequency Interference (RFI) The unwanted non-deliberate interference due to a radio transmitter causing difficulty in receiving a wanted signal. For example, an amateur transmission affecting a nearby television receiver because the television does not have sufficient rejection of the amateur signal.

Radio Frequency Choke (RFC) An inductor designed to be used in radio frequency circuits. Often used for feeding the DC supply to a point in the circuit at RF voltage.

Reactance (X) The opposition to the flow of AC signals presented by either an inductor or capacitor. The unit of reactance is the Ohm (Ω).

Receiver (Rx) An item designed to convert electrical signals from an antenna and present an output in an intelligible form.

Resistance The opposition to the flow of DC. The unit is the ohm (Ω). Circuit reference R.

Resonant Frequency The natural frequency of a complex network when the reactive components cancel. See also Series and Parallel Resonance.

Selectivity A measure of how well a receiver is able to reject a strong off-channel signal. It ultimately depends on how narrow the IF bandwidth is and the slope of the filter. Normally given by the attenuation at some frequency offset from the centre frequency. For example -70dB at ±15kHz.

Self Resonant Frequency (SRF) The frequency at which a component becomes resonant due to parasitic stray capacity or inductance effects.

Sensitivity In receivers, the level of the smallest signal it is able to successfully detect. Normally stated as the signal level in µV or dBm for a given Signal to Noise Ratio (SNR), for example 0.25µV for 10dB SNR.

Series Resonance The resonance in a circuit where maximum current flows in a network and the voltage amplitude is minimum. See also Parallel Resonance.

Shot Noise In an amplifier, the name given to random bursts of noise which sound like lead pellets falling on a metal plate, caused by electron movement. In audio amplifiers it is also known as Popcorn noise. Shot Noise is the basis for the Barkhausen effect.

Side Band The signal generated adjacent to a carrier when it is modulated. See also AM and SSB.

Single Side Band (SSB) A modulation method where only one side band is transmitted. This can be either the Lower Side Band (LSB) or the Upper Side Band (USB).

Signal to Noise Ratio (SNR) In a receiver, the ratio of wanted signal to the background of receiver or sky noise. Normally expressed in dB.

Shape Factor A measure of how quickly the selectivity of a narrow band filter approaches the ultimate. The -6dB and -60dB BW points are commonly used to quantify a filter. For example, an SSB filter of 2.5kHz BW (at -6dB points) and 5kHz BW (-60dB points) would have a 2:1 shape factor. (A perfect rectangular shaped filter - 'brick wall' - would have a shape factor of 1).

Slotted-Line An item of test equipment consisting of a short section of rigid coaxial line that can be probed via a narrow slot cut in the outer wall. It is the modern version of a Lecher Line. Used at mainly UHF and microwaves to measure frequency or VSWR. Normally a crystal detector diode that can be moved along the line is used to measure the voltage peaks and troughs.

Stop Band In filters, the frequency where the filter attains the minimum required attenuation. Any frequency further away is considered to be in the Stop Band.

Speed of Light (c) Speed at which light waves and radio waves travel in a vacuum. The accepted value is 300,000 km/s (≈186,000 miles/s).

Suppression The elimination, or greatly reducing the level, of signals. Often performed by filters. Also a term used when tackling RFI by fitting components known as suppressors. The common use is on spark plugs in vehicles to prevent RFI to receivers.

Transmitter (Tx) A radio frequency generator used to convey intelligible signals to a distant receiver by exciting an antenna.

Tetrode A 4-electrode valve often used for power amplifiers.

Triode A 3-electrode valve used for low noise receivers and power amplifiers.

Valve A thermionic device used as an amplifier, rectifier, mixer or oscillator. Today largely superceded by solid state devices except when high power output is required. The American name is a 'Vacuum Tube' or simply 'Tube'. The correct name is 'Thermionic Electron Device' or TED. Television and computer monitors (excepting Liquid Crystal Displays types) use a version for the display screen.

Volt (V) The unit of electrical 'pressure' we call voltage. Named after the French scientist Voltaire (real name Francois-Marie Arouet, Voltaire was a nom de plume used for writing scientific papers).

Voltage Standing Wave Ratio (VSWR) The measurement of how well a transmission line is matched to a load, e.g. an antenna. When perfectly matched the VSWR is 1:1. Also a means of establishing the impedance of an unknown load.

Watt (W) The unit of electrical or mechanical power. 1 Watt = 1 joule per second. Named after the Scottish scientist James Watt. Smaller and larger units are used to denote power levels, e.g. µW, mW, kW, MW, GW etc.

Wave-Guide A type of low loss transmission line used at microwaves to convey signals from one point to another, for example from a transmitter to the antenna. Normally a hollow metal tube of rectangular cross-section although circular and other shapes are commonly used for special applications. Available in rigid and flexible types for a wide range of frequency bands. Normally only used for frequencies above about 3GHz where the attenuation of available coaxial cable becomes excessive.

Wave-Guide Beyond Cut-Off A special property of wave-guide is that it behaves as a high pass filter preventing lower frequency signals from propagating along the wave-guide. Other transmission modes prevent higher frequencies from propagating. Used for filtering at microwaves.

Wavelength (λ) The physical distance between the peaks or troughs of an AC signal. The unit of wavelength is the metre, but at high radio frequencies the cm and mm are commonly used. A wavelength of 1m is a frequency of 300MHz. At optical frequencies the wavelength of visible light is approx 400nm (nanometres). Light waves have frequencies measured in THz (terahertz).

Wave Meter An item of test equipment to measure frequency.

Index